Hollywood Be Thy Name

The Warner Brothers Story

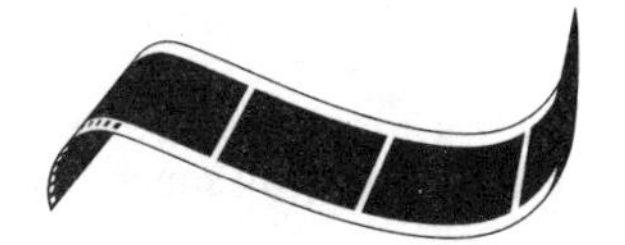

Cass Warner Sperling
and
Cork Millner

with Jack Warner Jr.

THE UNIVERSITY PRESS OF KENTUCKY

Publication of this volume was made possible in part by a grant from the National Endowment for the Humanities.

Scholarly publisher for the Commonwealth,
serving Bellarmine University, Berea College, Centre College of Kentucky, Eastern Kentucky University, The Filson Historical Society, Georgetown College, Kentucky Historical Society, Kentucky State University, Morehead State University, Murray State University, Northern Kentucky University, Transylvania University, University of Kentucky, University of Louisville, and Western Kentucky University.

Editorial and Sales Offices: The University Press of Kentucky
663 South Limestone Street, Lexington, Kentucky 40508-4008
www.kentuckypress.com

07 06 05 04 03 7 6 5 4 3

Library of Congress Cataloging-in-Publication Data

Sperling, Cass Warner, 1948-
Hollywood be thy name : the Warner Brothers story / Cass Warner Sperling and Cork Millner, with Jack Warner Jr.
p. cm.
Originally published: Rocklin, CA : Prima Pub., 1994
Includes bibliographical references and index.
ISBN 0-8131-0958-2 (paper : alk. paper)
1. Warner Bros. Pictures–History. 2. Warner, Harry Morris, 1881-1978. 3. Warner, Jack L., 1892-1978. I. Millner, Cork, 1931- II. Warner, Jack, 1916- . III. Title.
PN1999.W3S66 1998
384'.8'06579494–dc21 98-27317

This book is printed on acid-free recycled paper meeting the requirements of the American National Standard for Permanence in Paper for Printed Library Materials.

Manufactured in the United States of America.

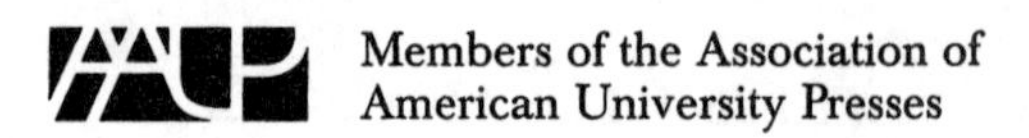

Members of the Association of American University Presses

Hollywood Be Thy Name

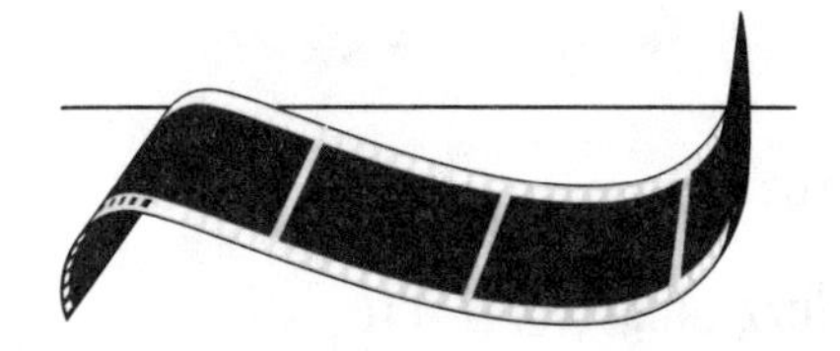

Contents

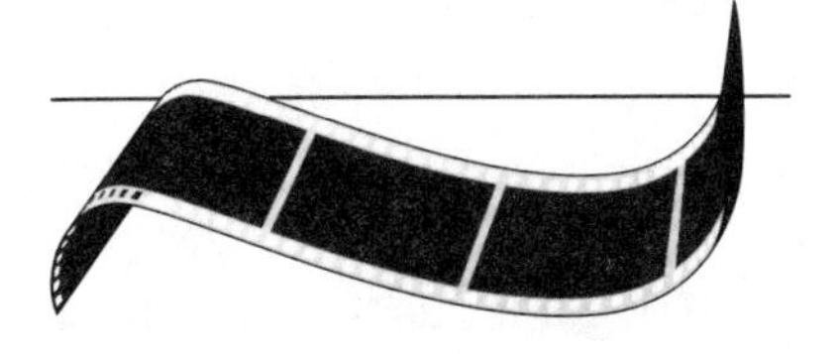

The Promise

A guard opened the iron gate at the entrance of an old, austere Bel Air mansion with big columns flanking the front door. My mother and I stepped into this huge mausoleumlike house with shiny marble floors.

Why was Grandpa Harry here and not at his beloved ranch?

I listened for his greeting, but it didn't come. Instead, I was led down a long hall into a dark, antiseptic-smelling bedroom. The person stretched out on the bed was like a statue, not the vital, spirited grandfather I knew.

I heard the nurse say he had eaten a little, but that he was still unable to speak. She bent down close to his ear. "Mr. Warner, your daughter and granddaughter Cassie are here to see you." He gave no response.

A mysterious force drew me to him as if he were a candle in the dark. It was a gentle force.

His eyes were open, and moved to take me in. He smiled. I watched his hand inch slowly across the sheet toward me. His hand found mine, and he tried to speak, but no words came. Instead, the look in his eyes transmitted his thoughts. His hand tightened around mine, sealing the message—something special was being entrusted to me. I squeezed back. A promise was made.

He tried to sit up. The nurse rushed over. I was being escorted out. I felt something tugging at me, telling me not to leave.

"Cassie."

I turned and saw him reaching for me. It was the last time I would see him.

I was ten years old when this happened, and it wasn't until years later that I would realize the magnitude and meaning of that moment. For years I had a feeling of incompletion, as if I was keeping someone waiting for me. As I did my research for this book, and learned more about what was important to my grandfather, I came to interpret the unspoken message that had passed between us as a responsibility that I had agreed to take on—a responsibility to convey to others his deep beliefs and ideals.

From a very early age, I was aware of my privileges. I had time to daydream, play, and take piano and dance lessons, while our gardener's son had to help his father tend to the yard every day after school. Often, I felt as if kids were being my "friend" so they could come see the movie stars who visited our house. Almost every weekend these celebrities would gather to watch films on our full-size living room screen.

Yet my family was different from those who were concerned only with how they looked, who they knew, and what parties they were going to. I was able to compare this vacuousness with values I was absorbing at home. Around the dinner table, my family discussed world affairs and civil rights. My grandfather and my parents were quick to point out injustices they saw. My father would always lighten things up with anecdotes about some situation that had happened at the studio or in the newspaper.

One of my favorite pastimes was sitting in my father's overstuffed leather chair in his home office and listening to him as he worked. As a screenwriter, he would often collaborate with another writer. There was always laughter along with their intense concentration. I felt like I was participating because he'd give me a script to hold on my lap and read. I could sense the exhilaration he felt in creating with another, and the pleasure he got from being a producer and putting all the pieces together.

I also watched my mother as she dedicatedly worked daily in her studio, practicing her detailed pencil sketches or spreading oil paints on her canvases. She taught me the value of being a mother as well as the importance of using one's artistic abilities.

There was nothing quite as special as going to the studio. I loved watching the cast and crew working together and seeing how much teamwork went into the magic of film.

Even today, when I go to the movies and see audiences uplifted and moved by what they're watching instead of bombarded with violence and desensitized by meaninglessness or superficiality, I have tremendous appreciation for the creative process and the choices made by those in charge. It's almost as if film casts a mysterious spell on those watching it, like the beat of a tribal drum that's subtly conveying beliefs and values. Despite the trends being what they are, I believe the wonderful art of drama and storytelling will endure. The Warner Bros. motto, "Educate, entertain, and enlighten," is a Hollywood legacy.

It's with great pleasure and pride that I give this recollection of the brothers' story as a gift back to my grandfather, and to family and friends who have contributed to my well-being, optimism, and understandings. I've tried to give a truthful and full account, as they would have wanted.

This book is in honor of and dedicated to dreamers and determined, caring souls like my grandfather, Harry Warner. May their hopes and visions guide humanity toward a better tomorrow.

My promise has been kept.

Cass Warner Sperling

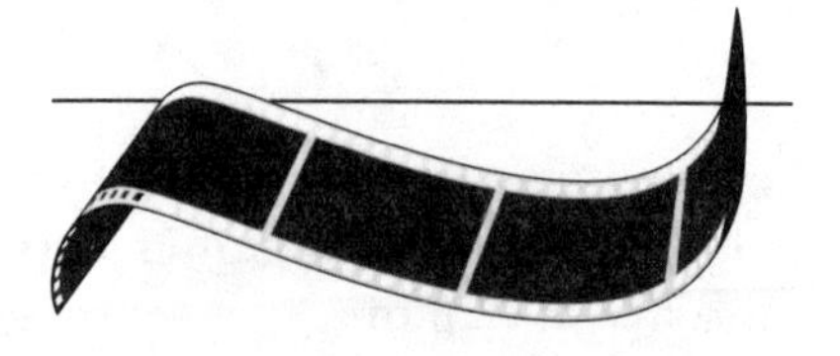

Foreword
by Jack Warner Jr.

> Someone may place a wreath on your grave, but the headstone will remain unmarked, for only your children can write the epitaph. What mementos will you leave to record your passing?
>
> —*Los Angeles Times,* April 13, 1967

It is said that Georgie Jessel, long noted for his many eulogies at funerals of the Hollywood famous, was so certain that he and only he could handle the oration at his own service, he prerecorded a long and glowing tribute to himself. No doubt others have written the epitaphs to be carved on their headstones, such as the one attributed to W. C. Fields: "Frankly, I'd rather be in Philadelphia," and the one more to the point by an unknown decedent which read, "See, I told you I didn't feel well!" All this shows that contrary to the quotation above, it is possible to bypass your children in the writing of your epitaph—but such was not the case with the Warner Brothers.

In a certain sense this book is the epitaph written by their children and grandchildren, about unique forebears who left a bittersweet legacy of triumph and tragedy.

It is difficult to think of these men as having been brothers, so different were they one from the other. Harry, the oldest, so serious, moralistic, hopeful that movies would be used primarily to educate and uplift humanity. Sam, full of fun and ever seeking to win friends and be where he could make things happen. His early death, and that of Harry's son, Lewis, changed the brothers' course. Abe, easygoing and calm in storms, was perhaps the brother who most enjoyed life. Henry, David, and Milton were all struck down before they could leave a firm imprint.

Then there was Jack, the most complex and confounding of all the brothers. For years I have tried to find the keys to the labyrinth of my father's mind, but it remains now what it was throughout most of his lifetime: boxes within boxes, rooms without doors, questions without answers, jokes without points, scenarios based on contradictions, omissions, and deceit. His was the anguished story of a man driven by fear, ambition, and the quest for absolute power and control—the little brother telling the big boys he saw as his tormentors to go to hell.

If ever two people were born on a collision course, Harry and Jack L. Warner were those men. Some time at the midpoint of their lives their basic moral outlooks took violently opposite turnings, and the most talented scriptwriters could not have created a more tragic plot. This almost biblical drama of brother against brother and how it played out in its final lamentable episodes is the story told here.

You will read here of the sons of immigrants at first closely bound together, achieving impossible dreams as they began to produce films on what Hollywood then called Poverty Row. Through the years of absolute trust in each other, blind faith that they would achieve what they went after, and the brutally hard work it took to achieve those dreams, they finally assembled one of the world's great motion picture production, distribution, and exhibition enterprises, whose crowning achievement was the marriage of sound to image—the first commercially successful sound picture.

Then, engaging in the most terrible kind of warfare, the battle of brother against brother, they transformed the great company they had created and nurtured into the betrayal of a dream.

While Harry Warner, like his brother Abe, seemed to be one man, devoted to family and friends, charity, and honest relationships, Jack L. Warner was many different men. He could be a likable fellow, telling jokes to relax and entertain an audience, then suddenly, as he grew in years and power, turn unbelievably cruel in debasing those absent or unable to respond. He would mouth maxims and philosophies counseling great deeds, the triumph of good over evil, and the furtherance of a beneficent moral humanity, sounding even at times like Benjamin, his father, whose warmth of heart and soul were never questioned. Then, without warning, from the dark recesses of his psyche, he would shock listeners with vulgarity and foul language he would never permit from others in his presence.

Jack's ego, which he pretended didn't exist, was actually enormous. When he parried arguments with his famous question "Whose name is on the water tower?" he was ramming home the fact that he and he alone ran the studio and that the symbol of his name high in the sky was like holy confirmation. Although his name was indeed on the water tower, his older brother Harry, president of the company, had the corporate position and power to overrule him, and it was his name too on the water tower, though he never mentioned it. This was the seed of discord that would grow into a choking vine of disaster.

Now they are gone, but the great company they started lives on, reincarnated in another body. It's a pity that Jack L. Warner did not live to come into contact and conflict with those who eventually took what had once been Warner Bros. Pictures Inc. and threw it into the pot with other ingredients to meld and merge into Time-Warner, the world's greatest media corporation. No water tower exists large enough to hold the names and egos of these men. I can almost smell the smoke rising, feel the heat and flame of that imagined confrontation, with J.L. screaming "Time-Warner! What bastard gave us second billing? Warner-Time—that's the way I want it! Reshoot the main title!"

Many years ago Warner Bros. Pictures adopted a motto lifted out of a *New York Times* review: "Combining Good Citizenship with Good Picture Making." The tragedy of their story is that a more descriptive motto for the company and its founding brothers might better read "United they stood—divided they fell."

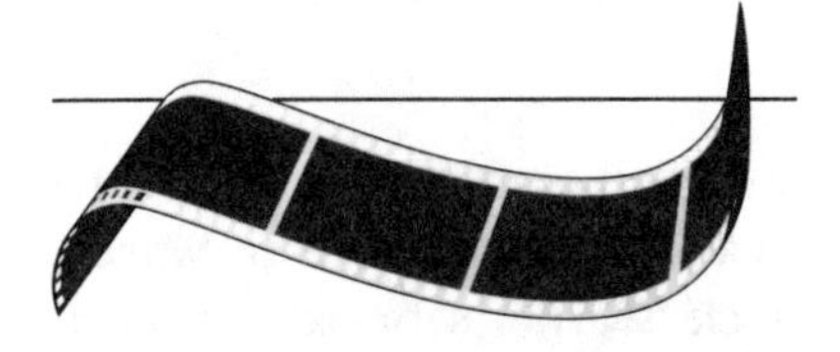

Acknowledgments

This book is the result of many people contributing in many different ways. I hope that those who helped will regard this mention and the book itself as a form of my appreciation. I'm especially grateful to my children (Tao, Cole, Vanessa, and Jesse) for their six years of patience and understanding; to my father and mother for having faith in this project, sharing their experiences, and inspiring this all along the way; to Jack Skinner for encouraging and being a part of the birth of this book; to my collaborator, Cork Millner, for hanging in there when things got tough, and doing a brilliant job of bringing this to fruition; to cousin Jack for all his personal contributions and interest in telling the definitive story; to Estuardo Rojas for his incredible calorcito; to Howard and Anne Koch for their friendship and artistic support, and Howard for the title of the book; to Dick Dorso for his insights and friendship throughout; to Barbara Warner a most warm thanks for the beginning of a wonderful friendship and for breaking the family curse and sharing her personal family photos so willingly; and to Doris Kearns Goodwin for her special hand holding. Special thanks go to Leith Adams, archivist for the Warner Collection at the University of

Southern California, and Ned Comstock in USC's Department of Special Collections for their years of assistance; Ann Warner's estate for permission to quote from Jack Warner's book, *My First Hundred Years in Hollywood*; Warner Bros. Studio for allowing me to dig around and use their photo files.

Others who aided the cause in a variety of appreciated ways: Bill Schaefer, William T. Orr, Charles Yoss, Eric Sherman, Gregg Orr, John and Lewis Steel, Viki King, Stanley K. Sheinbaum, Ann Sperber, Ronald Reagan, Richard Gully, Lina Basquette, Sam Halper, Vicki Hiatt, the late Ruby Keeler, Ian Bernard, Anne Francis, Paul Lazarus, Dennis Hopper, Lita Heller, Shirley and Jim Lavine, Linda Janklow, Lois McGrew, Kate Horowitz, Jonathan Kirsch, Dorothy Wall, Mike Weltman, Dasheng and Guilan Pan, Karen Sperling, Hilbert Lee, Ray Castellino, Bill Cleary, Daniel Frank, Steve and Ann Gilbar, Barry Klein, Abigail Albrecht, Tony Cohan, Carlynn and John McCormick, Joke and Frizell Clegg, Gay Ribisi, Lynda Millner, Tom Klassen, and the Princeton Theater Collection.

Prologue

Camelot

An amber blade of light slices through the darkness.

The blare of trumpets breaks the silence, punctuating the single word on the screen:

Camelot

In my mind's eye, the proud old man sighs heavily, rolls an unlit cigar between his fingers, feeling its silkiness, and settles into the cushioned seat. The movie credits roll:

Richard Harris, Vanessa Redgrave, Franco Nero . . .

The man mumbles, "Yeah, and not one of them could sing a lousy note," but his words are lost in the chorus of voices performing the title song and the void of the studio screening room. There is no one to hear him.

The song increases in tempo as the names of the composers and music director flash on.

Then, in bold letters, the credit:

PRODUCED BY
JACK L. WARNER

The old man stares at the name—*his name*—etched across the screen like a page from a medieval illuminated manuscript.

Reflected light washes over Warner's face as the credits dissolve into a wooded scene of snow and ice. Richard Harris, as King Arthur, is emoting: "How did I blunder into this? When did I stumble? When did I go wrong . . . ?"

The voice of Merlin the Magician breaks in: "Think back, Arthur . . . think back to the day you met Guinevere . . . the day she came to Camelot. That is the beginning . . ."

Jack Warner spits out a shred of tobacco and grumbles in the darkness, "Some beginning." His eyes glaze, dissolving the color on the screen; his ears shut out the sound. A window of his mind opens onto the flickering images of a black and white film, the first "feature" from the early silent era: *The Great Train Robbery.*

The film is lashed with jagged black scratches and has been spliced together so many times it jumps erratically from scene to scene. Yet the audience leans forward on the edge of the benches in the darkened space, their emotions audible as the deadly shootout unfolds magically on the bedsheets pinned to the canvas wall. They scream and duck as the menacing black locomotive rushes toward them.

Sweating inside the cramped projection booth, Sam Warner, seventeen, cranks the Kinetoscope projector, coils of celluloid piling up around his knees. Outside the tent his brother Abe, nineteen, sells tickets to the line of people waiting for the next show to begin. He passes the coins to Harry, the eldest brother at twenty-three, who drops each nickel admission into a cigar box. Jack, the youngest boy, now thirteen, waits to sing his ballads between showings, a practice the brothers have discovered quickly clears the room.

"Some beginning," Jack Warner grumbles again. "But it worked. *We damn well made it work.*"

The shadowy sound of another voice, from another place and time, intrudes: "Together, together you are strong!" Their father, Ben Warner, big-shouldered, square-jawed, is standing before a kitchen table on which is piled a stack of wooden sticks. Harry, Abe, Sam, and Jack watch as their father takes a stick, grasps it in both hands, and snaps it in two. "Alone, the wood is weak, it breaks easily," Ben says. He takes two sticks and cracks them apart. Then three; harder this time. Finally, he picks up four sticks, gathering them into a tight bundle. This he cannot break. He looks hard at his four sons and repeats, "Together you are strong."

Jack Warner chews on the end of his cigar. "Sure, sure, always together. Well, nothing lasts forever." He pushes out of the seat, staring at the image of King Arthur on the screen and hisses, "Damn you, Harry, *damn you!*" Moisture glistens on his eyelashes.

King Arthur sings:

Don't let it be forgot,
that once there was a spot,
for one brief shining moment, that was known as Camelot . . .

"Camelot? It was Warner's lot! *Jack Warner's lot!*" He lashes out at the face of Arthur now filling the screen: "I showed you, Harry. Took the whole goddamned thing . . . right out from under you. Jesus, you trusting son-of-a-bitch. *You fell for it!*" He reaches back to lower himself into the seat.

The towers of Camelot dissolve into a watery blur.

Part I

1890s & 1900s

1

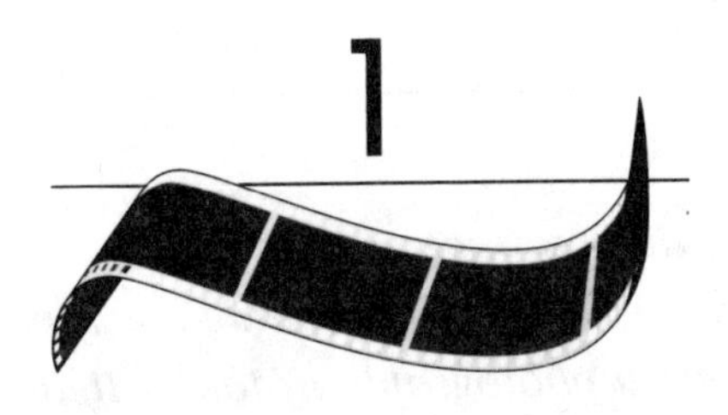

Two Nickels and a Handshake

Summer 1955

I called it the "Mystery Room."

I was seven years old when I started exploring in Grandpa Harry's ranch house and found it. It was a dark place with paneled walls, a big wooden desk, and a curtain that was always drawn shut. On the walls were photographs. I had to stand on tiptoe to see most of them. Many were of my grandfather smiling that warm smile of his while standing next to important-looking men whose names I didn't know. (I later learned these men also had important names: Franklin Delano Roosevelt and Harry Truman; "Hap" Arnold and "Ike" Eisenhower; Albert Einstein.)

There was also a huge glass case filled with trophies, plaques, ribbons, and silver cups. I wished I knew how he got all these ribbons and things. Was Grandpa an Olympic runner?

Then one day while I was snooping around in the room I heard Grandpa softly say my name. "Cassie?" He came over to the showcase I was staring at.

"Grandpa, why do you have all these?"

He put his hand gently on my shoulder, as if he realized it was time that I knew, took a photograph off the wall, and led me to a dark maroon leather couch in the center of the room. On the way, he opened the curtains: light streamed into the room.

"This was our first theater—the Cascade," he said, tilting the photograph so I could see it better. His fingers traced an outline around one of the figures standing in front of the theater's arched entryway. "That's your great-uncle Sam."

"He died," I said.

"Yes, many years ago." My grandpa sighed. "But not before he gave us, and the world, the gift of music and the spoken word on film." He sat there for a few moments, his fingers still touching the photograph. Then, with great care, he hung the photo back on the wall and came back to the couch. He opened a huge scrapbook lying on the coffee table in front of us. "Here, this is Sam describing how it all began." Very slowly he read from a yellowed newspaper clipping:

> *Harry and I were walking on Smithfield Street in Pittsburgh in 1905, when a sign attracted us which read:*
>
> NICKELODEON
>
> *We put two nickels on the plate and Harry and I walked into the place. We sat through three shows, until we were shown the exit sign. When we got outside, we stood for a time watching the people go in and out of the nickelodeon. We then and there shook hands and said:*
>
> *"We're in the motion picture business."*

Spring 1906

Black smoke churned in angry belches from the idling car's exhaust and gritted the air, burning seventeen-year-old Sam Warner's eyes. He tilted his head and lifted his shoulder to his cheek, trying to blot away the sweat, careful not to slow his cranking of the Kinetoscope projector. With his other hand he fed the film strip into the machine's take-up spool, feeling the gears click as metal claws poked into the sprocket holes to move the film forward frame after frame.

On the white sheet sewn to the side of the unventilated black tent, shadows came to life, revealing the panorama of Yosemite National Park. The six people crammed into the back seat of the touring car were enthralled with the scenic ride. Why, it was just as if they were there, in the mountains, passing a cascading waterfall, smelling the lush green foliage . . .

"George!" Sam Warner tried to attract the attention of the other man hunched behind the car. In the flickering shadows Sam could see his colleague pushing up and down on the car's bumper, then rocking it from side to side as he tried to simulate the car's bumpy journey through Yosemite. "George!" Sam repeated louder. "We gotta cut the engine . . ." Sam coughed, his throat raw.

George Hale looked fiercely at his young projectionist and growled, "Shaddup!"

Sam closed his eyes and sucked air in through his nose, hoping that would filter out the acrid fumes. His arm felt like lead from cranking the wooden handle of the projector. *Great, now I'm in the motion picture business,* Sam thought, as he remembered that day a year earlier when he and Harry had strolled down the Pittsburgh street and had been attracted to the NICKELODEON sign.

Sam smiled at the recollection as he wearily fed the last bit of the travelogue film through the projector: clickety click, clickety click . . . The screen went black. George stopped rocking the touring car, switched off the ignition, and opened the flap to the tent as Sam helped the passengers to step out of the back seat and onto a wooden box, then to the sawdust-strewn ground.

"Great trip, young fella," a portly man said, fingering the watch chain stretched across his belly. The man puffed heavily on a cigar, trailing a cloud of gray smoke that mingled with the lingering black fumes.

Stepping past the tent flap, Sam filled his lungs with fresh air and grabbed a can of water he had left in the shade of the ticket taker's bench. Swallowing a mouthful, he glanced up at the banner which proclaimed:

HALE'S TOURS
Latest Lifesize Moving Picture Tour
Yosemite National Park
Moral and Refined
Pleasing to Ladies, Gentlemen & Children

Sam flexed his aching fingers as he watched George climb back onto the little platform in front of the sign, grab his megaphone, and announce to the passing parade of fairgoers:

"Lay-deez and genn'men, welcome to the miraculous scenic tour of one of America's wonderlands. Enjoy it in the privacy and luxury of your own touring car . . ."

Sam poured the last of the water over his face, then, pushing his shirtsleeves higher on his arms, went back into the tent to rewind George Hale's "miraculous scenic tour."

Even though Sam hated returning to the hot tent, he loved working with anything mechanical. A year earlier (about the time he and Harry had walked into the nickelodeon) he had read, enthralled, of the Wright Brothers' flight at Kitty Hawk. Sam laughed when people scoffed at airplanes as they had at autos in the '90s. He was clever beyond his years and was always burrowing into some engineering book or other that he couldn't quite understand.

At six feet he was solidly built, with a square jaw and a thick head of rust-colored hair. He liked to laugh to his friends: "Everyone thinks I'm Irish, not Jewish." His deep blue eyes had a faraway quality in them that saw beyond the confines of Youngstown, Ohio.

Grandpa turned the pages of the album carefully, smoothing each page with the palm of his hand until he came to a brown-toned photograph of a football team. The young men were in uniforms thickly padded with cotton, each of them staring seriously into the camera. Grandpa pointed to a figure on the back row, the largest player in the group.

"Abe," he said. "Your great-uncle Abe."

Three years older than Sam, Abe was making a few bucks of his own. Barrel-chested like his father, with huge hands and solid, stern features, Abe had stayed in school longer than any of his brothers, entering Youngstown's Rayen High School in 1900, where he played quarterback on the football team, before deciding to quit and take a job in a steel mill. He quickly discovered that the cavern of forges spewing rivers of white-hot metal was a terrifying place to work. Cascades of hot sparks shot sparkling fingers of fire at him, and coal cinders scratched his eyes. Even worse, on his first day at the mill a worker was crushed to death when a huge ingot fell on him. On his second day, Abe saw a man's leg smashed by a metal beam.

His mother, Pearl, could see the dark lines of worry on her son's face when he came home from work that second day. After he told Pearl of the accidents, she spoke to him in Yiddish, the language she favored, having yet to completely master English: "You stay home from work."

Abe, his face blackened by sulfur and soot, answered, "Mama, I can't stay home. The job pays well."

"No!" She grasped one of his huge hands in hers. "Life is more important than money."

The warning was enough for Abe, who started looking for other employment. When a representative from Swift and Company in Chicago visited the family grocery store, he talked Abe into becoming a salesman. In Chicago, Abe, to his dismay, found himself a failure as a soap salesman.

At night, in his hotel room, he would line up the different kinds of soap on the bureau, then walk up and down in front of them practicing his sales pitch. He even carried a bar of soap in his pocket to take out and present to the storekeeper the minute he got in the door. Still he couldn't get an order and was thrown out of many places. Then Abe got an idea: He'd raise the price of the soap and throw in one free box with every six he sold. Of course, the customer was paying exactly what he would have paid for seven boxes, but the excitement of getting something "free" brought in the orders. Without realizing it, Abe had invented discount trading. In a year he was the company's top soap salesman.

It was while he was working in Pittsburgh selling soap that he saw his first motion picture—at the same theater Harry and Sam had seen theirs.

"In 1903 a nickelodeon had opened on Smithfield Street, and the pictures shown were changed twice a week, Monday and Thursday nights," Abe recalled later. "I was first in line to see each new picture and I began to figure out the attraction. If moving pictures had so much appeal that I never missed one, then it ought to be a pretty good thing to be in. I thought about it some more and decided to get into it some way or the other. So I resigned as a soap salesman, and with no definite idea of how to get into the movie business, I started home. When I got back to Youngstown I discovered that my brothers Harry and Sam had had the same idea."

"Strange . . . the three of us thought about it at the same time." My grandfather began turning the pages of the album and paused at

a page showing Jack hamming it up in a strip of five snapshots: one with a fake cigar stuck in his mouth; the others feigning laughter, surprise, anger, and pensiveness.

I heard my grandpa say under his breath, "Jack should have been an entertainer." I could see that he was bothered, and he slowly closed the book, put his arms around me, and gave me a big hug.

Jack Warner had his own ideas about getting into "pictures." He loved them—as long as he was the subject. In 1903, when he was eleven, he'd hang around the Banner Studio, where they made photos-while-you-wait at a penny apiece. "He became a pest around the studio," the owner, Bill Stanton, recalled years later. "When the camera wasn't working well, we'd use him for test pictures."

When Jack wasn't clowning in front of the studio camera, he worked as a delivery boy for his father's grocery store, a task he did well, albeit reluctantly.

But it was Harry who had borne the major burden of the family workload since he began to labor as a child in his father's Baltimore cobbler shop in 1888. Ben Warner had shown his young son how to salvage bent nails by straightening them with a short claw hammer. He taught him how to hold new ones in his mouth so he wouldn't have to fumble for them on the bench. Together they worked long hours building their reputation and living up to the store-window sign that read SHOES REPAIRED WHILE YOU WAIT.

By 1890, when Harry was nine, Ben thought his son was capable enough to give him the job of managing the shop.

Although by 1906 the Warner boys knew they were interested in the picture business, none of them had figured out *how* to get into it. Sam still had his job at Hale's Tours, but he didn't want to slave the rest of his life cranking a projector for someone else. He wanted one of his own.

A Youngstown friend, George Olenhauser, who ran small machine shop, had taken in one of Edison's Kinetoscope projectors for repair and, knowing Sam's interest in anything mechanical, called him to help work on it. Sam looked at the brass label on the projector's box that read "Edison Kinetoscope Deluxe Model AA Projector" and could hardly wait to take it apart.

When he lifted the brass and copper machine from its box, turned the latch, and opened it to see the myriad gears, wheels, sprockets, and the mirrored carbide lamp, he felt as if he had truly touched a magic lantern. All he had to do was rub its gleaming surface to loose the genie inside.

A few months after Sam had learned how to take apart and put back together George Olenhauser's Kinetoscope ("I can do it blindfolded," he bragged), the opportunity for the brothers to get into the motion picture business presented itself.

A woman who ran a theatrical boardinghouse near the amusement park where Sam still worked for Hale's Tours confided in Sam one day that her son, Joey, had just returned, broke, from touring with the Kinetoscope he owned.

"Sam, you maybe know somebody . . . wants to buy it?" she asked.

Sam swallowed, trying to stay calm. "How much do you suppose Joey wants?"

"He says two hundred fifty."

Sam's shoulders sagged. George Hale had told him that was what he had paid for the machine he used for his Yosemite tours. A thousand bucks. A lot of nickels.

"Joey's go a film, *The Great Train Robbery*, goes with the machine," the woman said. "There's rolls of tickets, blue and gold, he had printed up.

The Great Train Robbery, produved by Edwin S. Porter in 1903, was 800 feet of red-blooded drama in which desperadoes rob a passenger train and end up chased by a posse. Audiences would yell out "Catch 'em! Catch 'em!" as puffs of smoke erupted from six-shooters. The most terrifying moment came when the villian, in a close-up, aimed his revolver directly at the camera and fired. Women swooned in their seats.

Pearl Warner, a robust woman of forty-six, whose waist had widened from twenty years of childbearing,stood over the kitchen stove stirring a heavy pot of soup with a wooden spoon as she listenend to her boys talk to their father. It was a cool, brisk day late in October 1906, but the warmth emanating fromthe black wood-burning stove and the aromas drifting across the room from Pearl's boiling pots made the kitchen a cozy, warm place to be.

"It's a real bargain," Sam was saying.

"Pops, like Sam says, it comes with a film, rolls of tickets . . . " Abe said, leaning back from the table while his older sister, Annie, pushed a clean plate in front of him. Across the table, fourteen-year-old Rose was setting knives and forks by Harry's plate. Two of the younger Warner children, David, eleven, and Sadie, nine, were in the small living room playing. Milton, only seven, the youngest, lay in the bedroom next to the kitchen with a cold, his chest greased with mentholated oil.

"It's a big opportunity," Sam added, nervously twisting a button on his jacket. *He had to have this projector.*

Ben Warner leaned back in his chair and crossed his arms across his barrel chest. He was a big man with a square jaw and a bold black mustache, and when he spoke his children listened. In broken English, he said, "If it's such a big—how'd you call it?—'bargain,' how come this Joey, he want to sell?"

"The machine just needs a little fixing," Sam said, pressing his father. "I know how to do that."

Ben looked at his oldest son. "And what do you think?"

Harry hunched forward, elbows on the table. "Pop, we all talked about this before. This movie business, it's new. It's not like Sam taking odd jobs all over the country or Abe off selling soap while I help you run the store. This is something we can all work at *together.*"

Ben Warner nodded. He liked that—working together, helping one another. Yes, that's the way his boys would succeed. How many times had he told Harry that, as the oldest, it was his responsibility to keep his brothers together. *As long as you stand together you will be strong.*

Just then twelve-year-old Jack burst into the room, avoided his father's look, and, encircling his mother's ample waist, sang:

> You're as welcome as the flowers in May,
> And I love you in the same old way.

Ben Warner slipped his gold watch out of his vest pocket, opened it, and tapped its face. "My son, he doesn't know we eat as a family?" Ben sighed. What would he do with this rebellious boy, who had quit school after the fourth grade only to hang around with toughs, and to sing?

"It's work, Pop." Jack reached into his pants pocket. "Two dollars! Just for one week's work."

"You give it to your mother," Ben said, wondering how his son could make such money singing in an opera house.

Jack warbled lyrics, such as the one he had just sung to his mother, from the corner of the stage at the 180-seat Dome Picture Palace in Youngstown as tinted color slides of outdoor scenes flashed on the screen. Jack would sing "Dear Old Georgia I Long to Roam" while pastoral chromes illuminated a line of magnolia trees leading to a Southern mansion. These illustrated ballads, inserted between the vaudeville acts, gave the ushers time to nudge people out of the theater's seats to make room for the next group waiting in line outside.

"Bubeleh," Pearl said, "maybe you sing for your supper?" She brushed a lock of hair from his face, then started to ladle the soup onto the thick white plates.

Sam scooted his chair closer to his father's. "Pop, the projector . . ."

"Projector!" Jack said. "You mean we're going to do it? Open a nickelodeon?"

Harry said, "Sam's found a Kinetoscope. We would need a theater . . ."

"Wow! Rose could play the piano," Jack said excitedly. He slipped into his chair, his nose hovering over the steaming broth laden with chunks of meat from their butcher shop. "And I can sing!"

Now Ben began the *broche,* the blessing.

"Baruch ato adonoy eloheynu melech ho-olom . . ." Blessed art Thou, oh Lord our God, King of the universe . . .

Ben's eyes moved to Abe, who never quarreled or bickered with his brothers, the one who was built a lot like Ben, tall, bullish . . .

"Ha-motsi lechem min ho-oreets . . ." Who bringest forth the bread from the earth . . .

He glanced at Sam. Sam was the farsighted one, the smiling, bright, blue-eyed son, the boy who looked to the future, the one who had big dreams.

"Bray preeh hagolfen . . ." Who givest us the fruit of the vine . . .

Once again Ben looked at Harry. Solid, serious Harry, who at twenty-three had earned his father's respect as the most responsible of his sons and had secured his place as future head of the family.

Oh, Lord, take these my boys and make them strong, each in his own way . . .

After dinner, as the girls cleared the plates away, Ben lit a hand-rolled cigarette, savored the smoke as it filled his lungs, then leaned over and put his hand on Sam's shoulder. "Now, my son, let us talk about your great machine."

In the next hour it was decided. The family would take the chance on the new invention. The money? The boys pooled all their cash assets and the total came to $75.

"It's not enough," Sam said sadly.

"We can sell the store!" This from Jack, who bounded out of his chair.

"No!" Harry said. "We must keep the store until we see what happens."

"What will we do for more money?" Sam asked.

The brothers looked at each other blankly. The only sound in the room came from the coffeepot bubbling on the stove.

Finally, Ben slipped his fingers into his vest pocket and pulled out his gold watch. As his sons watched, he coiled the chain in his hand and laid the watch on the table. Lightly tapping the shiny curved surface, he said:

"I have this."

2

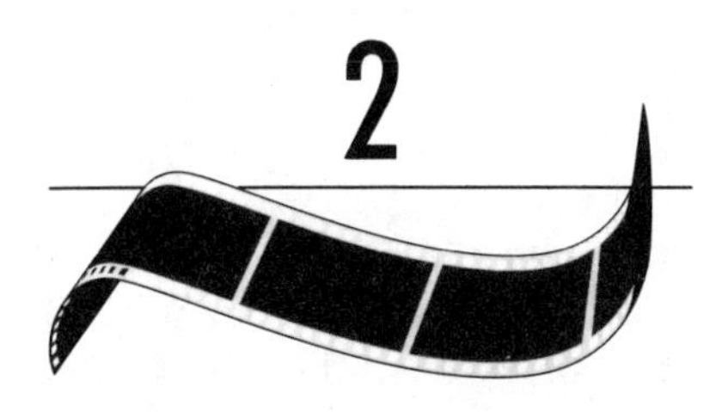

The American Dream

My great-grandfather, Benjamin Warner, brought his prized gold watch to America in 1883 when he fled from the tiny village of Krasnashiltz, now in Poland, then in czarist Russia. He had left the village, the home of his ancestors for 300 years, for one reason: so he and his family could survive.

Krasnashiltz, Poland, 1857

Benjamin had grown up in a hostile, threatening world where the sound of horses' hooves heralded the alarm, "The Cossacks are coming!" The call sent daggers of fear through the shtetl, the Jewish section of Krasnashiltz. The village was part of the Pale of Settlement, one of the twenty-five provinces of Russia that had been established in 1772. Jews needed special permission to live outside this Pale. There wasn't a Jewish settlement in the land that hadn't felt the wrath of Czar Alexander as he periodically unleashed his marauding Cossacks to massacre this "unwanted" race of people. The strong young men of the village were rounded up to work in labor camps, never to return. It was into this terrifying world that Benjamin was born on July 10, 1857.

Benjamin grew up to be a tall, powerfully built young man, who in his teens sported a wide, bushy mustache. When he walked down the streets of Krasnashiltz he drew wistful glances from the village maidens. Although he was a striking man, he had little else to offer; his pockets rarely jingled with more than a zloty or two, the meager saving from what he earned as a cobbler. His only possession of any value was the watch his father had given him, an heirloom that had been passed down through several generations.

Like his father and his father before him, Benjamin had no formal education, having been barred by the Russians from going to school. What little he learned was taught in secret by the village rabbi.

"We had to hide to learn," my grandfather Harry told me. "Perhaps this need to steal to gain knowledge was the seed that made me strive to make movies the way I did. I didn't want to just entertain, I felt the need to educate."

Many years later, Jack Warner would say of the ghetto families in Krasnashiltz:

> Like conspirators in crime, they herded their children into the largest stable available, and there the rabbi taught them the lore of their religion and race. . . . There was always a lookout posted in a loft where he could cry alarm at the approach of the village police. . . . The children, long trained in the arts of escape, crawled through a crude earth and rock tunnel into the Christian cemetery. There, because tradition made it a sacred little island where the law dared not trespass, the Jewish children huddled in a cave beneath the gravestones, and were safe.

When Benjamin was nineteen he met Pearl Leah Eichelbaum, a full-figured girl six months his junior. Ben and Pearl were married in 1876, and less than a year later their first child, Cecillia, was born, followed by another daughter, Anna. In 1881 a son arrived. Benjamin and Pearl named him Hirsch.

Worried about the ever increasing threat of pogroms to his growing family—there was a real fear that his daughters might someday fall prey to Cossack rapists—Benjamin began to look at the German border not many kilometers from his home as a route of escape. Would his family's life be enriched across that border? He knew of one person who had fled to Germany, a young man named Waleski who had been known as the village idiot.

"I will go to Germany, then work my way to Hamburg," the scatterbrained youth had excitedly told Benjamin before his departure. "I go where the ships depart like . . . like great flying birds and sail to America!"

America. In Benjamin's imagination America was only a misty fantasy, a land shrouded in mystery.

One day while he was working in his tiny cobbler shop, he was surprised to receive a letter from a place called Baltimore. It was, he quickly realized, a city in America! The letter, wrinkled and worn from its long journey, was almost illegible.

"Come to Baltimore . . . riches . . . earn two dollars a day . . . the streets run with gold." Benjamin looked at the envelope: there was an address scrawled on the back. He folded the letter and put it in the pocket of his stained leather apron. It was something to think about. After all, Waleski had added as a last thought: "Everyone in America wears shoes!" And Benjamin *was* a cobbler.

"Gold in the streets"? That sounded like the babbling of an idiot. Yet if Waleski *had* found riches in America, why not he? Benjamin worked harder, saving every zloty.

In late 1881, several months after the birth of Hirsch, Benjamin and Pearl's first child, Cecillia, died of an infection. She was only four years old. Although it was expected that many children in each family would die, Benjamin wanted his children to be given the opportunity to grow strong, to have children of their own, to live somewhere other than this place of death and terror.

Thus, in 1883, the year of his twenty-sixth birthday, Benjamin said good-bye to Pearl and his two children. Cupping his wife's face in his hands, he said, "If Waleski is right and America is a land of hope, I will send for you."

Before he left, Pearl sewed a secret pocket in the waist of his pants. "It is for your watch," she said, "to keep it safe."

Castle Garden, New York, 1883

This no-man's-land temporarily held the hordes of immigrants before they funneled into New York and the rest of the country. (Ellis Island would not be opened until ten years after Ben's arrival.) There, inspected by the cold eyes of health officials, the immigrants awaited their fate. A hacking cough might mean tuberculosis, a facial sore an infectious disease; either was enough to send the ashen-faced man or woman back home.

"Name?" the final inspector asked Benjamin as he stood straight and tall before the wooden table. Not understanding English, he looked dumbfounded.

The clerk repeated his question, this time in a mixture of broken Russian and Polish.

Benjamin nodded and said his name. The clerk cocked his head, trying to translate—as he had done countless times before—the sound of the name into English. Was it Verner? Waner? Varner?

Finally he wrote "Warner."

Ben never told anyone the original family name, no doubt because he wanted to leave that part of his life behind him. Perhaps the name was something like Varnereski. More likely it was Varna, which meant blackbird in the language spoken around Krasnashiltz.

I sometimes wonder how "Varna Bros." would have looked on the screen.

Ben Warner made it to Baltimore only to learn that Waleski was not at the address scratched on the letter. Carrying his leather bag of clothes and a worn blanket, Ben began his search for the crazy man who had told him to come to America. He walked endlessly, confused by the clamor of the hostile new world he had been thrust into. It was the hard part of winter and the Baltimore streets were covered with soot-blackened slush. Hardly the river of gold Waleski had told him to expect.

He went from shop to shop, shoving a card with WALESKI crudely printed on it in front of dozens of faces. When Ben did find him, ten days later, Waleski was working in a cold cellar, cutting soles for shoes from thick pieces of leather. In the candlelight, Waleski looked at the towering man who appeared before him and shuddered, not from the cold, but from fear.

"You lied to me!" Ben roared. His great sweeping mustache glistened with the melting snow.

The little man shrank back into the shadows of the cellar.

Then Ben reached out his arm and grabbed Waleski, and gave him a great, friendly bear hug. "It is fine, *luntzman,* my countryman. I am glad you lied to me." He laughed, his breath misting the cold air. "Had you told me the truth, I would not be here."

Ben, more determined than his countryman, opened his own shoe shop at Pratt and Light streets in Baltimore, a hole in the wall with a cobbler's bench and a chair for customers. He called the place "The Baltimore Shoe Repair Shop" and placed a hand-lettered sign in the window that read SHOES REPAIRED WHILE YOU WAIT.

Most customers had only one pair of shoes and readily waited in Ben's secondhand chair, toes wiggling in stocking feet, while this giant from Russia nailed on new soles. Ben was soon taking in three dollars a day, far more than he had ever made in Krasnashiltz. In Ben Warner's eyes, the soot-streaked streets of Baltimore had taken on a golden hue. In less than a year he was able to send for Pearl and the two children.

Winter 1883

The huge door of the Customs Barn opened; Pearl, Anna, and Hirsch emerged into the sunlight. Bundles swinging at their sides, they ran to Ben. After practically crushing them with joy, he looked at Hirsch, who had grown so much! Only then did he notice his son was shirtless. A shawl of Pearl's had been wrapped around his shoulders, yet the boy was shivering from the crisp wind blowing off the water.

"What is this?" Ben asked. "My son, you come to your new country half-naked?"

Hirsch looked steadily at his father. "That man . . . he took it!" The boy said it as if a great injustice had been done.

Pearl explained, "The man inside said it had lice. . . from the ship . . . it had to be burned. It was the only shirt I could bring."

Ben laughed and slapped his son across the shoulders. "My son, he comes to America without a shirt on his back!"

My grandfather loved to tell that story. "I arrived in America without a shirt on my back!" And he'd laugh in his soft, low way, his happy eyes alive. As a seven-year-old I would smile back at him with compassion, as if I understood the story, but at the time I didn't know why he thought it was so funny.

One of the first things Ben did was to Americanize his son's name, changing it to Harry. He also changed Anna to Annie. Ben decided that, except for Jewish holidays, when Yiddish was to be spoken, the family would learn to speak this difficult new language, English. One of Ben's goals was to become a naturalized citizen. The next decision he made was to add to his family. Albert, nicknamed Abe, was born in late 1884; between 1886 and 1891 the next four of the twelve children Pearl bore arrived: Henry, Sam, Rose, and Fannie. Henry, a frail child at birth, died in 1890 when he was only four.

Not long after Henry's death, my great-grandfather Ben had a family portrait taken. It was as if he had decided that, from that moment on, he would preserve his family on film. In the picture, Ben, in a black suit with his watch chain looped through his vest, is seated in the middle of his family, while Pearl, tightly corseted and wearing a flowered hat, rests her forearm on his shoulder. Annie, the oldest child at thirteen, stands next to her mother. She wears a striped dress, her hips as wide as Pearl's. Rose, age one, holds on to her father's knee for support as he cradles the newborn, Fannie, on his lap.

The three boys, Harry, Abe, and Sam, are grouped together at the right of their father, each dressed in dark, knee-length suits. Everyone except Harry and Sam stares blankly at the camera. Sam appears to be contemplating his baby sister—or a faraway dream—while Harry stares intently at his father as if cementing the link between them.

What were their actions after this stiff, formal portrait was taken? Did they laugh? Go someplace special for a treat? Had they come from the synagogue in their Sabbath finery? Were the boys' outfits rented for the occasion? More than likely, after the camera preserved their image, they did what all immigrant families did: they went back to work.

Although work was as natural to Ben Warner as his religious devotion, he began to weary of cutting leather and sewing battered shoes. His cobbler shop provided a modest, steady income, but hardly the kind of success he had dreamed about. "This is a land of opportunity. I must find new work," he told Pearl.

He had heard from a Polish friend that a new railroad was being built in Bluefield, Virginia. That gave him an idea: Railroad workers got paid, and with money in their pockets they needed supplies, pots, pans, food items. Ben bought a horse, which he named Bob, and a wagon, which he packed full of goods. Leaving Harry, only ten, in charge of the cobbler shop, Ben headed for the railroad crews. The peddler business was better than he expected, so much so that he sent for Pearl's two brothers, Hyman and Berrill, both newly arrived from Krasnashiltz. The three of them did well until Berrill ran off with the week's take.

During his travels, Ben met a man who told him there was money to be made trading supplies to Canadian trappers for furs. The man said he would act as his partner, open a store in Montreal and exchange the furs for cash. "We can both be rich," the man enthused.

Ben returned to Baltimore and told Pearl, "We're moving to Canada." The next day, the Warner family, with all their belongings stuffed into the wagon, headed north. For the next two years they lived like nomads, traveling the rutted roads from trading post to trading post, bartering supplies for pelts. Ben would occasionally unload furs and ship them to Montreal, confident that in that far-off city his fortune was growing.

With a wagon full of the last few pelts he had collected, Ben pointed his horse, Bob, and the wagon in the direction of Montreal. When he arrived at the warehouse where he had been told the furs were stored, he discovered his partner was gone. Slashing open one of the boxes, Ben was sickened to find them stuffed with paper strips, not furs.

As Ben headed back to Baltimore, the leather reins grasped tightly in his strong hands, his heart felt a great weight. Not only had he failed his family, but on the trip he had lost his beloved daughter Fannie—the one he had held so tenderly in his lap in the family photograph. She had not been able to withstand the harsh Canadian winters. To Ben, children were security against the future; to Pearl, they were a great joy.

No one child had brought her greater happiness than her new son, Jacob, who had been born in 1892 in London, Ontario, while they were still gathering furs.

This youngest son grew to hate the name Jacob and in his early teens changed it to Jack. He added a middle name, Leonard, because it sounded "classy." Jack admitted he got the name from an actor in a traveling minstrel show that had played in Baltimore. In his mind "Jack L. Warner" had a theatrical ring to it.

Between 1893 and 1896, in order to provide enough money for the family to survive, Harry, Sam, and Abe found jobs while their sisters worked at home. One of them always managed to supply the money while the others were trying to find their way. The word "mine" was never heard in the Warner home. Everything from a child's toy to a winter coat was referred to as "ours."

Ben's "all for one, one for all" philosophy was driven home the day a gang of tough youths tried to bully the Warner brothers off a street corner where they had set up a shoe-polishing business. To get customers to stop at their stand, the Warners had hand-lettered a sign to read 5 CENTS A SHINE, which was two cents less than their young competitors were charging. A brawl ensued. The brothers, fighting side by side, were able to chase the toughs off.

It took many fights for the Warner boys to hold down their shoeshine stand. Late at night they would cut the price of a shine from a nickel to three cents in order to coax enough customers to their stand to make the two dollars that Harry had set as the goal to bring home each night. To add to this fund they also sold papers, but at a penny each the profits were small.

On a lucky day, when their take exceeded two dollars, they treated themselves to donuts and coffee at Harvey's restaurant on East Baltimore Street.

One of the whistlestops in William McKinley's presidential campaign of 1896 was Youngstown, Ohio. Harry, who was now fifteen, wanted to see McKinley, a future president, but he had another reason for traveling: to see whether Youngstown offered better work opportunities than Baltimore.

Harry took a few dollars from his shoe repair earnings and bought a train ticket to Youngstown. Upon arrival, he was surprised to see, not McKinley, but a town swarming with Poles. A new steel mill had opened and there were thousands of Polish workers wandering the streets.

With that much shoe leather being worn out, Harry decided to open a repair store. He had his father send him a shoe jack and tools, then rented a small vacant store and set himself up for business. Because he couldn't afford to put out a sign, Harry stood in the store's window, hammer in hand, mouth filled with nails, repairing shoes. He was an instant sidewalk attraction and became so overwhelmed with business that he wrote his father, urging him to come.

Ben packed his family (which had grown larger by three: David, 1893; Sadie, born on the first day of the year 1895; and Milton, the last child, in 1896) in the wagon and headed to Youngstown.

With three Warners now standing in the window (Abe, now twelve, had also taken up the trade), the crowds grew. Harry began to wonder how he could cash in on this gaping audience, and that's when he suggested to his father that they make the back of the shop into a grocery store and a butcher shop. Ben, feeling the enthusiasm of his oldest son, happily agreed.

With nine children to care for, housing was a problem. They lived in four rooms above the grocery store, the boys crowded into one room, the girls into another. The kitchen, with its large table, doubled as a family meeting room.

Jack later remembered the crowded living conditions:

> We were jammed in there like bait worms in a can . . . Worst of all, though, was the leather. Because there was no room anywhere else, my father had great slabs of leather piled up in one corner of the room, and these hides, most of them fresh from the tannery, could have walked away by themselves. I have a very unpleasant memory of sleeping on a blanket spread out on top of the stacked leather . . .

Ben and Pearl finally settled their brood into a large but run-down house on West Federal Street in the Polish section of the city. It had an outhouse and a water pump in the back yard, only a few steps from the porch, where Pearl filled a galvanized tub and bathed the children each Saturday night.

Jack recalled:

> Mom hung a curtain around the beat-up sitz bath and the girls soaped up first. When they were covered with foam up to their ears, Mom would put them into the tub, one at a time, and douse them with a huge pitcher of cold water. This system was fine in summer, but when there was snow on the ground . . . you could hear the Warner kids gasping and yelling two blocks away.

With the grocery store, butcher shop, and shoe repair shop in full swing, the Warners' finances began to improve. In 1899, Ben took out an ad in the Youngstown *Vindicator.* It read:

> BENJAMIN WARNER, Propr THE BAL SHOE REPAIRING CO, a full line of new shoes from $1.00 up, hand made shoes, $2.25, men's half soles & heels 50 cts, ladies, 40 cts, while you wait.

By the turn of the century, the grocery store was thriving. It was the family custom for the boys to turn over to Pearl all of their weekly earnings—except fifty cents. The rest of the boys spent their half dollar in record time, usually in one day, but Harry stretched his. He had developed an intense interest in girls, and loved to invite them to the house. To save money, he made sure the girls didn't live far from his home so he could walk them both ways. Practicing with his sister Rose, Harry became a smooth dancer and together they won several waltz contests.

In late 1899, Harry and Abe went to a bicycle shop one day, intending to buy two bicycles. They cost $30 each, a tidy sum for the day, and double what the brothers had. Harry offered to pay cash for one and pay off the other in thirty days, but the owner refused to give credit. Harry, his mouth set in a firm line, told Abe he had just decided to get into the bicycle business. He opened a shop that sold bicycles "on time," then profited by the repair work.

The bicycle craze was sweeping Youngstown and the business flourished. Abe worked in the shop, renting bicycles for fifteen cents an hour. He also got interested in bicycle racing and entered local events.

One day Jack brought home a stack of sepia-toned photos mounted on thin cardboard stock which he had got from the penny photo studio he liked to hang around. Each photo showed a horse running at full stride on a racetrack. Eadweard Muybridge, a British photographer, had taken the series of photographs in 1872 by setting up a battery of cameras along the racetrack and tripping the shutters in sequence as the horse raced by. Muybridge's experiment proved that the hooves of a running horse left the ground all at one time. Jack knew nothing of the Englishman's test, but he was mesmerized by the photographs. Grasping the stack at the bottom, he flipped the pictures at the top, and—wow!—the horse seemed to be running. The effect was crude and jerky, but it created the illusion of movement. He showed the stack to Sam, who flipped through them.

"So?" Sam said, handing the photos back to his younger brother.

"Hey, look! They move!" Jack thumbed the stack under Sam's nose, disappointed that his brother wasn't crazy over his new find.

Seeing his brother's crestfallen look, Sam said, "Edison, the inventor, has already made an electric machine that makes pictures move."

"You saw one?" Jack asked, incredulous.

"Read about it in the newspaper. The machine's been around for five or six years now." Seeing Jack's blank stare, Sam added, "Don't you ever read anything?"

"Naw. Got no time to read."

Sam explained: "You peek at the pictures through a machine, something called a Kinetoscope. In New York they've even shown them in theaters on a big screen."

"Suppose we'll ever see one in Youngstown?" Jack said.

"Sure. Someday."

"Whoever does it first is going to make a pile of nickels."

Sam grabbed Jack around his waist in a bear hug, lifting him off the floor. "That, my dumb little brother, is the smartest thing you've ever said."

For a long time after that, Sam thought about the running horse, an image that had been frozen on a photographic plate, yet appeared to move—an image that had danced like shadows before his eyes.

And now, in 1906, Sam and his brother had the opportunity to buy the Kinetoscope projector that would make those shadows blend into moving images on the screen.

Unfortunately, even with their father's offer of the family watch to secure the balance needed to buy the projector, they would still be more than a hundred dollars short.

Ben told them that the balance would come from ol' Bob.

"Sell our *horse*?" Jack asked.

The boys looked at him, stunned. Sell the horse that had taken them to Canada and back? The gentle horse that for years had pulled their meat delivery wagon through the streets of Youngstown? They couldn't sell Bob.

Harry looked at his father intently. "You mean hock him."

The pawnbroker offered $25 for the watch, but was reluctant to accept ol' Bob. After an hour of haggling, he added another $75. "I'm only going to feed the animal for two weeks," the pawnbroker warned as the brothers walked away with the money.

They were still short $100.

Sam had no choice but to go back to the landlady and offer what they had. He rushed inside the boardinghouse and spread the money out on Joey's mother's dining room table. "Nine hundred and fifty, that's all I have," he blurted.

She stood there, hands crossed over her ample bosom, and looked down at the crumpled bills fanned out on the table.

Sam waited. He could hear a grandfather clock ticking in its large case in the hallway. Joey's mother picked up the bills and slowly counted them. The clock ticked away the seconds, each one louder in Sam's ears.

Finally, "Ach, the machine is no good to Joey. The money is enough." As she handed the projector box and canvas bag of film and tickets to Sam, she looked at him like he was crazy.

Walking outside, hugging the Kinetoscope case to his chest, the bag tucked under one arm, Sam Warner felt like Aladdin. He held the genie that made pictures move.

3

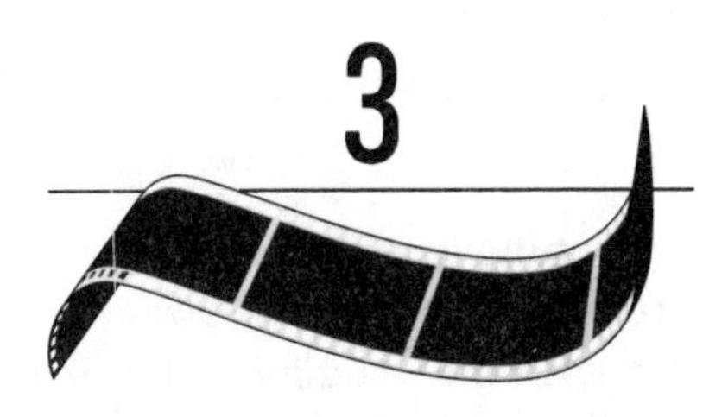

The Cascade Theatre

Excitement!

The supper dishes were done and Pearl put out a special treat, dishes of latkes, potato pancakes topped with large spoonfuls of applesauce and sour cream. Sam had set the Kinetoscope on the kitchen table and threaded the film. Now, with the family waiting around the table with anticipation, he pulled the string to the ceiling light and clicked it off. The projector began to whir . . .

For an instant the sheet on the wall was illuminated by white light as the first clear frame of film clicked through the projector. A black web of scratches wiggled across the screen.

"Oh, Mama, look . . . bugs," Sadie said, pointing a finger at the sheet.

"Shush," Pearl said, her eyes as wide as her child's.

Sam continued cranking the wooden handle. A jumble of blurs appeared on the screen and he twisted the long brass lens tube with one hand, first forward, then backward. The image cleared a little, enabling the gathered family members to make out the fuzzy letters of the title:

THE GREAT TRAIN ROBBERY

Sam slipped the lens tube backward again. The title blurred more. Forward. It was no better.

"What's the matter with the machine?" Ben Warner asked.

"A little focus problem with the lens, Pop," Sam answered, feeling the heat from the hissing carbide lamp. He tried the lens again, then watched the screen in dismay as the image faded to a dull gray. Then to black. The hissing stopped. Sam let go of the wooden handle.

The room was quiet.

"Rose, turn on the light," Sam said.

The kitchen ceiling light went on. Everyone seated around the long table turned in their chairs and stared at Sam.

"It's, uh, there's a problem with the focusing barrel, and, uh . . . well, the lamp . . ." Sam grinned lamely as he berated himself for not testing the machine before trying to show the movie to the family.

Sam's disarming grin didn't placate Harry. "You can fix it," the older brother said. It wasn't a question.

"Just needs some adjustment. Maybe ten minutes."

"Okay!" Jack jumped from his chair. He bowed deeply. "Before the film you will all now have the wonderful experience of hearing and seeing the star of the Dome Picture Palace, Leon Zuardo!"—he strutted a few steps, thumbs jabbing at his chest—"who will sing his latest ballads." Jack gestured toward the piano. "Rose, if you please . . ."

Rose sat at the piano in the corner of the room while Jack warbled, "Sw-e-e-e-t Adeli-yan—my Adeli-yan," in his prepubescent soprano. The three younger children, David, Sadie, and Milton, clapped their hands in delight.

Sam loosened the screws on the right side of the Kinetoscope and looked at the gleaming parts inside. Harry and Abe sat across the table with their father, watching grimly as Sam lifted the carbide lamp assembly from the projector, almost burning his fingers, and quickly set it on a dinner plate.

Abe looked at the smoking contraption. "Maybe this little Joey who took our money . . . he's a thief."

"His mother said the machine never worked," Harry added, worried about the family's investment.

"No, it's all right," Sam said, peering into the opening at the metal sprocket claws. "Joey, he's no mechanic . . ."

Ben Warner absently fingered his watch pocket, feeling the empty space. He sighed. "So much money . . . that I can see. The picture? I wish I could see."

"Pop, it's just the focus . . . ah, see, this film-feeding mechanism is loose, out of line . . ." Sam made adjustments with a dinner knife, then, satisfied that it was working properly, turned his attention to the carbide lamp. He filled it with white powder from a tin box, poured water into the lamp's tank, then pumped up the pressure unit. Something was wrong; it wouldn't hold the pressure. He felt a tiny stream of air on his hand. Yes, that was it, the tank was leaking around the base seal. Sam clamped it tightly together, crimping the edges with the knife, then hung the lamp assembly on its brackets in the machine. Releasing the valve, he struck a match to light the gas coming out of the lamp. It worked! As the blue-white light illuminated the inside of the projector, he threaded the leader of the film through the slots and gates, then announced, "It's ready."

Moments later, the Warner family ducked and screamed as the puffing locomotive rushed toward them, then clasped their hands to their ears as gunshots from the great train robbery felled the train's conductor. Yet the only sound in the kitchen was the click-clack of the projector and the hiss of the carbide lamp (which did sound like steam from the locomotive).

In the flickering light, Harry turned his face from the screen and watched the excited family audience as they viewed the magic shadows dancing on the sheet. *If this is what it does to my own brothers and sisters, and my parents, then think what it can do to others.* He imagined himself selling tickets to long lines of people waiting to see this new wonder on the wall come to life.

Yes, this was the business he and his brothers had thought about for almost a year, the one they could work at together. What they lacked in school education would be balanced by their individual talents, talents that had been honed—as Harry was to say many times later—by the school of hard knocks. Abe had developed into an excellent salesman; Sam, though a dreamer, had natural mechanical ability; and Harry had become adept at financing businesses. Each of their strengths complemented those of the others. And Jack? He was too young to be any real help. But, with his energy, he'd make a good errand boy.

To test their new enterprise, the brothers set up a makeshift tent in their back yard and invited neighbors and customers of the grocery store to sit on the lawn and watch the Western melodrama. The enthusiasm of the audiences confirmed that the enterprise could be a success. All they needed was a permanent place to show the film.

Sam and Abe heard that a carnival was opening in nearby Niles, Ohio, and proposed that they pack up the Kinetoscope and the print of *The Great Train Robbery* and find a place to show it near the carnival grounds. Harry agreed, saying he'd stay in Youngstown, working at his bicycle shop and the family grocery to ensure some money coming in.

After looking around Niles for several hours, Sam and Abe found a vacant store which they converted into a theater by hanging a sheet on one wall and scrounging up whatever benches and chairs they could find. Most of the customers would have to stand. The two brothers tacked up handmade posters all over town, and when the carnival opened they were ready for business.

Sam ran the projector and Abe sold tickets. Occasionally the film would break at a moment of great excitement—something the old print did with increasing frequency—and Sam would flash a slide on the screen that said TWO-MINUTE INTERMISSION. While the audience shuffled and whistled, Sam spliced the film's brittle edges together. At the end of the show, Abe would call out in his bass voice, "All who have not seen the entire performance will kindly remain for the next—this way out, please."

When the carnival reached the end of its week's run, the two brothers each had a bulging pocketful of coins.

Abe and Sam packed up the Kinetoscope, climbed aboard a train, and took the show on a circuit of the small towns near Youngstown, such as Steubenville and Warren, and ventured across the nearby Pennsylvania line into Sharon, Meadville, and a growing steel town named New Castle.

Abe recalled: "We started writing managers of opera houses in Akron and other neighboring towns offering to provide entertainment in the intervals between the appearances of Shakespearean companies and road shows. Although movies weren't particularly popular with managers at the time, the plan was accepted."

In each town it didn't take long to use up the available customers. When opera-house managers closed them out, they'd rent a small hall or empty storefront. Once the receipts slowed, the Warner boys moved on. Their traveling show paid off until winter came to Ohio and Pennsylvania, and with it a violent blizzard that swept in from the icy lakes to the north. Customers stayed away from the Warners' unheated storefront theaters.

"At the last town, we found that we had money enough to buy us a cup of coffee—at three cents a cup—and streetcar fare home," Abe recalled.

To Harry's way of thinking, failure was an incentive to take the next step forward. Although Abe and Sam would state their opinion and offer options, Harry, at twenty-three, made all the decisions—and his brothers never questioned them. He firmly believed that to make a vision into a reality they had to take chances.

"We will lease our own theater," Harry announced to his brothers when they arrived back in Youngstown. He could see they were tired and discouraged, but he knew that there was a future in this movie business. They had to try again.

"Not here in Youngstown," Abe said. "Too many nickelodeons already."

Sam turned from warming his hands over the wood stove. "We did well in New Castle, before the blizzard chased us out. It has a growing industrial district. More steelworkers coming in all the time."

"I checked the town out," Abe said. "The only other entertainment is the Genkinger Opera House, which shows some second-rate vaudeville acts in the evening. On Sundays the workers can take their families to Cascade Park for a picnic in hopes of hearing a concert."

"Or a singer, like me!" said Jack, who had followed his brothers into the kitchen and had been listening with rapt attention to the idea of having their own theater. "I can knock 'em dead with 'Sweet Adeline.'"

Harry frowned.

Abe, sticking to practicalities, didn't pay any attention to Jack. "New Castle's only fifteen miles away. Sam and I can take the trolley from here."

"Hey, how about me?" Jack stuck his head between his two big brothers. "I can go, huh?"

Sam playfully rubbed the top of Jack's head. "Singers we don't need, little man. But if you want to rewind the film, sweep out the theater . . ."

Jack grimaced. "Some show business."

Abe paced the room, scuffing his big feet across the wooden floor. "Yeah, but even if we find a place, where are we going to get the money to rent it, fix it up?"

"We need to get some new films," Sam said. "I can't keep splicing this old one together. The locomotive's beginning to look like it hiccups along the tracks."

Harry stood up. "We'll sell the bicycle shop."

The brothers found a landlord who allowed them to convert rooms of a former inn in New Castle to a small hall. Harry, with part of the money he got for the bicycle shop, rented it. They chose the hall because the outside looked something like the way they thought a theater should look. A gingerbread arch, supported by pillars on each side, curved across the recessed entryway, and above that were a line of electric light bulbs. The boys hung poser boards fromt he pillars that would announce the daily attraction, and set two potted palms at each side. In the middle of the entrance was a phone-booth-size box office with ADMISSION 5 CENTS printed on the glass. Below that was a sign that read:

REFINED ENTERTAINMENT FOR
LADIES, GENTLEMEN AND CHILDREN
CONTINUANCE PERFORMANCE

None of them were schooled enough to notice their spelling error.

The Warner brothers named their new enterprise the Cascade Theatre.

A platform was built for the Kinetoscope and the opposite wall was painted white. It still looked as bleak as a cold storage warehouse. At Harry's insistence, a couple of modest oil lights were installed on the side walls. He had learned that most women wouldn't visit amusement houses because they were not lighted. The lady patrons didn't want to accidentally sit in a stranger's lap while finding a seat in the dark.

The theater had one major drawback. It didn't have any chairs. The brothers had spent most of the money from the bicycle shop on the building rental and sprucing up the place. What was left was needed to rent new films. They couldn't afford seats. But that didn't stop them.

A few doors away was a funeral parlor, and the brothers made a deal with the undertaker to borrow ninety-nine chairs. They had to hold the seating to less than a hundred because a local ordinance stated that anything over that number required safety features, such as special exits, fire extinguishers, and a projector set up on a balcony rather than the ground floor. There was no balcony in the Cascade.

"What if this undertaker has a funeral the same time we're running our picture?" Abe asked.

"Don't worry," Sam said. "Nobody's going to die on us."

"Yeah, I hope you're right," Abe replied, worried.

The Cascade Theatre opened for business February 2, 1907. For the first showing, Ben Warner invited the Youngstown gossips and many of his friends in free to see the novelty. The front row was specially reserved for the Warner family. When the chairs were full, Sam closed the black curtain around the Kinetoscope. Before starting the movie, he clicked on a slide projector and a black and white sign flashed on the wall:

PLEASE READ THE TITLES TO YOURSELF
LOUD READING ANNOYS YOUR NEIGHBORS

There were a few grunts of approval from those who had been in a nickelodeon before as Sam switched to the next slide:

LADIES!
KINDLY REMOVE YOUR HATS

Several ladies plucked the long pins out of their wide-brimmed hats as the last slide flashed on the screen:

GENTLEMEN!
PLEASE DON'T SPIT ON THE FLOOR

Sam started cranking the Kinetoscope as Abe fed the film into the projector. The screen lit up with a train rushing down the tracks: The people on the benches gasped and several got to their feet in wordless panic. The scene flickered to a blond girl tied to the tracks, squirming to get free.

A man in the audience stood up: "Save her! Someone save her!"

"Shush!" several patrons hissed.

The hero arrived and carried the heroine away as the locomotive sped by. The show ended with a flash of white as the tail of the film passed through the gate of the projector. Sweating in his enclosed cubicle, Sam clicked on the slide projector and a title flashed on the wall:

ONE MOMENT PLEASE
WHILE THE OPERATOR CHANGES REELS

Abe fished the coiled film from the barrel and started the laborious process of hand-winding it onto a reel as Sam turned off the hissing lamp, added more water, pumped up the pressure, then threaded the next film. Sam's big fear was fire. He and Abe were knee-deep in celluloid film which even the tiniest spark from a cigar could turn into an inferno. Sam turned the handle and once again the beam of light stabbed out from the hole in the curtain. And once again came the shared gasps from the audience as images flickered on the screen.

After the second showing, Ben Warner escorted his friends proudly to the door. When he returned, he saw the empty coin plate at the box-office counter and the blank faces of his children looking at him. A smile crossed his face. "There's tomorrow. You'll see."

The next day, Harry, Sam, and Abe walked slowly down the streets of New Castle en route to the theater, talking among themselves, when Rose rounded the corner and almost crashed into them. "Come quick! Come see what's happening!"

In the stark light of the mid-afternoon sun, a crowd of jabbering men, women, and children were standing four abreast on the sidewalk in front of the theater, waiting to get in. As the Warners had hoped, the steelworkers of New Castle were flocking to the Cascade.

After collecting the nickels, Ben and Pearl, who had closed the grocery store for the event, stood proudly at the back of the darkened theater watching the picture as Rose played accompaniment on the piano.

The show ended—but no one got up to leave.

Harry turned the house lamps up, bathing the room in an amber glow. Still, no one left.

Panicked, Harry, Abe, and Sam ran up to their parents.

"What'll we do now?"

"They're not leaving."

"We have a line of people outside waiting to get in!"

Pearl leaned over and whispered something in her husband's ear. Ben smiled: "Don't you think it's time to let Jack sing?"

Sam laughed. "If that doesn't clear the house, nothing will."

"All right, we'll do it." Harry turned to Abe. "Tell Rose to stand by at the piano."

The ruse worked. Jack's voice, skipping octaves from tenor to baritone, sounded like ice cracking from a glacial floe and drove the customers from the place.

"You were great, kid," Sam said, coming out of the projection booth.

Jack was dismayed. "They all left."

"Yep, we couldn't have got them out of the theater without you."

At that Jack brightened. "Hey, you're right. I guess I've got a job!"

Harry called all the brothers and sisters to the entrance to see the mound of nickels they had collected. Scooping up a handful, he proudly handed each of them a shiny buffalo nickel as a memento. "Something to remember our first theater," he said.

The next day, Harry got ol' Bob and his father's watch out of hock.

Memorial Day, 1907. The boys were prepared for their biggest crowd. There would be hordes of people three abreast, nickels clutched in sweaty palms, waiting to get in. Then, what they had feared would happen happened.

The old-fashioned wall telephone in the Warners' kitchen rang. Sam lifted the earphone off the cradle and bent his six-foot frame to the mouthpiece. "Hello." He jerked back as a volcano of words poured into his ear. He listened for a moment, then said, "Wait . . . Abe, don't put the sign out yet."

"Sam," his exasperated brother was saying, "the damn undertaker isn't going to give us the chairs. He says he's got a big memorial service. Mr. Harmond, the big-deal school superintendent, died. We're going to have to put up the Standing Room Only sign."

"We'll lose too many customers if they have to stand." Sam thought about the money they would lose on the special—and expensive—1,000-foot film they had rented. "I'll think of something, some way to fix it."

"Chairs just don't fall out of the air."

"The chairs will be there!"

Abe felt relieved. This was too much for him to handle. Sam was the idea boy; maybe he could think them out of this mess. He held the earpiece to his ear and waited.

"Abe," Sam finally said, "you just go tell Mrs. Harmond that she and her kids can come to every show free until Christmas, *if* she'll put off her husband's service until tomorrow."

"I sure hope they got him stored in the ice house." Abe heard Sam's laughter and added, "But she's got eight kids and they'll come twice a week."

"You want the show or not?"

"Okay, Sam, okay. I like your idea, but you sure she'll do it?"

"I'll bet you a new buffalo nickel the grieving widow puts the funeral off and we get our seats."

Abe did get the seats and the show did go on—and kept going on.

Although serious-minded about making the Cascade a success, Harry had not lost his love for dancing, and still tried out all the new steps. It was at a New Castle dance that he met a black-haired, brown-eyed Jewish girl named Rea Ellen Levinson. Rea was wearing a high-collared dress so tight that Harry could see the stays beneath. Proud of her tiny waist, Rea had cinched it to the smallest possible measurement, which in turn emphasized her bust. She looked smart and fresh. Harry was intrigued by her figure, her natural elegance, her smile, and her quick, attentive mind.

Rea had arrived with her parents in New Castle from Manchester, England, a few years earlier. The family was educated, intellectual and cultured. Rea's mother, whose maiden name was Kreiger, was originally from Germany; as German Jews, Rea and her family considered themselves part of an elite group and looked down on Jews from Russia. Yet there was something about Harry's face, in which were written lines of strength, defiance, good nature, and humor, that attracted her. His features gave him a striking solidity. She also had to admit she liked his deep blue eyes and dark blond hair.

Not long after they met, Harry (then twenty-six) proposed. They were married August 23, 1907.

Harry was the first of the boys to get married (Annie had wed a Youngstown man, Dave Robbins, five years earlier) and to have a home where his brothers could get together. Sam, Abe, and Jack would hand Rea the money they made and she would put it into kitchen jars to save it rather than depositing it in a bank. She knew that if she got the money quickly, she'd keep the brothers from gambling it away, spending it on girls or fly-by-night schemes. They learned to depend on her steady, no-nonsense approach to their behavior. In the years to follow, she would repeatedly remind Harry, "If it wasn't for me, you wouldn't be where you are now."

Abe met a Jewish girl named Bessie Krieger in New Castle and married her shortly after Harry wed Rea.

Sam, with his strong chin, broad shoulders, full head of rust-colored hair, and snappy wardrobe, was a great hit with the girls. He savored the laughs and good times he was having and liked to brag that he wouldn't get married until he was forty.

Jack was still treated like a "little brother" by Harry, and that was a constant sore spot, one that would continue to fester throughout the years. Yet he worked diligently at the Cascade, doing whatever chore he was assigned, from sweeping out the theater to mending broken film. As long as Harry let him sing and joke with the audience at the end of each movie, he was content.

Getting audiences to attend the Cascade was no problem. Acquiring good films was. Harry carefully studied the catalogues of various film companies, looking for features to attract audiences. D. W. Griffith movies like The Redman and the Child *and* Rescued from the Eagle's Nest *were the best. He would order a short reel from one of the eight companies that he knew produced films and get "a picture of the Flatiron Building on a windy day with skirts blowing. We would send it back and the next day the same picture would be returned to us. And the cost of renting films from producers, even if you could get a good one, had become excessive."*

There were now thousands of theaters across America, popping up in every abandoned store, hall, and barn. This proliferation of theaters demanded a great deal of film. In small towns, a new show might be needed every day to attract customers who went to the "moving pictures" as religiously as they attended prayer meetings. Because the demand was so great, film rentals escalated to $100 a week.

Harry, along with several of the other theater operators around New Castle, decided they could take the idea one step further by exchanging movies. Instead of paying $100 for a four-reeler, they could trade films, reel for reel, and the price would average out to $15 or $20. This exchange concept necessitated introducing a distributor into the system. The distributor—who was soon referred to as the Exchange—would buy a film from the movie producer and rent it to individual movie theaters.

Abe remembered the Warners' first introduction to a distributor: "One day a man came around and told us he was getting twenty representative theater owners to pay him a hundred dollars each for ten weeks' supply of films—twenty reels. Well, we gave him the hundred dollars and got our films. It was a good plan and it worked. Everyone was satisfied, you see, and at the end of ten weeks the man still had his twenty reels of film that he could take to another territory and begin all over again."

The reality of the situation had quickly become apparent to financially minded Harry Warner: The big money was not in showing films, but renting them out. Why take in money from two theaters when a smart operator could dip into the profits of thirty, perhaps a hundred theaters?

Harry recalled, "I gave a young fellow a two-thousand-mile railroad pass and a hundred dollars, with the instructions to go out and count the theaters in Pennsylvania to whom we might distribute film.

"He found fifteen."

Harry sent Sam and Abe to New York to ferret out some movies while he kept operating the Cascade. In New York the two brothers met with magnate Marcus Loew, who owned a string of picture theaters. Loew never liked renting films, figuring he would save money by purchasing them outright. Once the movies had made the circuit of his theaters, Loew had them packed in a trunk and stored under a stage. When Marcus Loew told Sam and Abe that he had "trunks full of two-, three-, and four-reelers," the boys felt like they had discovered a mother lode. Paying Loew $500, Sam and Abe carted away three trunks of film. As they left the office, they happily shook hands and said, "Looks like we're distributors now."

That was the start of the Duquesne Amusement Supply Company.

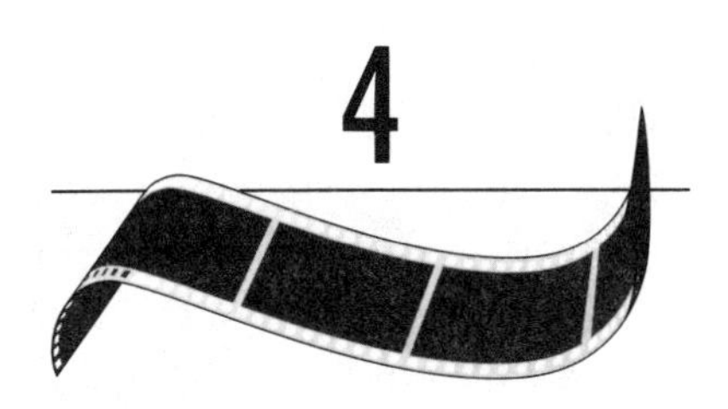

4

The Duquesne Amusement Supply Company

Spring 1907

Harry and his brothers named the Duquesne Amusement Supply Company after a college located in a suburb of Pittsburgh. They felt the name added a touch of class to their new enterprise.

Early in 1907, Harry sent Sam and Abe to Pittsburgh to manage the exchange while he kept control of their New Castle theater for security. Jack was told to remain at home in Youngstown.

Pittsburgh was not many miles away, but fifteen-year-old Jack Warner was devastated when Sam, his favorite brother, left. "I worshiped Sam," Jack recalled later. "He was a dreamer who could grab a handful of clouds and weave them into a magic rug. Sam never talked big, but his mind was like a periscope. He could look around corners and see things coming."

Sam had even proposed "taking old Mr. Edison" and opening theaters in South Africa, Madrid, and Lisbon. Wandering the globe with his brother sounded wonderful to Jack.

Harry had other ideas. Pop Warner needed a son to help take care of the family store in Youngstown. So instead of seeing the world, Jack ended up dipping chickens in boiling water and plucking the soggy feathers. To get the terrible stench off his skin, he said, "I took endless baths, and I stole cologne from my sisters," but he still smelled like chicken feathers. Young girls,who were beginning to quicken his pulse, shunned him. The frustrated young man had to get out.

Jack's break cam in 1908 when the three older brothers decided the film exchange was going so well—the income had reached an astounding$2,500 a week—that they could sell the Cascade in New Castle. A group of buyers got together and offered a good price: $40,000,with payments over five years. The offer was accepted, and Harry, his mind conjuring ways to open up a second film exchange, decided to let Jack in on the act. Much to Jack's delight, Harry sent him along as Sam's assistant ot open an exchange in Norfold, Virginia.

The four Warner brothers, Harry and Abe in Pittsburgh, and Sam and Jack in Norfolk, were now joined in what would become a lifelong partnership.

"No form of entertainment has more quickly taken its place in all the cities and towns of the country than that of the moving picture show, and with them comes naturally a general demand that they be kept clean and free from atrocities and evil suggestions." Finished reading, Sam tossed the newly printed issue of *The Duquesne Film Noise* onto his desk. "Do you think we shoulda put in that 'atrocities and evil suggestions' stuff? Sounds kinda high-falutin'."

Jack picked up the four-page bulletin and reread the copy. "It's written the way Harry wants it."

"Don't get on Harry," Sam said, cautioning his younger brother. "He does what's best for the business—and the family. He knows what he's doing—we distribute clean, wholesome entertainment."

Jack rolled his eyes but didn't say anything. He looked at his picture on the front page of their film exchange bulletin—something he had done a dozen times since it had come off the press earlier that morning. The bust-size photograph showed the eighteen-year-old looking newly washed around the ears, the hint of a smile on his mouth. He wore a white celluloid collar high on his neck, and a watch fob dangling from the breast pocket of his dark suit. (The two brothers had only one watch between them and switched it for the photographs.) Under Jack's photo was his name, J. J. Warner, and title, "Asst. Editor and Manager Film and Supply Dept., Norfolk Office Duquesne Amusement Supply Co."

Jack looked at the caption again. "J. J. Warner! How in the hell did we make a mistake like that? I'm J. L. Warner."

Sam grabbed the paper and satisfied himself that he was listed properly as "S. L. Warner—Editor and Manager."

Even with the printing error, Jack was pleased with the portrait, figuring he looked like a big-deal entrepreneur. He read the lead item below his photo to himself: "Just received: Pathé's *Passion Play.* Three thousand and two hundred feet in length and highly hand-colored." To Sam he said, "Who colored all those frames? Can you imagine some guy with a little brush painting maybe half a million little pictures?"

"Maybe Pathé hires a bunch of elves," Sam said.

Whether the film was colored by elves or artisans, the brothers were grateful they had anything at all to distribute. When they opened their film exchange in Pittsburgh, there were less than a dozen companies making movies in America, lusty human dramas with such titles as *The Artful Husband in Distress* and *The Magic Skin.* Because the producers were unwilling to risk a great deal of money, most of the films were of poor quality, churned out to meet the increasing demand. The most accomplished films were being made in Italy and France. *The Red Spectre* from Pathé Paris introduced some of the first special effects. It was filmed on a huge set in which a fiery hell was created, complete with backgrounds that moved and shook in the tumultuous flaming cavern. Pathé's Passion Play and its hand-colored frames was another example of emerging European epics.

Once, when one of the films received from a French producer turned out to be worthless because it was a negative print, which projected black as white and white as black, Jack rented it as "The first picture made by colored actors." He pocketed an extra $25 a week for three weeks for the rental fee.

The two brothers continued to spread the word about their exchange's films with their news bulletin, *The Duquesne Film Noise.* The young men weren't the best proofreaders. "Duquesne Film Exchange" was sometimes printed as the "Dequense File Exchange."

Yet the hype (typos included) sold films:

IT IS OUR POLICY
TO PURCHASE FROM THREE TO FIVE CEPIES OF EVERY
GOOD SUBJECT MANUFACTURED
MAKE SHIPMENTS IN AMPLE TIME TO AVOID DELAYS
AVOID REPEATERS
SHIP NO JUNK
GIVE WHAT WE PROMISE

Sam and Jack would laugh when a client pointed out the mistakes. After all, what did a few misspellings matter? They were making money and having a ball doing it.

The two brothers lived in a lively boardinghouse on Granby Street. Jack later recalled: "As in many guest houses, nearly all the boarders were young bachelors avoiding the trap of matrimony, or unattached ladies waiting for some Lancelot to take them away from the gas plate and the washboard."

Harry and Rea, meanwhile, had bought a nice home in Wilkinsburg on the edge of Pittsburgh, not far from the exchange at the Bakewell Building, and on October 10, 1908, a son, Lewis, was born. Harry, looking in awe at the baby boy, realized that he now had what he wanted most: an heir.

The Warner brothers were benefiting handsomely from the proliferation of movie houses calling for their films. They had established a solid business to support their families and their new lifestyle. But even more, they had learned a lesson that would stay with them the rest of their lives: In a competitive field they had been able to survive—and prosper. They were, in Jack's words, "gliding down easy street."

Then came the Edison Trust War.

In 1908, Thomas Edison, concerned that too many film exhibitors were making millions from his invention of the Kinetoscope without paying royalties, decided to organize film producers into a single company, to be called the Motion Picture Patents Company. Few were inclined to join until a federal court in Chicago held that independent producer William Selig had violated Edison's camera patents. With that ruling, Edison enlisted Biograph president J. J. Kennedy. The heads of the other studios (Selig, Essanay, Kalem, Lubin, Vitagraph, Gaumont, Pathé and Méliès) quickly joined. They would pay royalties and be the exclusive licensees of Edison's equipment. Producers who weren't members of his little club were forbidden by the courts to use the equipment.

Edison now had a monopoly on the films produced by major companies in the United States and France.

Thus was born the monopoly that was nicknamed the Patents Trust, then the Edison Trust, and soon thereafter, simply the Trust. The Trust sent out an edict:

1. All motion picture producers are required to have a license from the patents company to make films.
2. All exchanges are required to have a license to distribute those films.
3. A fee of $2 a week is to be levied on theater owners for each projector used in showing those films.

To fight the Trust, several small independent producers, Rex, Yankee, Actophone, and Tanhouser, made films in hole-in-the-wall studios. Others headed west. Some bigger producers tried to fight back. The strongest threat to the Trust was Carl Laemmle, who owned a series of exchanges. In 1909 he decided to create a movie production company to satisfy his own needs, the Independent Motion Picture Company.

On April 10, 1910, the Trust acted promptly to quash this new enterprise by organizing the General Film Company, whose purpose was to buy up all the independent exchanges. If the exchange operator did not agree to sell, the General Film Company simply cut off their film supply.

Harry watched and listened grimly as one independent exchange after the other fell to Edison's ruthless takeover. The Warners had no other alternative but to try and buck the Trust. They offered theater owners colored lithographs and banners to advertise the few pictures they could get their hands on. To distribute these films it was necessary for Warners representatives to approach each theater secretly with a can of film stashed under a topcoat, or between the pages of a newspaper, or hidden in a suitcase. The Trust hired thugs to thwart this distribution method. The Warner brothers realized they couldn't hold out much longer; their day was coming.

Sam told the story this way:

> One day in late 1910 a man walked into our office in Pittsburgh and said he was buying us out that day. We said, "But we don't want to sell."
>
> Then this man, who said he was from the General Film Company, said he would not supply us with any more film. We said, "Okay, we'll sell."

> He said he was going to give us a fair break and allow us $52,000 for our exchange. He handed us $10,000 in cash, $12,000 in preferred stock and the rest in four payments to stretch over a period of three years.
>
> Here we were— $10,000 among the four of us, not enough for a first-class peanut stand.
>
> We asked him, "How about a job managing our own office in Pittsburgh?" The man told us to come to New York and he would fix us up. The next week we arrived in New York and J. J. Kennedy, president of the General Film Company, said, "I'm very sorry, but I have just completed an arrangement to have someone else manage the Pittsburgh office."
>
> "Then we don't get the job?"
>
> "No."
>
> Abe, who was with us, said, "Very well, Mr. Kennedy, we'll stay right here and put you out of business."
>
> Kennedy replied, "Personally, I wish you the best of luck, but in business, I hope you break your neck."
>
> Then he opened a box of cigars and didn't even give us one.

The Duquesne Amusement Supply Company had ceased to exist. Harry was still convinced that America was the land of opportunity, but had learned that under free enterprise, if one man had too much freedom, you had to have twice as much enterprise. To try to balance the scales, Harry immediately joined forces with Carl Laemmle's independent company to distribute their films in the Pittsburgh area. The Warners' slogan became "What Warner Bros. Promise, Warner Bros. Deliver." But the Patents Company stepped up pressure on the independents and the legal tangles were endless. The pictures that were available to the Warners were also of terrible quality. By 1912, Harry realized it was useless to continue.

He took some of the money from the sale of their film exchanges and bought a large grocery store on Federal Street in Youngstown for his parents. "We'll stay here and go into business on a big scale," Harry told his brothers. "I believe this is the best way to make money, a steady income from regular customers who are satisfied with the service." Although the "super" market amazed the town with its huge size, Harry was far from satisfied. Still determined to fight the Trust, Harry and his brothers opened the Rex Theater and showed the films they had saved from the exchange. After the supply had been exhausted and the audience had tired of reruns, the theater was closed down. But like Abe, Sam, and Jack, Harry kept looking for a way to return to the film business.

Realizing they couldn't jump back in immediately, the boys scattered, each searching for work of his own. While Harry stayed in Youngstown and ran the market, Abe went on the road again as a salesman, and Jack wandered from city to city trying to find a job in film exchanges, finally ending up in New York. Each day Jack read the trade papers—*Billboard,* the *New York Dramatic Mirror,* and the *Moving Picture World*—looking for some kind of work in the film industry. He was finally hired as a film splicer in an exchange. During his off hours he went to the theaters at Fourteenth and Herald Square and watched Sarah Bernhardt and Lillian Russell perform.

His favorite pastime was visiting the vaudeville theaters, where he envisioned himself on stage, singing and clowning for the audience. He saw every moving picture show that was exhibited in the city, mesmerized by the scope of *A Tale of Two Cities,* the historical detail of *The Life Drama of Napoleon Bonaparte and Empress Josephine of France.* He wondered how they had been filmed: how were the crowd scenes directed; what were the production costs; why weren't the scenes that slowed the dramatic action cut from the film? Little by little, Jack was learning the techniques that he would later put to use as a producer and editor.

Sam wasn't about to get out of the entertainment field. To him, the odor of celluloid was like the smell of greasepaint to a performer. He too decided to go to New York and check out the film exchanges, see what new movies were being offered. To his surprise he discovered a hand-tinted five-reeler that had been filmed in Italy and was not under the jurisdiction of the Patents Company. It was *Dante's Inferno,* an extravaganza based on the classic poem. Sam watched in awe as hordes of writhing bodies were consigned to the flames.

Once they'd decided to take *Dante's Inferno* on the road, Sam and Jack hired an out-of-work, perennially drunk actor, named him "Professor Bush," and supplied him with a podium from which to read the scenes from Dante's poem. Sam set Jack up with a wind machine and a sheet of metal that he could rattle like thunder. They opened in Hartford, Connecticut, and, as Sam had hoped, audiences were enthralled. The sound effects, the booming voice of the Professor, and the tinted film with its huge cast dazzled viewers who were used to black and white fare with five or six actors. Sam and Jack took the show from town to town until the Professor keeled over one night in a drunken stupor. The brothers closed the show shortly thereafter, took the $1,500 profit they had made and treated themselves to a sumptuous five-course dinner, then blew most of the rest of their cash on a crap game. They had enough sense to save a few bucks for the train trip home.

Yet they were happy. They had accomplished what they wanted; for a brief time they were once again in the motion picture business. But now what?

Tired of everything being controlled by the Trust and unable to get good movies from independent producers, they were once again motivated by someone trying to stop them. Harry decided to take the next step.

A step called Warner Features.

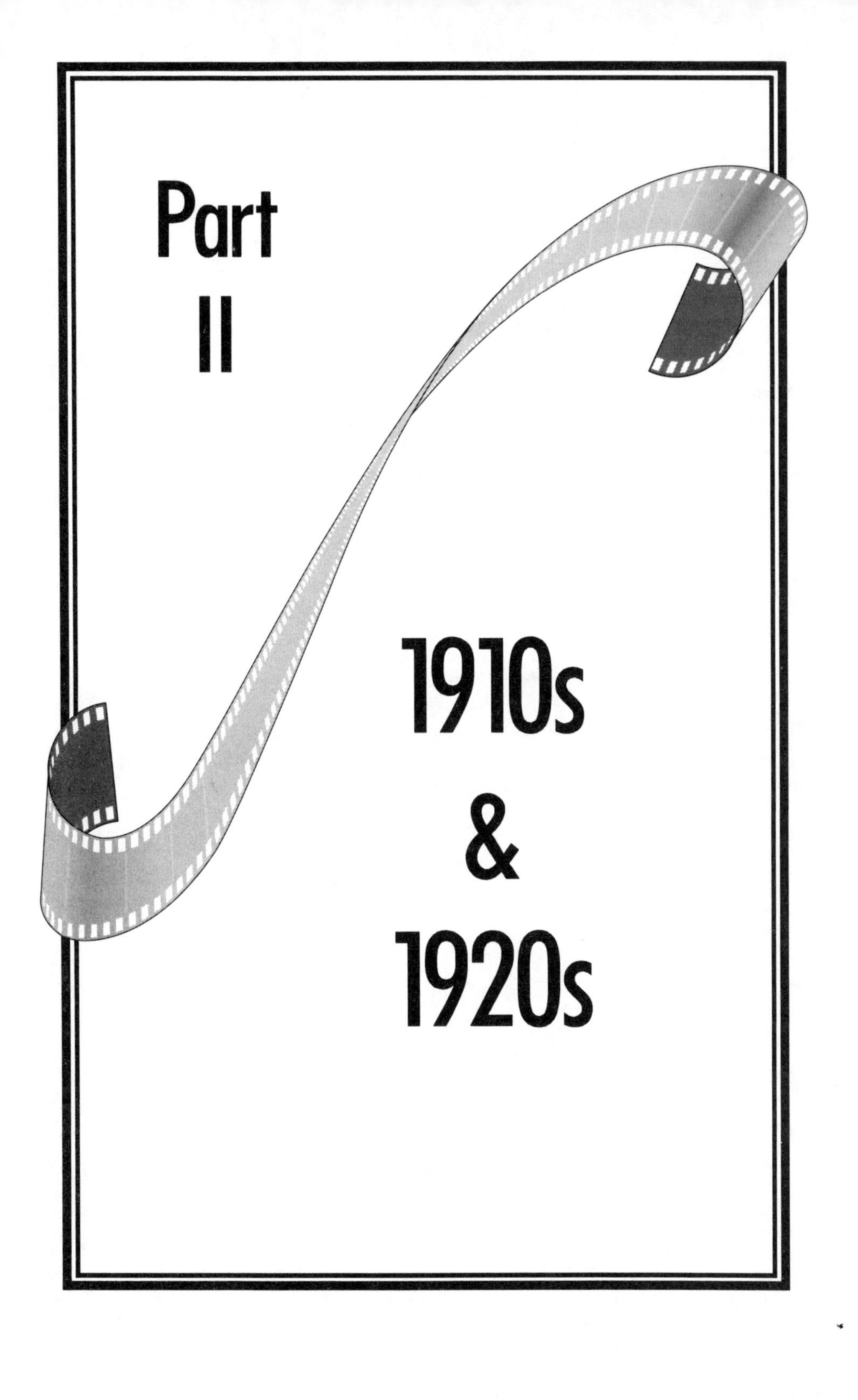

Part II

1910s & 1920s

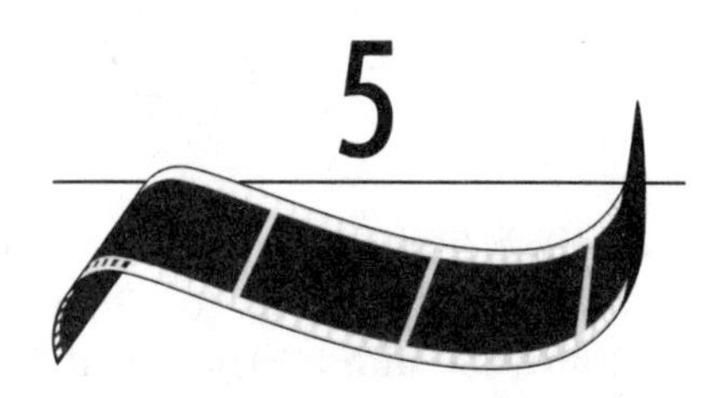

5

Warner Features

Summer 1912

"Sam, Jack, why don't you make a couple two-reelers?" The two brothers glanced sharply at each other, as Harry added, "*Cheap* two-reelers."

Jack grinned and nodded eagerly at Sam, then at Abe.

"How do we make a cheap anything?" Abe said. "Producing movies is something we don't know anything about."

"So what did D. W. Griffith know before he started cranking out films?" Sam said. "All you need is a camera."

"And a story." This from Jack, who was no doubt visualizing his name on the title credit:

PRODUCED BY JACK L. WARNER

And:

WRITTEN BY JACK L. WARNER

Abe, always pragmatic, asked, "How could we do it with the Trust breathing down our necks?"

"I was in St. Louis," Harry said, "and there's an old abandoned foundry that has enough space to use as a studio." He looked at Sam and

Jack. "You'll need money for film, a camera, and rent on the foundry. Between the four of us we should be able to dig up the dough."

"How do we distribute the films?" Abe asked. "The Edison group's got the exchanges locked in."

"Yeah, and us locked out," Sam said.

Jack didn't want to hear all these objections. *He wanted to make pictures.*

Harry said, "Remember what Pop says: As long as we all stick together, we can do anything."

"Okay, let's do it," Sam and Jack said in unison.

"What are we going to call this new production company?" Abe asked.

"We ought to have our name in it," Jack put in. "The Warners . . . something-or-other."

Harry nodded. "How about . . . Warner Features?"

This was the first time the brothers used the family name in one of their enterprises. It would be over fifty years before they would sell the name.

Sam and Jack rented a camera and boarded a train to St. Louis. En route they devised a scenario in which a "bunch of bloodthirsty Indians" threaten a group of settlers as they cross the plains in covered wagons.

The two brothers rented the foundry, and with Sam directing and Jack producing (which meant rounding up actors, as well as jumping into some of the filmed sequences, first as a cowboy, then as a red-painted Indian), they began to crank out their two-reel quickie, *The Covered Wagon.*

Dot Farley, a pretty young girl with—as Jack said—"auburn hair and soulful brown eyes who could ride like Paul Revere," portrayed the heroine. Dot could ride all right, but she couldn't act her way out of a celluloid barrel. The local Missouri National Guard was conned into service as westward-bound immigrants, blasting away with their (1910 Army issue) carbines at the marauding Indians, fording the Mississippi.

Sam later recalled: "When we made *The Covered Wagon* we only had three wagons, and it hadn't rained that year so the river was smaller, but it was the same story we released under the title *The Peril of the Plains.*"

While they had the camera in place and the background available, they made a second film, *Raiders on the Mexican Border.* Of the two movies Jack later wrote, "It was obvious to [us] that Sam and I were not going to put the great D. W. Griffith out of business." As far as he and Sam were concerned, both films were turkeys. Warner Features' future looked bleak.

The boys returned to Youngstown and screened the two-reelers for their brothers. Harry and Abe watched and in the darkened room shook their heads in dismay. After the last image flickered from the screen, Abe said, "I can't sell that stuff."

Sam rewound the film on the reel. "C'mon, Abe, you always said you could sell anything."

"Soap I can sell. Pots and pans I can sell. Junk I can't sell."

"It's not that bad," Sam said defensively.

Jack, his anger rising, looked at Harry. "How in the hell are we supposed to make a good movie on the shoestring budget you gave us?"

Harry shot back, "How do we get more *shoestrings* to make better movies if we can't *sell* the first ones?"

Harry knew the two films weren't marketable, as he had seen some of the innovations being produced by Griffith and filmed by his cameraman, Billy Bitzer. The 1912 drama *An Unseen Enemy,* with its evil faces photographed in close-ups and its guns blasting at the screen, terrorized audiences, but kept them coming back for more. The 1912 Italian film *Quo Vadis,* an eight-reel epic with a huge cast, was impossible to compete with on the Warners' small budget. The picture that intrigued Harry the most was *Queen Elizabeth,* starring Sarah Bernhardt and photographed directly from her stage performance. Although the single-set scenes were too static, Harry saw it as an important union of entertainment and education. It was a lesson he would not forget.

The four brothers sat in silence. Finally, Harry said, "Well, our films aren't that good, but they're done. As Warner Features, we will distribute them ourselves."

Harry's three brothers looked at him like he was crazy. They had already been bought out by the Trust, which still maintained a stranglehold on independent filmmakers and distributors.

Harry, always the entrepreneur, had a new idea. He had decided to go along with Carl Laemmle and his Independent Motion Picture Company and fight the Trust, taking the risk and reentering the exchange business. But not in the east, where J. J. Kennedy and his lawyers and thugs could get at them. The Warners would join the exodus to California.

Once again the big problem was money. They were broke. Sam liked to joke, "We called in the old-clothes man to pay off our debts. When our monetary worth was totaled, we found we had a dollar sixty-five." Ben and Pearl didn't have a lot of money saved, but when they learned how broke their sons were, they gathered together $400 and gave it to the boys. Harry would later say: "Whatever we have today was built on the four hundred dollars our parents gave us at that time."

In late 1912, Sam was dispatched to open an exchange office in Los Angeles, near a small community of movie producers called Hollywood. (Only a year before Sam's arrival, the Nestor Film Company had rented an abandoned tavern for $30 a month and set up the town's first studio. Carl Laemmle, as part of his continuing struggle against the Trust, bought Nestor a year later and established what would eventually become Universal Studios.) Sam opened an office at a cost of $16: $10 for rent and $6 for furniture. He hired a girl who, Sam said, was "the booker, winder, cutter, and all the rest of the help."

While Harry and Abe opened an office in New York, Jack was sent to San Francisco. It was his kind of town. The Barbary Coast with its gambling tables and flashy ladies was a long way from Youngstown, Ohio, for a fresh twenty-year-old. Jack later wrote of the venture, "Sending [me] to a city like San Francisco was as risky as having Don Juan run a girls' boarding school. Not that I was Don Juan. But I was more interested in bright lights and broads than I was in squeezing out nickel-and-dime profits on pictures."

Jack did as Harry had told him to do and opened the Warner Exchange at 217 Taylor Street, not far from the bars and bordellos of the city. In the daylight hours he worked hard and the cash began to flow into the exchange. At night he could go to the special restaurants, the Mansions, on Nob Hill, where on the first floor customers could sit and have dinner and on the second they could lie down and have dessert.

The largest percentage of the money the two exchanges took in was sent to the brothers' central fund, but Jack (with a small loan from Sam) was able to save enough to buy a little theater, the All-Star on Fillmore Street. It was with great joy that he stood outside his movie house and watched patrons plunking down cash to see films that he had paid just a small fee to rent.

With the Warner exchanges thriving, Harry and Rea welcomed the arrival of their second child, a daughter, on September 13, 1913. They named her Doris. Although Abe had married shortly after

Harry, he and his wife, Bessie, had not had any children. Sam and Jack were still enjoying the bachelor life. Jack, however, had fallen in love.

A year after Jack arrived in San Francisco, a new friend, young Sid Grauman (who in 1927 would build Grauman's Chinese Theatre in Hollywood), introduced him to the blond teenage daughter of a well-to-do German Jewish family. Irma Claire Salomon's parents had come to San Francisco soon after the Gold Rush, but made their fortune with the City Lottery, which prospered until just after the disastrous 1906 earthquake. Irma said of her blind date with Jack, "He was a young man with shiny patent leather shoes, a high collar, and a striped suit." Moving fast, Jack wooed Irma relentlessly until she consented to be his wife. They were married on October 14, 1914. At first, their union was not well received by the bride's family, but the brash, bright, dapper young man from Youngstown finally managed to impress them with the force of his personality. The marriage was a good match for Jack: Irma's parents' money would come in handy in the operation of the exchange.

The Warner exchanges on the West Coast continued to flourish, helped by the demise of the Trust. On August 15, 1912, the Justice Department had filed an antitrust suit against the Motion Picture Patents Company, and in this single act the Trust lost its legal legitimacy. With the demise of the Patents Company, independent distributors were now able to open offices anywhere in the country. The Warner brothers had finally fulfilled Abe's threat to Patents president J. J. Kennedy: "If you put us out of business, then we will put you out of business."

In the glow of this triumph, Harry and Abe decided to produce feature-length photoplays: four to six reels filmed directly from theatrical plays. The concept of filming live theater fit right in with Harry's emerging principle of producing quality entertainment of educational value. Exhibitors, however, were wary of any "intellectual" property longer than two reels. They wanted fast-paced shoot-'em-ups, or teary-eyed-damsels-in-distress "flickers." Harry and Abe couldn't give away their photoplays.

That didn't deter Harry; he wanted to stay in the production end of the business. Still committed to doing serious films, he bought the rights to a poem, *Passions Inherited,* a heartwarming story with strong family values by the popular writer Ella Wheeler Wilcox. Instead of having his brothers film it, Harry scraped together $15,000 and hired a British director, Gilbert P. Hamilton. The site selected for the filming was a hundred miles north of Los Angeles, in a desolate area that had been used by others for making Westerns.

Sam and Jack thought the whole project was a mistake (even Abe was wary), and argued with Harry that the hearts-and-flowers plot was not the kind of story moviegoers would pay to see. The filming got off to a bad start when Hamilton asked for another $5,000. Harry paid. When two weeks had passed without Harry hearing a word from his director, he sent Jack to find out what was going on. Jack drove the 300 miles from San Francisco to discover Hamilton bedded down in a tent with the movie's heroine; a new car (purchased with Harry's money) was parked in the dust outside. Jack threatened a lawsuit and was told that the movie had been completed, except for the ending, and that the first five reels had been shipped off to Los Angeles. Jack got hold of the film, duplicated a lovers' embrace from the middle, and spliced it onto the end. *Passions Inherited* was released—and bombed. Once again, the brothers were broke.

After such adventures the average person would have abandoned the infant film industry, but instead of becoming disillusioned, the Warner brothers became more enthusiastic about the future of the movies and were determined that no matter how frequent the disappointments, they would become a significant part of it.

This determination attracted the attention of several major film organizations, notably the General Film Company. Sam recalled, "We were getting along fine when Lubin offered us the job of general manager at a salary of $50,000 a year, plus a bonus of $120,000 at the end of the year, if we would sell out our business to them for $250,000."

"What about it, boys?" asked Harry. "Do we stay on our own, or do we work for them?"

"We stay independent!" said Abe, Sam, and Jack in unison.

They wouldn't sell their name. The brothers were big-stakes gamblers. They had to have a dream or there was no direction for the future.

A few newspaper items were beginning to be written about them. One noted: "The career of the Warner Brothers reads like a combination of the Revolutionary War and Ireland's struggle for freedom." Another said: "You can make brains do a lot of tricks with a good bank roll, but one thing is certain—you can't monopolize brains when they are coupled with nerve and the desire for independence. Or, to be brief, 'You can't keep a good man down.' The Warner brothers have certainly proved that."

The brothers were living on the edge—and pushing it one step further.

The Warner family experienced one great tragedy during this hectic period. Milton, the baby of the clan, died of a ruptured appendix on June 13, 1915, at the age of nineteen. Too young to follow his brothers into the movie business, Milton had shown great prowess as an athlete, particularly in baseball. A four-letter man at Rayen High School in Youngstown, he was known to have a pitching arm "like a flame thrower." The New York Giants, Milton's favorite team, sent a telegram offering him a job at $150 a month. Disappointed by what he thought was an unduly low amount, he rejected the offer. Just before his death, he learned that it had been a misprint and should have read $150 a *week*.

On March 27, 1916, Irma Warner gave birth to a son, whom Jack proudly named Jack Warner Jr. Naming his son after himself violated Jewish tradition, which required naming a baby after the most recently deceased family member. Jack did give his son the middle name Milton.

At the age of one year, Jack Jr. was taken to Youngstown to be presented to Ben and Pearl. Irma went along, and as Jack Jr. recalled years later, "My mother went into deep culture shock when she found the language of my grandparents indecipherable and the kosher cooking indigestible. She blanched when introduced to matzo balls, potato kugel, and blintzes and borscht." The parents of the groom referred to their new blue-eyed blond daughter-in-law as the shiksa. The three sisters were like a Greek chorus: Sadie was a critic, Rose a dominant personality, and Annie a quiet lady. In spite of these basic differences (and the weeklong quarantine of the family brought on by the baby, who had imported a case of measles), the marriage was working out well. Irma learned how to make matzo ball soup and to stop asking for butter at the dinner table.

By 1916, Sam and Jack were still prospering in the exchange business, mostly by buying "state's rights" from producers, which meant they had exclusive rights to rent their films to movie houses in that area. Their biggest success was with a European import titled *War Brides,* which starred the exotic actress Nazimova. They paid $50,000 for the California-Nevada-Arizona rights and reaped a huge profit.

With the success of *War Brides,* the brothers decided to distribute a picture produced by one of filmdom's pioneers, Col. William N. Selig, a Civil War spectacle that was reputed to be as grand as *Birth of a Nation.* It was titled *The Crisis.*

The Crisis was aptly named. The movie opened April 5, 1917, to excellent reviews. Jack rubbed his hands together gleefully. The show was going to be a smash. The next day, the United States declared war on Germany.

Jack stood forlornly outside the theater, watching as a trickle of people bought tickets for the matinee. The evening performance looked like a miser's funeral—nobody showed up. No one was interested in a make-believe story of a war that had happened fifty years earlier, not when the country was involved in a real war. The brothers' investment was a total loss.

The military felt it could put the Warners' experience in the motion picture business to good use in the war effort. Sam and Jack were assigned to work with the Army Signal Corps making educational films for servicemen. The first script they received was titled *Open Your Eyes.*

"Must be something about getting up in the morning, you know, reveille," Sam said, tossing the script to Jack.

Jack thumbed through the pages. "It's about the clap."

Sam sat up straight. "No kidding. What the hell does the title, *Open Your Eyes* . . . what's it supposed to mean, then?"

Jack laughed. "Look before you leap?"

After getting in a huddle with Harry, they decided to produce the film themselves using Signal Corps equipment and signed a contract stating that the Warner brothers had rights to show it commercially after the war. Harry especially liked the idea because parents could take their sons to see it as a warning.

The documentary was filmed at the Biograph studio in the Bronx. Sam and Jack, wanting to make sure that the movie was done right and realizing that neither of them had the expertise to direct it, reluctantly hired Gilbert P. Hamilton again. They first informed him that there would be no money for new cars and not one penny for bedding down starlets.

In *Open Your Eyes,* a quack doctor is shown selling a soldier a bottle of worthless medicine. The soldier was played by Jack Warner; he said it was his only featured role in a film. The Army liked the finished product and showed it to servicemen before they went overseas.

With the war at its height, few commercial films were being made in the United States, so Harry tried to import whatever foreign films he could find. He heard that a Mr. Zeigler of Anderson and Zeigler, an importing company, had just returned from Paris with several films. Sam later recalled:

> Zeigler had brought back two negatives; one was titled *Redemption,* and the other *The Glass Coffin*. He wanted $30,000 for the two prints. We looked at both pictures, liked *Redemption,* but didn't like *The Glass Coffin.* Mr. Zeigler said if we bought them, we had to take them both—and that was the beginning of "block booking."
>
> We succeeded in selling 50 percent of the country on *Redemption,* but practically nobody wanted *The Glass Coffin.*

One West Coast theater owner, having received a wire from Harry that the film had been shipped to him in care of the local postmaster, went to the post office and asked, "Did you receive *The Glass Coffin*?"

"No," said the postmaster, "not even a wooden one."

"This isn't an ordinary coffin. It's the glass one. Warners shipped it to me from New York."

"So our California undertakers aren't good enough for you?"

"You don't understand . . ."

"Ha, you patronize a New York slicker and let our fellow citizens starve."

"You still don't understand . . ."

In 1917 the Warner brothers decided to make a serious bid for success in the rapidly growing movie industry. Jack was sent to Los Angeles to operate the film exchange, but mostly to be closer to the movie action that appeared to be centering in the little town named Hollywood.

During the day, Sam and Jack attended studio screenings, looking for new films to distribute, and at night went to any party they could wangle an invitation to, talking to producers and directors. It was fun and they loved playing the role of movie executives on the rise, but in fact they were floundering. The brothers knew they could make it really big if only the right project came along. They knew how, they just didn't know what.

It was a warm day early in the summer of 1917 when Sam left the Alexandria Hotel, a well-known melting pot of agents, exhibitors, out-of-work actors, hookers, and movie-house bookers, in downtown Los Angeles to walk to the Warners' exchange on Olive Street. Passing the window of a bookstore, Sam saw—as he later said—"a poster in the window of the Kaiser in a web of spiders with diplomats grouped around him, advertising *My Four Years in Germany,* by Ambassador James W. Gerard. Stacks of books advertising Gerard's bestseller surrounded this display."

Intrigued, Sam went inside the store, thumbed a copy of the book, then bought it. Sam had always read voraciously, trying to improve a mind that had had little formal education. Arriving at the exchange, Sam pushed aside the stacks of papers that covered his desk, propped his brown oxfords on an open drawer, and started to read. Soon he was sitting straight in his chair, absorbed in the dramatic story. His mind began to dissolve the words on the page into images on a screen:

> Ambassador Gerard tries to negotiate with Kaiser Wilhelm II to avoid war by stopping the U-boat sinking of Allied ships . . .
>
> Hindenburg orders the disposal of the conquered populace of Belgium: "Healthy ones to the farms, use your discretion with the young and old." . . .
>
> The ambassador speaks with a young woman in a camp. She cries, "Are we slaves?" . . .

Halfway through the book, Sam called Harry in New York. "Harry! This is our chance. This book is great! I mean, this book has . . ."

"Whoa, Sam, what book?"

"*My Four Years in Germany,* Ambassador Gerard's book. It's about this big-deal guy's experiences trying to stop the country from getting into the war with the Germans. The story's got the Kaiser, Hindenburg, war speeches in the German Reichstag. It's got . . ."

". . . a great wartime title," Harry interrupted, beginning to see his brother's vision.

The phone line crackled as Sam went on: "Look, Harry, this could be big stuff. This is our big chance."

"If it's that good, some studio's got an option on it already," Harry said.

Sam was silent. That problem had occurred to him. Finally, he said, "Suppose I send this Gerard a wire—he's an ambassador, so he's got to be in Washington—ask him how much he wants for screen rights? I'll lay it on pretty thick, about how we're big studio guys, how we'd spend a lot of money on the picture . . ."

"If we had a lot of money," Harry sighed.

"That doesn't worry me none. You're a wizard when it comes to money."

Harry smiled to himself. If there was one thing he was getting better at, it was talking banks and investors out of cash. He breathed deep. "Okay, let's do it."

Sam sent the telegram to James W. Gerard and got an immediate reply. The ambassador was willing to listen to their offer. Harry caught a train to Washington to make his pitch.

"Mr. Ambassador," Harry began, "film is the great founder of peace. When people understand each other, they need not fight. I feel it is our patriotic duty to the thousands of Americans who cannot read to make a motion picture of your book."

Ambassador Gerard said that several producers, including Lewis J. Selznick and William Fox, had made offers for the movie rights to his book. "In fact, this very morning I received a letter from Mr. Fox containing a check for seventy-five thousand dollars." Gerard picked up the letter from his desk. "Seventy-five thousand is a lot of money."

Harry admitted to himself that it was indeed a great deal of money: It was ten times more than his brothers possessed at that moment. Trying not to look at that awe-inspiring check in Gerard's hand, Harry stood straight and said, "I believe, Mr. Gerard, that I am in a position to make you a better offer."

With that, Gerard motioned Harry to a chair, saying, "I really don't believe a successful film could be made of my personal experiences in Germany."

Harry remained standing. "Ambassador, the movie I make can carry to the American people, and to the world, your stirring warning about the menace of the German military threat. The picture will help arouse the world at large as to why we must fight for civilization." Harry stopped. He had come on very strong. But he had meant every word.

A faint smiled crossed the ambassador's face. "That's why I wrote the book. Perhaps it could be made into a movie, as long as it gets in the hands of the right studio." Gerard spoke slowly, articulating each word carefully.

Harry knew he was far less educated than the ambassador, and he realized he couldn't con him into a contract with a bunch of fast talk.

Instead, Harry told him how his parents had come to America thirty-four years earlier to escape the oppression of the czar, how his brothers had thrown themselves into the movie business, and how they had prospered, only to lose it all and have to begin over again—and again. "But one thing, Ambassador," Harry said. "The Warner brothers will become a major force in the movie industry. Today, we are not a large company. Perhaps that is for the best, as we can provide the intense, personalized effort your book needs."

Harry took a deep breath and went on. "Sir, we can't offer a lot of money. Frankly, we don't have it. But we can assign you a good percentage of the profits."

"And how much is your offer?" Gerard asked.

Harry thought about the cash he and his brothers had on hand, less than $2,000. He gazed steadily at the ambassador and plunged into the deep water: "We can pay fifty thousand, plus twenty percent of the profits."

Gerard nodded thoughtfully. He didn't tell this straightforward negotiator—whom he was beginning to like—that before Harry had walked into the office he had decided to accept the William Fox offer. "I would have to see the—what do you call it?—the scenario."

"Yes, sir, I'll have a scenario showing how we would film your story to you later this summer. I'll need your consent to do this."

Gerard pulled a sheet of paper out of his desk drawer, scribbled his consent, and signed it. Before he handed it back to Harry, he said, "Mr. Warner, in a month I have to go west to visit some relatives in Montana. Suppose we adjourn this meeting and see each other again in your studio in Los Angeles."

Harry's heart almost stopped. There *was* no studio in Los Angeles. True, Sam had rented a $60-a-month barn and he and Jack were trying to get together enough capital to shoot a couple of quickies, but their entire equipment consisted of a small camera they had used to film the covered-wagon flop.

"Perhaps we could meet at your place in Montana," Harry suggested. "Los Angeles is terribly hot this time of year . . ."

"But I do want to go to California," Gerard said. "I'll meet you there and read the scenario. If I like it, it's a deal."

Harry's first problem was to find a suitable studio for filming Gerard's story. He called Sam and Jack to New York and had them rent the Biograph in the Bronx. While the boys set up outdoor locations on a New Jersey farm, lent by Hearst editor Arthur Brisbane, Harry and Abe scouted New

York for a benefactor with $50,000. Jack wasn't worried. He had always said, "Hell, Harry could sell swimsuits to South Seas natives."

Harry found his film backer: Mark M. Dintenfass, whose name was strikingly similar to the German word for "inkwell," *Tintenfass.* Dintenfass was—as he liked to say—"always in the black." The backer gathered together some cash and delivered it to Harry and Abe's office in a shoe box. It turned out to be only $28,000, all Dintenfass said he could scrape together. Harry and Abe had to scrounge up other investors before they could assemble the $50,000 advance and have enough left to cover production costs.

One month later, when Gerard boarded the train to travel to Montana, he was joined by Harry and screenwriter Charles Logue with a finished script. As Harry paced, looking blankly out the window at the passing countryside, Gerard read the script. The clickety-clack of the metal wheels on the rails pounded like hammers in Harry's head.

Turning the last page, Gerard took off his glasses and said: "Mr. Warner, I give you my consent to make the picture." They signed a formal contract on the train and had it notarized at the next station stop. Harry wrote a check for $50,000 and caught the next express to New York.

Later Gerard told the story that Harry had offered to toss a coin to see who paid the contract's fifty-cent notary fee. "I lost," Gerard recalled.

Harry had said, "Don't worry, you'll get it back."

Gerard laughed. "I know I will."

Filming began under director William Nigh, most of whose previous experience had been with Keystone Kops two-reelers. Gerard was played by an unknown actor named Halbert Brown, who, with his stocky build, neatly trimmed mustache, and thinning hairline, looked startlingly like the ambassador himself. To save money, documentary footage was inserted into the film, a technique that hadn't been tried before.

These sequences of violence and brutality, which were to have a great impact on the audience, lent authenticity to the film. The Warners advertised the film as "fact, not fiction." Unknown to them at the time, they had created the kind of picture that would become their trademark. *My Four Years in Germany* was good entertainment that at the same time performed a public service: letting Americans know what was happening in another part of the world.

But the Warners' troubles were not over. As they made plans to exhibit the movie, they discovered that powerful competing studios were prepared to do anything, even threaten violence and personal harm, to keep them from breaking into the big leagues. The major studios, such as Paramount and First National, were able to coerce the owners of Broadway movie houses into not showing the Warner feature.

One night while the brothers were huddled in their New York office trying to come up with a distribution plan, a visitor dropped in. He told them he was a man with important connections. "It is my intention," he said, "as well as that of the interests I represent, to offer a cash settlement to buy the completed screen version of *My Four Years in Germany.*"

The brothers just stared at him as the man offered less money than it had cost them to produce it. "You can take my offer," the man added, "or else." The threat was not lost on the Warners. The "interests" he represented would see to it that the film was boycotted by theater owners.

Abe rose slowly from his chair and hovered over the smaller man, growling, "I give you exactly three seconds to get the hell out, or so help me, I'll throw you out the window!"

The man fled.

Although the brothers laughed at the man's hasty exit, Harry later said, "I shall never forget that night. That was the nearest I ever came to committing suicide."

Surprisingly, a week later, William Fox offered the brothers $375,000 for the movie. They turned him down!

The brothers had put too much effort into developing and filming *My Four Years in Germany* to sell it off. This was their chance to become established in the industry as producers. They knew they had eight reels in their vault that had the earmarks of a smash hit. All they needed was a company to distribute it.

"We finally let J. D. Williams have it for First National," Sam later recalled. "They were having hard sliding; their pictures had been disappointing and their franchise holders were dissatisfied. Williams contacted us and said he'd distribute the film through his theaters for a small percentage of the profits. We quickly agreed."

December 1917

Opening night in New York was a nail-biter for the four brothers. Ambassador Gerard relaxed in a box seat and watched the events of his four years in Germany unfold: his interview with the German foreign minister; the inspection of German prison camps; his chat with a British soldier locked in a barbed-wire cage; the witnessing of a Prussian officer tearing a young girl away from her parents with the intent to commit rape.

It was inflammatory footage that intensified anti-German sentiment, striking just the right patriotic chord with the war-conscious American public. When the movie ended, the audience rose to applaud, not for Ambassador Gerard, but for the men who had made it happen: the Warner brothers.

My Four Years in Germany grossed an impressive $800,000. After Gerard got his 20 percent and First National and the investors and the production costs had been paid, the Warners were left with a net profit of $130,000. Sam said at the time: "When our share was split up between the four of us, it wasn't much for the work and worry we put into it."

The real value was in the new status that Harry, Abe, Sam, and Jack had achieved. They were not just four ragtag film distributors, they were moviemakers. Just like Zukor and Goldwyn and Laemmle.

They had arrived—in Hollywood.

6

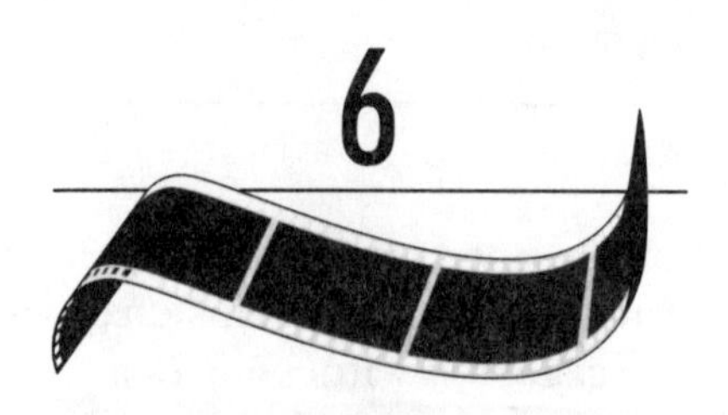

Hollywood Be Thy Name

Hollywood.

Silent-film actress Agnes De Mille described early Hollywood as a place where "the streets ran right into the foothills and the foothills straight into the sagebrush. Wild, wild hills where you could gather wildflowers by the armful." She went on to say of the first filmmakers:

"Hollywood residents used a disparaging word for these film people. They called them movies *because they moved around a lot. They were really outcasts, but the locals found them amusing, because they beat everybody on the heads with rubber clubs."*

November 1918

"Harry, we got the sun out here, all year long. We got mountain backgrounds, the desert . . ." Sam began.

". . . a big city, a seaport, open range for Westerns," Jack interjected.

Sam's lips touched the black mouthpiece of the telephone. "Harry, we can film anything we want right here in our back yard."

"And film it cheap." This from Jack again. "Besides, I froze my ass off stomping around in chicken shit and cow piles filming the 'Kaiser.'"

"Jack said he froze his ass off in New York . . ."

"My ass *and* balls!" Jack added, loud enough for Harry to hear in the New York office.

Harry laughed and pressed the telephone to his chest and said to Abe, "We're going to set up a studio in Los Angeles." Abe nodded in agreement. Their second try at boarding the patriotic bandwagon, *Kaiser's Finish,* a fictitious World War I story, had had far less impact—and far less monetary return—than *My Four Years in Germany.* The Armistice, November 11, 1918, put to rest any further plans for flag-waving films. Besides, all the production companies were moving to this new place called Hollywood.

Hollywood got its name from Daeida Wilcox, who, with her husband, Harvey Henderson Wilcox (a wealthy Prohibitionist from Kansas), had bought the tract of land on the edge of the growing community of Los Angeles in 1883 (the same year Ben Warner arrived in America). On the train trip to California, Daeida had talked to a woman who owned a summer home in Chicago, a home she had named "Hollywood." Daeida Wilcox thought the name sounded just right for the community of God-fearing, sober citizens she and her husband wanted to establish on the West Coast.

Cecil B. De Mille put the little community on the map when he filmed *The Squaw Man* in 1913 in a rented barn for the Jesse L. Lasky Feature Play Company. D. W. Griffith was at work on some vast, mysterious epic called *Intolerance.* For it he had built a spectacular set on Sunset Boulevard (a dusty tin-lizzy trail that curved through canyons to the Pacific) and had hired four thousand extras (at the unprecedented rate of $2 a day) to perform in it. No one knew exactly what was going on, but everyone knew that Griffith was pushing filmmaking into another era—that of the movie spectacle.

By 1914, most of the major producers—Zukor, Laemmle, Goldwyn, Fox, and Mayer—had established studios in Hollywood. By late 1918 when the Warners decided to set one up in Hollywood, the city's population had grown to 35,000. (By 1935 it would reach 130,000.) Douglas Fairbanks started the exodus to Beverly Hills when he built a house called Greyhall, then erected Pickfair on adjoining property after he married Mary Pickford.

Before the year 1918 ended, Sam and Jack had rented space at Eighteenth and Main streets, not many blocks from downtown Los

Angeles. The property had a few crude outdoor stages and an open-ended wooden shed—referred to thereafter as "The Studio." They laid several planks across the grass and dirt and grandly referred to the area as "The Lot." Nearby was the Selig Zoo, run by producer William Selig. The brothers made a deal with the Colonel to use the animals and shot a succession of films in a special movie cage, with Sam coproducing and Jack in floppy hat and black boots as the director.

Irma's cousin Doc Salomon, who had been a salesman for the Warner exchange in San Francisco, left the city by the bay and became the fledgling studio's first employee, serving as janitor, office boy, night watchman, and eventually the prop man. Doc also organized crews, acted as assistant director and stuntman, and once when a tired old lion escaped from its cage and ambled toward Jack and Sam beside the camera, Doc rushed for a rope which was tied to nothing and jumped off a platform to land on his behind in front of the escaped king of beasts, who sat down and gently licked his face.

Sam and Jack didn't want to make animal pictures, but they could use the docile old jungle beasts to cash in on the current craze for Perils-of-Pauline serials. These shorts, each ending in a cliff-hanging scene that left the audience wondering how the hero or heroine would escape, lured audiences back to the theaters week after week. Sam and Jack scribbled out a couple of story lines in which a beautiful girl was menaced by vicious beasts, and hired a curvy blond named Helen Holmes to play the heroine. Two serials were filmed, *The Lost City* and *The Tiger's Claw;* in the latter, audiences practically climbed out of their seats when studio lions romped through the hillside chaparral after the wild-eyed Helen Holmes.

Jack later recalled:

> Serials were very big at the box office during this period, and the cheapest possible villains were gorillas, tigers, lions, wild elephants, and other dangerous beasts. . . . The last scene [could show] the gorilla dragging the half-naked heroine into a cave, and the fade-out title CONTINUED NEXT WEEK left the audience in Freudian frustration. As a matter of fact, they really had something to worry about, because the biggest apes were always young guys dressed up in gorilla suits, and people knew very well they weren't going into the cave just to talk about heredity.

A year later, when the lease was up on the studio, Sam and Jack moved to what was known as Poverty Row, an uneven line of wooden buildings with movie-set facades that disguised the ugly warehouse studios. Poverty Row (or Busted Acres or Shoestring Alley, as the fledgling filmmakers variously called it) was an area on Sunset Boulevard between Gower and Beachwood Drive where dozens of would-be producers had set up shop.

"Most of them guys down there aren't going to make it," Jack had said before they moved.

"We will," Sam answered.

It wasn't going to be easy. If anything, the new studio was worse than the first one. From the front it looked like a residence, with wood-shingled sides painted white and a cement porch. A couple of rooms had been converted to offices, and out the back door was an open space.

"Looks like the city dump," Jack said. In one corner of the lot, a pile of dented oil drums dripped fluid, staining the earth black. Several wrecked cars, their windows smashed, rusted in the sun. A green lizard slithered out of a broken headlight and flicked its tongue at the two strangers who had invaded its home.

"We can always make a dinosaur epic," Sam added dryly.

Jack shook his head at a small stage that looked ready to collapse. "With the cash Harry's giving us, we'll be lucky to film an ant, let alone an anthill."

The boys cleaned it up as best they could but were so ashamed of the place that they deliberately didn't hang a sign to identify it as the home of Warner Pictures. In this dismal setting they filmed a comedy, *His Night Out,* written by a pair of young brothers, Kenneth and Howard Hawks. The star was an Italian comedian by the name of Mario Bianchi, who had been discovered by Fatty Arbuckle. Arbuckle had changed Mario's name to Monty Banks. Jack thought Banks was as good a comedian as Charlie Chaplin.

Jack Jr. remembers the stories his father told him about those first magical days in Hollywood.

Other Voices

Jack Warner Jr.

Those were happy days, lighthearted, insecure, and exciting. The next day's payroll was today's dream, and tonight's vision would become tomorrow's shooting script. Jack and Sam were like joyful kids turned loose in a toy factory. They dreamed up plots, picked up crews, called in friends and family as extras, and stole comics from producers who left them around loose.

Al St. John, a tall, lanky, sad-faced comic, led the chase through the Hollywood hills until the light got bad. Next day it would be an imitation Charlie Chaplin breaking in a pair of Sam's shoes, then came Mario Bianchi, rechristened Monty Banks, a little Italian comic, complete with top hat and spats.

Jack and Sam ground out one- and two-reel comedies for Harry and Abe to book into theaters and use as collateral to borrow more money to make more movies. My father during those days of stress, pratfalls, chases, and custard pies was a wonderful guy with a remarkable sense of humor, genuinely crazy about this looney business.

Fragments of memory flicker back from those relatively happy early times of beginning when my mother, father, and I lived in a little bungalow on South Berendo Street, within walking distance of what passed for a motion picture studio. My father found time to be home for most of his meals and a night of sleep, and enjoyed taking me to work with him early in the morning when I had no school. Often while we walked he sang songs and performed little tap dances remembered from the Cascade Theatre days.

When we arrived at the run-down studio just off Gower, with its sagging warp-floored stage, property and scene storage bays, and several little offices scattered off a dark hall, my father would sit down with Sam and review the scraps of paper covered with ideas scribbled between shots the previous day, which would now become the shooting script.

My father had this tiny leather notebook and on the first page was stamped the grand title "Warner Feature Film Co." On the pages he jotted down ideas for comedy scenes. One page listed several slapstick ideas for showing off a girl's leg:

Dropping handkerchief, fellow picking it up, observing limbs.
Seltzer bottle on girl's leg.
Picking up hat with cane while stooping over, picks up skirt and hat at same time.
Girl lighting match, looking at limbs.

Some days began with a round of telephone calls to family, friends, and passersby, who would get rides to the Pacific Palisades, Echo Park, or nearby Larchmont Boulevard, where the power poles and trolley tracks ran right down the middle of the wide street to make the chase scenes look especially perilous. The crowd of extras, whose pay would be a few dollars plus a box lunch, ran in circles around the hand-cranked camera mounted on a creaking truck.

I was usually in the front rank and when we raced behind the camera and out of the scene, we would switch coats and hats to reappear instantly as brand-new characters. The real veterans brought along different wigs, mufflers, beards, and umbrellas so they could be in more scenes and become established Hollywood actors. It was all quite informal, very spontaneous, and it was obvious that the director-producers, Jack and Sam, were having the most fun of all with their gang of pals making movies.

Finally, Sam could stand the dumpy quarters on Poverty Row no longer. "Let's get out of this sleaze pit," he told Jack.

"Where we gonna go that's as cheap as this place?" Jack asked.

"I've been looking at a ten-acre lot on Sunset and Bronson. It's got a building on it we can use for offices and a barn for a studio."

"I'm game."

With a little haggling, Sam and Jack got the property owner, William Beesemeyer, to agree on a total price of $25,000, nothing down and $1,500 a month. For the first few months the brothers couldn't make the payments and would go empty-handed to Beesemeyer. He would say, "Don't worry, boys, pay me when you can, next month, whatever you want."

Between 1919 and 1920, the Warner studio was on the edge of collapse. Motley Flint, a young executive with the Security Bank of Los Angeles who had financed their serial *The Tiger's Claw,* had first met the Warners on a train trip to New York and had seen something in them that other bankers had missed. Flint felt the Warners were "going in the right direction," and told them, "I don't worry about your debts; I know you'll make it." He became the Warners' biggest benefactor in those early years and helped them borrow a million dollars to carry them through.

It was a good thing the brothers had Motley Flint, because undercurrents of anti-Semitism ran through the banking community. Producer Joseph Schenck was called a "kike" by a bank officer who refused him a loan. Years later, Schenck went back to the same bank officer and said, "The kike wants to borrow one hundred million. My security is Twentieth Century Fox." The bank officer said, "I'll be happy to do business with you." Schenck responded, "Fuck you!" and walked out.

The Jewish issue was becoming a problem in the minds of anti-Semites. Henry Ford's *Dearborn* (Michigan) *Independent* reported in 1921 that the film industry was

> Jew-controlled, not in spots only, not 50 percent merely, but entirely; with the natural consequence that now the world is in arms against the trivializing and demoralizing influences of that form of entertainment as presently managed. As soon as the Jews gained control of the movies, we had a movie problem, the consequences of which are now visible. It is the genius of the race to create problems of a moral character in whatever business they achieve a majority.

Meanwhile, on May 4, 1920, Harry and Rea had become the proud parents of their third child, Betty May Warner, named after two of Rea's best friends, both of whom had died in the flu epidemic.

The 1920s started slow for the Warner brothers. Sam and Jack were able to make only one film, *A Dangerous Adventure,* a serial in fifteen episodes starring a traveling circus of animals and two sisters wandering half lost through Africa searching for treasure. Sam and Jack, who thought they had a hit on their hands, codirected. It was a disaster: the two heroines nearly tore each other's hair out with incessant fighting. When not acting as referee, Jack and Sam had to fend off attacks by an over-excited monkey and a foraging elephant. The series was a failure, so Jack edited the thirty reels down to seven and released it as an ordinary feature. That version proved to be successful and brought a few needed dollars into the Warner till.

In 1921 Harry decided to make a series of dramas that dealt with social themes. One of the first ideas the brothers came up with was *Ashamed of Parents,* which had autobiographical overtones (although the title did not express the brothers' true feelings toward Ben and Pearl). The main character, Silas Wadsworth, bore a striking resemblance to Ben Warner: he was a cobbler who worked long hours to be able to send his son to college. Just as the four Warner brothers were making good in the movie industry, Silas's son made good as a football hero and got a college degree. A second film, *Your Best Friend,* told the story of a Jewish mother who found herself snubbed by her son's gentile wife. Both pictures bombed.

Undaunted, Harry borrowed money wherever he could, many times at outrageous interest rates. Jack knew how Abe felt about these transactions: "As treasurer of Warners, Abe was a very suspicious guardian, and when he used red ink on the books, he was meaner than a cornered coon." Even with banker Motley Flint's support, Harry complained, "Most of our time was spent obtaining money from loan sharks," at interest rates as high as 40 percent.

To start off 1921, Harry chose to make *Why Girls Leave Home,* a morality tale about the corrupting effects of big-city life. The film got bad reviews in New York. Sam later recalled:

> After it was done, nobody wanted it. They argued the title was bad, nobody wanted to see a picture of that kind, so it looked like we were up against it again. We had a conference and decided to go over to Atlantic City, rent a theater, and exhibit our picture. We did, and packed the house for three weeks. After that, everybody wanted it. The picture cost $45,000 to make, and we grossed over $750,000.

With the success of *Why Girls Leave Home,* the Warners decided to open a theater in Niles, Ohio, where they had shown *The Great Train Robbery* almost twenty years earlier. In 1922 they held the "Grand Gala Opening of the Warner Theater." The "Proprietor and Manager" was listed as "Mr. Benjamin Warner." A special program was printed announcing the movies to be shown: *Miracles of the Jungle* and *Why Girls Leave Home.* On one side of the program was a photograph of Pearl with the caption THE BOSS OF THEM ALL—"MA" WARNER. On the opposite side of the page was Ben's photograph surrounded by those of his boys, Harry, Abe, Jack, Sam—and David. In the photograph, David looked sternly at the camera through pince-nez glasses. He appeared vibrant and alive. In fact, he was slowly dying of encephalitis lethargica, an inflammation of the brain, commonly called sleeping sickness.

David had been the silent, introspective brother, the one the family had hoped would be the first son to graduate from college, and for a while he did attend Western Reserve University. Harry had him run a film exchange for a while in Cleveland, where he married a young woman, Mary Ginsberg (described as charming and classy, a Jewish social butterfly in the posh Cleveland suburb of Shaker Heights), and had a daughter, Shirley. Harry was concerned about his brother because he appeared to put so little energy into running the exchange. The truth was that the degenerative disease was destroying his life.

As the Roaring '20s got under way, Hollywood found itself to be a boom town. It wasn't just movies that were being produced in this sunlit city—but stars. Names like Charlie Chaplin, Mary Pickford, Harold Lloyd, Pola Negri, and John Barrymore were becoming legends. Rudolph Valentino, a young bit actor who had played villains with slicked-back hair, had been catapulted to stardom in the 1921 film *The Four Horsemen of the Apocalypse.* Stars were making an impact on the American public's way of thinking and dressing. The flapper hairdo, a straight bob with bangs, was copied from a Japanese doll by Colleen Moore, who played a flapper in *Flaming Youth.* Moore's flat-chested, close-cropped look changed the concept of ideal beauty from big breasts, wasp waists, and hair cascading over the shoulders.

Movie producers discovered they had more power than a king: not only could they make actors rich—they could make them *famous.* No one else in the world could do that. The casting couch became the first stop on the way to stardom, but it wasn't always the casting director or producer seducing the young actress; sometimes the actress would be offering herself to anyone who could make her a star. Within six months of being "discovered," an actor or actress could walk down the street and people would fight to get an autograph. Fans would hysterically climb on a star's chauffeured car, block the entrance to studios and theaters, anything to see, to *touch* that shadowy image they had seen so many times on the screen.

The power of the studio moguls was enormous. They reigned over city hall, the police, the media, *everything.* They could make news; they could control news. They could entertain royalty. Where did the king of Spain want to go when he visited the United States? He wanted to go to a studio and see stars.

The film stars of Hollywood didn't realize they were establishing a new art form. They were happy just to be part of an exciting growing family. Not only was it fun, there was a lot of money to be made—and spent.

On Labor Day, 1921, Fatty Arbuckle, the butterball comic actor, decided to throw a party to celebrate his new three-year, $3 million contract with Paramount. He invited a crew of entertainment people to the party at the St. Francis hotel in San Francisco, including a showgirl named Virginia Rappe. Arbuckle took her to his bedroom and allegedly had sex with her, shortly after which the girl died of peritonitis. Arbuckle was charged with manslaughter. After two trials with hung juries he was acquitted, but the scandal destroyed his career. A year later, director William Desmond Taylor was found shot to death in his home, and although stars Mabel Normand and Mary Miles Minter were implicated, the mystery of Desmond's violent death was never solved.

These scandals made lurid reading, and branded Hollywood the "Babylon of the West." (As early as 1917 an actress appeared nude in a movie, an event that shocked the sensibilities of middle America.) Of the Fatty Arbuckle affair, a journalist wrote: "It was then that the finger was pointed at Hollywood as being this terrible sin city."

Movie attendance began to slump and more and more banks refused to finance pictures. There was a growing resentment toward the movie colony's high living. Shouts of "Clean up Hollywood or close it down!" began to echo in producers' ears.

Director King Vidor recalled, "I think it was Louis B. Mayer who said, 'If this keeps up there won't be any motion picture industry.'" Deeply concerned that their new art form would self-destruct, the major producers decided they would have to police their own product. They formed the Motion Picture Producers and Distributors of America Inc. and turned to Washington for someone to head it. President Harding appointed Will H. Hays, a former Postmaster General.

To satisfy the public's demand for reform, the first thing Hays did (reluctantly, he admitted later) was bar Fatty Arbuckle from the screen. The scandal-shadowed actor's movies couldn't even be shown in prisons because they might be a bad influence.

Hays quickly became known as the "czar of all the rushes" and the "Hays Office" established a moral code with a long list of "Don'ts and Be-carefuls," such as:

Don't have any suggestive nudity.

Be careful not to film brutality and possible gruesomeness.

Don't ridicule the clergy.

Be careful—the sale of women or a woman selling her virtue will not be filmed.

Don't hold a kiss longer than three seconds.

Be careful not to show women drinking.

Don't show excessive violence.

Be careful not to show a man and wife in the same bed.
Don't dare mention illegitimacy.

Hays's actions quickly reassured Americans that the moral standards of Hollywood would remain under close scrutiny. He asserted, "What the world needs is more human and heartwarming pictures."

Making movies that were "human" fit right in with Harry Warner's philosophy. He announced a series of "Classics of the Screen" that would drastically improve the quality of Warners' output, a step that was necessary if they were ever to reach the top in Hollywood. Included in this new plan were Sinclair Lewis's best-selling novel *Main Street* (the movie rights to which Harry paid $60,000 for) and *The Beautiful and the Damned,* F. Scott Fitzgerald's lost-generation novel portraying the youth of the early '20s as a hard-drinking, devil-may-care crowd. Harry also bought Charles G. Norris's *Brass* and George M. Cohan's *Little Johnny Jones.* The major producers scoffed at the Warners' ambition. "They'll never get their money back," they said. "They're crazy!"

To publicize the classics, a truck loaded with books taller than a man, the titles emblazoned on the spine, crossed the continent from New York to Hollywood. The publicity paid off: Audiences flocked to "Classics of the Screen."

Harry next got an appointment to talk to David Belasco, the great Broadway impresario. Other film producers had tried for eight years to purchase the rights to Belasco's plays, but he always responded with a firm no. Harry, with his sincere approach about making better movies for the American people, talked the showman into a $250,000 contract (plus a percentage of the profits) for three of his plays, *Deburau, Daddies,* and the hit *The Gold Diggers.* Released by Warners in 1923, *The Gold Diggers* was a success. The *Los Angeles Times* wrote, "The effect of Mr. Belasco's allying himself with the silent drama will be profound among producers of motion pictures. The magic of 'David Belasco Presents' flickering in a motion picture title is expected to pave the way to greater and better things in the industry."

Even Belasco approved of the picture. Writing to the Warner brothers, he said:

> My dear friends,
>
> After reviewing the final cutting of the print of "The Gold Diggers," I am deeply grateful, as it assured me that it was possible to faithfully portray a legitimate comedy upon the picture screen.

> It made me realize there was much already accomplished in the way of better productions to bring the screen nearer the theatre.
>
> You have indeed made a fine picture. I compliment you.

Harry borrowed money from Motley Flint and paid off the debt to Belasco for his three plays, then built a $250,000 rambling structure on the brothers' property on Sunset Boulevard to house executive offices, workshops, properties, and costumes. An elegant facade with white columns was constructed and above the colonnades block letters were added that proclaimed WARNER BROTHERS WEST COAST STUDIOS.

They spent another $50,000 on a big stage, its roof supported by fifty trusses. Jack referred to the naked interior as "The Barn." Although Jack preened when his name was listed as "Producer," the brothers decided they needed someone with experience to help out on the production end.

They added Harry Rapf, who had been Lewis Selznick's production manager, to the staff. Jack also took on a young publicity man named Hal Wallis. Other early department heads proved to be of great value. Frank Murphy ruled with an iron hand over the $500,000 worth of electrical equipment. Louis Geib started with Warners in 1918 as a technical director. In those early days he designed sets, and helped build them, paint them, and tear them down. With able employees and a decent facility to work in, the Warners were now in a position to turn out movies technically as sophisticated as any made in Hollywood.

Harry decided to take another bold step forward: He got his brothers to agree that it was time to form a corporation. On April 4, 1923, they announced the new company, Warner Bros. Pictures Inc., and stated that the four brothers would keep three-fifths of its authorized capital stock.

To celebrate, the brothers gave a buffet banquet at their Sunset studios, inviting family members and studio employees (who now numbered close to sixty). Long tables were laden with meats and cheeses, and huge punch bowls, generously fortified with bootleg gin, sparkled under the bright overhead lights.

Motley Flint addressed the assembly, saying he had first become interested in Harry, Abe, Sam, and Jack because he could see the deep love they held for their parents. "Further investigation proved to me that they were sincere and honest," Flint said. "I know that they are headed for a great success and I am glad that I am supporting them financially."

He ended by saying: "I intend to continue to support them anytime they need me."

When Flint sat down an unusual event occurred: Ben Warner got up, tapped his glass several times with a spoon to quiet everyone, then began to speak. It was the first time the sixty-six-year-old patriarch had ever stood before a gathering of strangers with any thought of addressing them.

"I do not come here to make a speech," he began in a loud voice that could be heard throughout the cavernous studio, "but now I have something I want to get off my chest.

"I hear two men . . . they talk about my boys. One of them I hear say, 'The Warners, they will never succeed in the film business.'

"'Why not?' the other man said. 'They know the business, and are hard workers.'

"'Because they are too honest.'"

Ben looked around the room. There was silence as he continued: "This has been the greatest moment of my life to hear Mr. Flint, the banker, say that he supported my boys because they are honest."

Ben stopped, thought about this for a moment, then added, "I think it would be nice for me to make one exception to this being my greatest moment." He touched the shoulder of Pearl, who was seated beside him. "It would be best if I put my wedding first."

Everyone laughed, then cheered. Before the evening was over, the punch bowls had to be replenished many times.

One note of sadness intruded a few months after this celebration: Abe Warner's wife, Bessie, died of influenza.

Now that they were incorporated, they needed to expand—and that meant borrowing more money, something that Harry was far from reluctant to do. He was not going to stand still. The studio needed capital for stages, props, actors' wages, and better literary properties.

Flint helped Harry make contact with several Wall Street bankers. One man, who would become as important to the Warner brothers as Flint, was Waddill Catchings of Goldman, Sachs.

When Catchings met Harry, the Warners were selling their pictures through franchise holders—powerful independent distributors who also advanced them money to make more pictures. Harry explained the high percentages he was paying to loan sharks, and Catchings replied that he could get far better rates for the Warners. Impressed by the movies Harry was making and by the frugal private lives Harry had forced the brothers to lead, Catchings took them in hand and said he could make sure that they got some real money. All Harry had to do was conform to what Catchings called a "master plan." After listening to Catchings's proposal, Harry quickly agreed.

The first part of the plan was to appoint Catchings to the Warners board as chairman of the finance committee. In turn, Catchings said, he would work to secure loans from banks, most of whom had made it a rule to never lend money to a movie production company. Catchings's concept was to provide Warners with a multimillion-dollar credit fund. This fund would be used to buy up other studio production facilities, exchanges, and theaters in which to show their films.

No one was more impressed with this last bit of creative financing than a vaudeville team who joked:

"Varner broders makes it a great success mit the movink pitchers, ain't it, Abie?"

"How's that, Mawruss?"

"Vell, ven dey come out here, dey got forty cents by their name, and now dey owe forty millions. Dot's success, Abie."

Jack enjoyed the joke, but he realized the serious financial situation the studio was in. "We had a continuing problem at our new Sunset lot to make ends meet. We often had to rent our studio equipment to the Wallingfords, who had bank accounts."

Something had to happen to get the studio on a solid footing. They needed a Charlie Chaplin, or Rudolph Valentino, or Gloria Swanson, someone the moviegoers would line up at the box office to see. They needed a *star.*

The star they finally got was not a baggy-pants clown, or a broad-shouldered hero, or a dark-eyed vamp.

It was a dog.

7

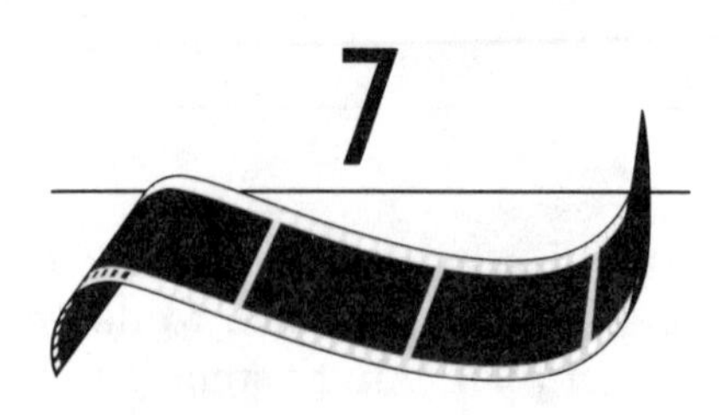

Building the Dream

Winter 1923

Harry Rapf had popped into Jack's office saying he had a great story, *Where the North Begins,* about a dog that's adopted by a pack of wolves in Canada and rescues a fur trapper. Rapf agreed there was only one problem with the story—they needed a performing dog. Rapf went on to say he knew a fellow, Lee Duncan, who owned a trained police dog.

Jack told his producer, "Hell, I've made pictures with gorillas, lions, and chimps, so this I gotta see, a dog that can act."

Lee Duncan, a tall, lanky man who had been a lieutenant in World War I, was ushered into Jack's office by a secretary. Following close at Duncan's heels was a large brown German shepherd. Duncan sat in a sofa chair and pointed his index finger at his right foot. The dog sat obediently, his eyes fixed on Jack.

"Nice-looking animal," Jack said. "Where'd you get it?"

"In France I found five puppies in a bombed-out trench," Duncan said, his voice soft. "Two of them I brought to California. Nanette didn't survive the trip. This one"—he rubbed the shepherd behind its ears—"was exceptionally easy to train."

"Does tricks?" Jack asked.

"Obeys commands," Duncan said.

"Okay, have the mutt do something."

Duncan made a quick motion with his hand. The shepherd immediately rose, bared his teeth, and growled menacingly at Jack, who involuntarily slid his executive chair back a few feet until it bumped the wall. "Okay . . . Okay . . ." He remembered the time a chimp had taken a bite out of his arm.

Duncan snapped his fingers; the dog relaxed. The trainer then pointed to a high-backed chair on the other side of the room and made a curving gesture with his fingers. The shepherd ran the short distance with effortless grace, vaulted the chair, and returned to its master.

Jack whistled. "Not bad. What's this wonder dog called?"

"Rin Tin Tin."

Jack signed Rinty—as the dog became known around the studio—for *Where the North Begins.* Rin Tin Tin was a great actor, following every command Duncan gave. Jack recalled that in the movie "the dog faced one hazard after another and was grateful to get an extra hamburger for a reward. He didn't ask for a raise, or a new press agent, or an air-conditioned dressing room, or more close-ups."

Where the North Begins played to sold-out houses across the nation, and Jack scheduled more Rin Tin Tin adventures. Ken-L-Ration put the wonder dog's face on every box of dog biscuits. The studio was deluged with fan mail, 12,000 letters and cards a week. Rin Tin Tin's pay went up to $1,000 a week, plus perks: a small orchestra for mood music; a diamond-studded collar, and T-bone steaks for dinner. Insuring against his demise, the studio trained no fewer than eighteen dogs who could double for Rinty.

Jack always referred to Rin Tin Tin as the "Mortgage Lifter."

Another of the talents Jack hired was twenty-year-old Darryl Zanuck, who had written comedy scripts for Mack Sennett and Charlie Chaplin. Zanuck came to the studio saying he had an idea for a Rin Tin Tin movie. Jack took one look at the scrawny kid and almost kicked him out of the office. Jack recalled that his first impression of Zanuck was of a "downy-cheeked youngster who looked as though he had just had the bands removed from his teeth so he could go to the high school prom."

When Zanuck offered to write a script, Jack scowled. "No, no, just tell me the story." If there was one thing Jack Warner hated to do, it was read. Zanuck began telling the story, then started acting out the scenes, playing each character and finally getting down on the floor on all fours and barking Rin Tin Tin's part. When Zanuck got to his feet, Jack told him, "Kid, go buy some glasses and grow a mustache. You'll look older. Oh, yeah, I like your story. We'll film it."

Zanuck was also a good investment and would quickly rise to the position of executive producer.

Jack Warner Jr., who was seven years old in 1923, remembers:

> Zanuck boasted he had been the prettiest baby born in Wahoo, Nebraska. Later somebody learned that up to then Darryl was the only baby born in Wahoo. Darryl read me scary scenes late in the evening when he and my father held script conferences around the table at our house, and he was the first person to introduce me to serious insomnia.

Jack Warner Sr. felt they needed a top director to work at the studio and wanted to add Ernst Lubitsch, a celebrated filmmaker from Germany, to the staff. Jack had talked to Harry about hiring Cecil B. De Mille (who he had heard was dissatisfied with working for Adolph Zukor at Famous Players-Lasky), but Harry squelched the idea by reminding his brother that De Mille was far too extravagant and demanding for their budget. Harry felt they could "handle" Lubitsch. In fact, Harry was so confident the director would follow orders that he signed a contract giving Lubitsch absolute control over the pictures he worked on.

Harry quickly learned that Lubitsch was at odds with his basic ideology about making socially important films. Lubitsch maintained that life was too drab. "The people vant happy pictures," he proclaimed. "They do not vant sadness, they do not vant failure!" Lubitsch, puffing on a cheap cigar (one of the dozens he chain-smoked each day while pacing incessantly), intoned, "I vill make delicious movies!"

Jack was mesmerized by the German's prestige and his Prussian directorial manner. But he was not amused by some of Lubitsch's on-the-set demands, such as the requirement that coffee and cakes, ordered from a special caterer, be served to the cast and crew every afternoon at four. Jack didn't like the idea of the whole crew idling on the set and canceled the coffee break. Lubitsch countered by sending everyone home and retiring to his room to play his cello. Jack heard of the delay and charged to the studio to confront the director. "What the hell are you doing?" he shouted.

Lubitsch stared haughtily down his nose. "No cakes, no coffee—no scene."

Defeated, Jack supplied the pastries.

But Lubitsch could make movies, debuting with *The Marriage Circle,* which became Warners' most prestigious production of 1924. (It was on the *New York Times* Ten Best list for the year.) Lubitsch also directed *Three Women, Kiss Me Again,* and *Lady Windermere's Fan,* based on the Oscar Wilde play. This silent movie was unique in that it was played without a single subtitle for Wilde's witty dialogue. After making *So This Is Paris,* Lubitsch, without saying a word to any of the brothers, resigned to accept a better offer from Metro-Goldwyn-Mayer. He left behind his young assistant, Henry Blanke, who would soon become one of Warners' most valued producers.

Warner Bros. had the studio and the stories, but what they needed was the stars to go with them. Rin Tin Tin and Monte Blue were kept busy making action dramas and comedies, but Warners did not have a big name to flash across the marquee.

A name like John Barrymore.

Sam and Jack got on the phone and called Harry and Abe in New York. "Harry," Jack began, "we've been hearing a lot about this guy . . ." He turned to Sam. "What's the moniker on this stage guy?"

"They call him the 'Great Profile.'"

"Yeah, Harry, the Great Profile guy. He's a big-deal dramatic actor in New York."

"You mean Barrymore," Harry said.

"Sure, the Barrymore guy. He's gotta have real acting talent." Hearing no reply from Harry, Jack added, "He's Broadway's top star, right?"

"Yes, but that doesn't mean he can act in movies," Harry replied, always the devil's advocate when it came to actors.

"C'mon, Harry. Women love him."

"There are stories about his womanizing," Harry said. "He drinks, too."

Jack ground his teeth, trying to hold his temper, and whispered to Sam, "Harry's worried that Barrymore will shtup his way through Hollywood. Hell, if we can get the guy, I'll set him up with every good-looking babe on the lot—and buy the champagne." To Harry he said, "You seen Barrymore act?"

"No."

"Just see him, Harry. Just go to the theater. See him."

Harry went to watch Barrymore play *Hamlet.* He sat through twelve performances before he was satisfied enough to offer the temperamental, hard-drinking actor a contract to star in one—and only one—Warner Bros. feature.

The film Sam and Jack had scheduled for their new star was *Beau Brummell,* with Mary Astor. When Barrymore, a famous connoisseur of female perfection, saw Astor, he whispered in her ear, "You are so goddamned beautiful, you make me feel faint."

Beau Brummell was so successful (it also appeared on the *New York Times* Ten Best list for 1924) that Harry signed Barrymore to a generous long-term contract. Barrymore got $76,250 per film, and if the movie didn't finish shooting in six weeks, he would be paid an additional $6,625 a week. The perks added to the contract were an actor's dream: leading-lady approval, choice of movies, a suite at the Ambassador Hotel, a limousine and chauffeur.

During 1923 the Warners advertised eighteen pictures, but produced twenty. As Sam Warner said, "Now the sky began to get smoky for us—the big producers began to knock us instead of boosting."

In a speech Sam gave in 1924, he told of his feelings during those early days of struggle:

> Things were so bad, whether it was intentional or not, I can't say, but we couldn't get a positive print for ninety days from a laboratory when our negatives were finished. Our films were delayed at film processing laboratories owned by the big companies, or lost in transit. My brother, Harry, came to me and said, "How soon can you put up a laboratory?"
>
> I told him, "Thirty days." This was a big order for a building 130 feet long, three stories high, conditioning rooms, etc. Well, I didn't do it in thirty days, but I did do it in ninety days so we could process our own film. We then had a plant big enough to do all the work turned out by Hollywood.
>
> There is only one real danger confronting this industry. If a large consolidation ever occurs, they wouldn't want so many theaters (half the theaters could take care of all the business), so the little companies would be forced out of existence. A general director would send you a notice of the picture you were going to play with its price and play date. Exhibitors would be at the mercy of a consolidated combine.
>
> We are the largest independent producers operating. We help keep up the courage of the other independent companies, but we are virtually forced out of "first runs" this year."

Then Sam paused and added the shocker:

> But Harry Warner is on his way east now with a guaranteed fund of ten million to build us a theater in the key cities.

In late 1924 Harry had begun to complete the plan that banker Waddill Catchings had proposed a year earlier. Part of the plan was to acquire movie theaters.

Distributing their films had become a real problem for the Warners. They had been working through a loose organization of independent agents who took large percentages of the profits to book their films into theaters. The agents believed the Warners, as well as other smaller family-owned independent producers, were at their mercy. Harry envisioned a Warner Bros. theater division to provide a guaranteed outlet for Warner films.

After conferring with Waddill Catchings and being guaranteed funds for construction, Harry began a tour of the major U.S. cities to select sites for a string of theaters in which Warner Bros. products could be given first-run presentation.

"It has never been our wish—or part of our plan—to engage in the exhibition end of the motion picture business," Harry said to a group of reporters who had followed him on his quest.

> Today we would be perfectly satisfied to go on producing photoplays if the exhibiting trade at large were in a position to give us what we consider a halfway run for our money.
>
> We are taking off our coats to do battle. We are going to fight the combinations which we are satisfied are out to ruin the industry for everyone but themselves. We will bring to the independent exhibitor the strength he needs to stave off destruction—and bring it just when he is beginning to feel the need of it most.

To prove he was serious, Harry confirmed plans for opening newly constructed Warner theaters in New York and Hollywood. The Hollywood theater would cost $1.25 million and seat 3,600. The *Los Angeles Times* noted that the theater would make "Warner Brothers the only organization to produce, print, and exhibit its pictures in Hollywood." Harry had virtually thrown down the gauntlet to Paramount, Universal, and Metro-Goldwyn-Mayer, the "Big Three" that owned and controlled chains of theaters. The Warners had, as a newspaper story said, "declared war to force their way to a place in the sun in order to insure first-run exhibition of their pictures in large cities."

"We wish we did not have to build theaters and could devote our efforts to the production of high-class pictures," Harry said to reporters, then added, "It is a Napoleonic desire for more power that is behind the movement of the monopoly. *Personally, I believe in letting the little fellow live.* I can make enough money out of producing pictures to satisfy me."

At a convention of 1,500 independent exhibitors in Milwaukee, Harry made an impassioned speech demanding that the Big Three "lay their cards on the table and admit to becoming a monopoly!"

The Big Three lashed back: "Warner's attack is just a case of publicity-seeking by an organization that hopes to attract attention to itself through attacking successful companies."

Harry countered by declaring that $500,000 would be spent for newspaper advertising in the campaign for independence. The same day, Harry heard the most welcome news since he had begun the fight: Mary Pickford, Douglas Fairbanks, and Charlie Chaplin had joined forces with Harry by pledging another $500,000 for publicity to support independent exhibitors against the Big Three. Harry's power play had worked.

Harry was now ready to take the biggest gamble of his career. Once again looking to Waddill Catchings's original plan, Harry set out to buy Vitagraph, one of the oldest motion picture companies in the country. Vitagraph had been formed in 1896 by Albert E. Smith and James Stuart Blackton, a reporter who had been sent to interview Thomas Edison about his new "continuity pictures" and became so intrigued by the inventor's Kinetoscope that by 1900 he and his partner had thirty machines projecting pictures in theaters throughout the country. By 1925, Vitagraph had a studio in Brooklyn, a twenty-acre studio in Hollywood, the largest motion picture library in the world, and, most important, over fifty distributing exchanges in thirty principal cities across the country, plus an elaborate system of foreign exchanges: four in Canada, ten in England, and ten in continental Europe.

Although their assets looked good on paper, Vitagraph was going downhill, deeply in debt—and ripe for a buyout. Vitagraph was a prize that Harry was determined to come home with.

He hurried over to the Vitagraph offices in Brooklyn and checked out the company's debt. The figure came to $980,000. Harry asked the company directors what they wanted for Vitagraph. They went into a huddle. The answer came at 3 A.M.: $800,000, plus enough to take over the $980,000 debt. Harry agreed. A paper was drawn up in longhand which they all signed. Harry wrote a check for $100,000, money he didn't have, for the biggest deal of his life.

The next day, Harry went to the bankers Catchings had recommended. Putting on a bold front, he told them what he had done. "If you want to get involved," he said, "it'll cost you a million and a half."

The bankers mulled this over, then said, "You'll need some capital, an extra million to act as a cushion."

Harry nodded his head.

The bankers pushed a few more papers around, then announced, "Just to be sure, we'll lend Warner Brothers four million."

When reporters later asked why bankers felt compelled to lend him money, Harry responded, "I don't always get the loan. I'm like the fellow who asks every girl he meets for a kiss. Occasionally I get slapped, but sometimes they kiss me."

I have many times wondered where my grandfather got the chutzpah to take on such odds. After all, he was the son of an immigrant cobbler, a man who didn't have even a grade school education, and yet he had no fear of approaching the most powerful, educated men in the banking industry.

On April 23, 1925, the contract to buy Vitagraph and all its subsidiaries was concluded, putting Warner Bros. in direct competition with the largest distributors and producers of motion pictures in the country. In the few short years since the brothers had arrived in Hollywood and set up their raggedy studio and filmed *A Dangerous Adventure,* they had become the largest family-owned independent producer of motion pictures in the industry.

On the day the newspapers announced the momentous purchase of Vitagraph, gossip columnist Louella Parsons noted:

> Albert Warner was married today at 11:00 A.M. to Bessie Levy Siegal in the home of her father, Mo Levy. Mo Levy the clothier. Immediately after the ceremony Albert and his bride motored to Atlantic City for a honeymoon.

Abe Warner, impervious to all the turmoil that had been unleashed by his brother, had gone ahead with his wedding to Bessie Siegal, an attractive, refined, and intelligent woman. Bessie's first husband had died around the same time as Abe's first wife. The two couples had been friends.

While Abe was honeymooning in Atlantic City, Harry began putting the Vitagraph holdings to good use by announcing that Warner Bros. planned on producing forty pictures in 1926, including a romanticized version of Herman Melville's *Moby Dick.* Retitled *The Sea Beast,* the movie would star John Barrymore, who had tired of playing "scented, bepuffed, bewigged and ringletted characters" and asked Jack to star him as Captain Ahab. Since Melville had neglected to supply a love interest, Jack got one added to the script and signed little-known Dolores Costello to play opposite Barrymore. (The actor's first choice, Mary Astor, was unavailable.) Barrymore called Miss Costello the "most preposterously lovely creature in the world," and their romantic scenes practically ignited the celluloid. Although *The Sea Beast* cost $800,000, more than any other Warner Bros. film, it was to become a huge, profitable success.

Moguls from the major studios continued to put the squeeze on these Poverty Row upstarts. Some of the dealings were so petty that the brothers loved to relate the stories. Jack took a young writer to lunch, saying he'd heard good things about his work and wanted to hire him. The writer said that he'd had lunch the day before with Louis B. Mayer, who'd also offered him a job.

"Don't do it, kid. I hear Mayer's about to lose his shirt."

"That's what he told me about you," the writer said. "Told me Warners was going under."

Jack had to smile at that. "So who do you want to write for?"

"Warners," the writer replied.

"Great, kid. Must be my charm that talked you into it, huh?"

"Nope." The writer wiped his lips with his napkin. "Mr. Mayer bought me chicken. You paid for steak."

One of the men Sam had hired as the company grew was an excitable Irishman, Frank N. Murphy, who became the chief electrician of the company. Both mechanically minded, Murphy and Sam began to take a look at an entertainment phenomenon that was sweeping the country: radio. In 1925 Los Angeles had two successful stations, KNX and KFI. Sam thought it would be a progressive idea to put up a third broadcasting station, this one at the Warners studio.

Besides, he told his brothers, "it would be a great way to advertise our movies. It would be the first of its kind on the coast. Think of it as another form of entertainment, one that could stand alongside the legitimate theater and motion pictures."

Jack, the frustrated performer, immediately liked the idea. Abe, who along with Harry was outraged at the rates charged for radio time in the Los Angeles area to advertise their new movies, thought it was something that should be looked into. Harry, seeing that his brothers were in agreement, said, "Buy one."

Sam discovered that the cost of setting up a new radio station was prohibitive. But he found an existing station, KWBC, for sale. One bright morning nine truckloads of radio apparatus and an engineering crew from Western Electric Company arrived in front of the studio.

Frank Murphy looked at the trucks, then at Sam. "What am I supposed to do with all that?"

Sam, sorting through the wires and cables, smiled at his electrician. "How long will it take you to install it?"

"How long? Hell, I don't know." Murphy conferred with the Western Electric engineers who had come along to supervise the installation. "Three weeks to a month" was their reply.

"I want it running in a week," Sam said, aching to get his hands on the equipment.

The engineers groaned. "Impossible."

The station was operating within the week. It was Western Electric's first run-in with Sam Warner, but it was one they would recall in less than a year.

Sam and Murphy, neither of whom had touched a vacuum tube before, were fascinated with the radio station. They spent hours each day studying the mysteries of amplifiers, microphones, and sound monitors. Sam translated the station's call letters, KWBC, as "Warner Bros. Classics."

The call letters were changed when Ben Warner dropped by the studio one day and told Sam he should rename the station KFWB. "Why, Pop?" Sam asked.

Ben smiled. "Because that stands for 'Keep Fighting, Warner Brothers.'"

On March 3, 1925, KFWB broadcast its first program. Jack later said, "This event was either gummed up or made memorable—depending on your point of view—by the singing of one Leon Zuardo, a former boy soprano whose voice had turned into what was more of a vodka baritone than a whiskey tenor. . . . Leon Zuardo and Jack Warner were one and the same. In spite of this, KFWB prospered."

Although Jack did get his opportunity to sing "When the Red Red Robin Comes Bob Bob Bobbin' Along," and relish the applause from a studio audience, the radio station's primary purpose was to advertise Warners movies. As part of the programming, a microphone placed on a studio set during the filming let listeners tune in to movies being made.

To Sam, the sound coming over the airwaves meant something more than it did to Jack. He realized that the public was becoming accustomed to radio as part of their daily lives. Each morning as he drove to the studio, the first thing he saw was the station's twin metal radio towers standing like sentinels, one on either side of the studio's colonnaded entrance. To Sam, the tips of the towers sparkled and crackled with the sound of voices.

Voices.

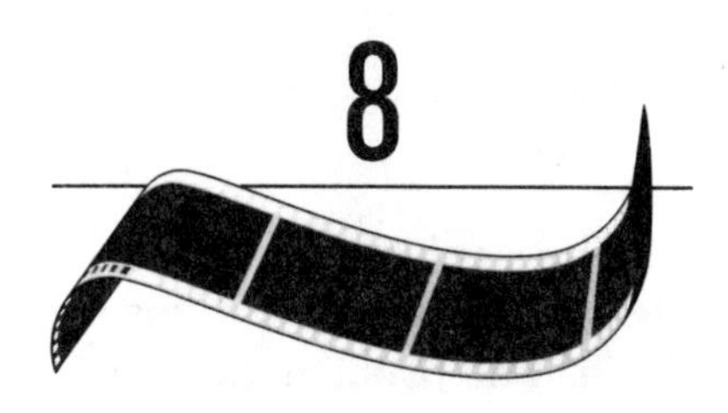

8

Sound and Fury

"Ch-ch-cheese c-c-crisps, cheese crisps."

In the small darkened room at the Bell Laboratories, Sam watched the screen intently as the man stuttered the words again. "Ch-ch-cheese c-c-crisps, cheese crisps." The man whistled a tune for a few seconds before waving his hand and telling the cameraman, "All right, you can shut the thing off now. Yeah, shut it off." He started to get perturbed. "Would you please shut the thing off!"

It was crude, but it was *perfectly synchronized,* and it sounded like a real person speaking.

On the screen a musician sat at a piano and smiled at the camera. Then, resting his fingers on the keys, he began to play. And Sam *heard* the music. Not just the music, but also the rustle of the musician's clothes, the scraping of the piano bench on the floor. The piano player was joined by another musician with a violin, then another . . .

Sam had been invited to the Bell Labs in New York to watch the sound demonstration by Nathan Levinson. Levinson, a representative of

Western Electric and Bell Telephone Laboratories, and a genius with electrical equipment, had worked with Sam for a year establishing the Warner radio station. He had no problem getting Sam, who greeted any new technical development like a kid opening an Erector set, to visit the Bell Labs.

After watching the demonstration, Sam was sold. The problem was to sell Harry and Abe. He remembered Harry stating: "All the major motion picture concerns have dismissed sound as foolish." Even Abe had said, "Talking pictures are the bunk!"

Harry knew about the sound tests on film being done by different individuals and companies, but he really wasn't interested. Ten years earlier he had sat in a small theater, eagerly awaiting a sound experience called Kinetophone that had been developed by Thomas Edison. Finally, a couple of costumed Romans appeared on the screen and began to speak; that is, their lips began to move. Then a thin metallic sound came forth—one of the men was speaking. But the voice was several seconds behind the lip movements. The film flickered and went out, but the voice continued, rising in a final outburst. It was a feeble effort, doomed by poor production and poor synchronization.

In early 1925 Harry had sat through another sound demonstration in which a man appeared on the screen, pursed his lips, and blew air. Whoosh, whoosh. "How's that? Can you feel the air?" The man scratched his eyelid with his finger. "I'd like to know if you can feel the air coming to you people that are sitting there watching the picture." He smiled, slightly embarrassed, then asked, "Can you feel it?" He squinted and looked upward. "Gee whittakers, those lights are awfully bright. Okay! You can turn them off!"

Even the great pioneer director D. W. Griffith had experimented with sound, filming a talking clip of himself introducing his 1921 movie *Dream Street.* In sepulchral tones Griffith pronounced: "Ladies and gentlemen, I want to say a few words concerning the play of dreams which you are about to see—*Dream Street*—and I wonder if there isn't a dream street running through the heart and soul of every human being in the world."

Griffith's dream did not bring sound to the screen and he quickly gave the idea up, saying it was "professional suicide." After all, only 5 percent of the world spoke English, and he would be depriving his films of 95 percent of their potential audience. A journalist of the early '20s wrote: "Silent films made us all one people all around the world with one language." Theater moguls agreed: They would stick with the silents.

The movie producers might have forgotten about "talkies," but the inventors hadn't, and they started tackling the problem from different, more radical angles. A determined inventor by the name of Lee De Forest developed a system whereby sound was recorded directly onto a strip of film on a narrow margin just outside the sprocket holes. It had one great advantage: It couldn't get out of synchronization, even if the film broke. There was one disadvantage: If there was one tiny scratch or blemish on the film, it made harsh static noises. De Forest went ahead and demonstrated his device at a New York theater in 1923 by matching music to a Pola Negri film, *Bella Donna.* Patrons thought it sounded like the musicians were in a deep orchestra pit.

Technicians and engineers at Western Electric, working in conjunction with Bell Telephone Laboratories, began developing sound-on-disc synchronization in late 1923. The idea called for recording the soundtrack on a 16-inch, 33⅓-revolution-per-minute wax disc that played from the center outward. During the filming, a series of gears connected to the camera ensured synchronization, then the same process was used to couple the disc to the projector.

Sam realized the opportunity to work with Bell Laboratories was coming at a turbulent time for Warner Bros. The Vitagraph deal was absorbing a lot of their energy; there were contracts to be signed, loans to be paid back; bankers were constantly at Harry's heels. At Harry's insistence, Jack had just announced they would make forty pictures the next year, and now Harry was crossing the country looking for theater sites in an attempt to break the Big Three's monopoly.

Sam knew he could never get his older brother to voluntarily attend a "talkie" demonstration, so he and Levinson contacted Harry and said they needed him at a meeting of Wall Street bankers. They told him that after the meeting they had planned a little reception—"for social purposes"—with some of the officials at Bell Laboratories. Harry agreed.

On the way to the Bell Laboratories "reception," Harry stopped at his bank to sign a few papers. He ran into Waddill Catchings and told him where he was going. Catchings, always formal in speech and manner, grasped the lapels of his dark blue suit and said, "My dear Mr. Warner, you can have all the money you want from this bank, but if you don't mind, I would like to have one of my partners accompany you to the Bell Laboratories." When Harry started to protest, Catchings, who knew Harry's penchant for plunging into any new deal, added, "It isn't that I mistrust you, but I'm afraid that, left to your own devices, you'd wind up buying the whole of Bell Telephone and Western Electric Company!"

Catchings assigned a fellow banker, a roly-poly man who looked like a German tuba player, to accompany Harry. Harry, figuring it would be a wasted afternoon, agreed to have the little banker come along. At least they could talk about investments.

The flickering light illuminated the screen in the projection room and a man appeared on the screen. Looking terrified, the man opened his mouth to speak—and stuttered:

"Ch-ch-cheese c-c-crisps."

Harry felt it was a rather silly demonstration of how speech, even impaired speech, could be synchronized with the motion picture. There were extraneous scratches as the voice came over the loudspeaker, but the quality was startlingly realistic. Harry was wary, holding his judgment until he saw more.

That was not the attitude of the banker who sat beside him. Eyes wide, his lips mouthing the stutterer's speech, the little man exclaimed, "W-w-w-wonderful. It's—it's wonderful."

Harry looked at the banker, then at the stutterer, and began to pay attention to what was happening up on the screen. If this little banker loved it that much, maybe others would too.

The scene on the screen changed to a short series of vaudeville acts, then a twelve-piece orchestra began to play. Each instrument seemed alive: Sound erupted from the brass . . . the siren pitch of a trumpet, the mellow growl of a saxophone . . . it was so real! Harry later said, "I could not believe my own ears. I walked in back of the screen to see if they did not have an orchestra there synchronizing with the picture. They all laughed at me. The whole affair was in a ten-by-twelve room. There were a lot of bulbs working and things I knew nothing about, but there was not any concealed orchestra."

Back in his fourteenth-floor office, Harry leaned back in his swivel chair and looked out the tall window at the stores on Eighth Avenue. After a few moments of silence he said to Sam and Abe, "I agree. We can use sound in pictures."

Sam grinned. "Great. Let's make a talkie. We take an actor . . . how about Barrymore?"

"Sam," Harry said, "maybe we shouldn't use the name 'talking picture.' The thought occurred to me that we quit the idea of a talking picture. Let's devote our energy entirely to music. All we'd be doing is taking what the audience is already accustomed to—background music with their movies—and making it easier for the theater owner. Think what it would mean to a small independent owner to buy his orchestra *with* his picture. Nobody'd have to have a piano or organ, or hire an organ player. He'd never have to put out any extra money for musicians."

Harry lit a cigar and continued, "Everyone's building five-thousand-seat theaters—motion picture palaces, they call them—and investing enormous amounts of money, and then what's everyone doing to keep the theaters up to standard? What? Hiring big orchestras and big acts, that's what. Just to keep an audience interested. And it's costing a lot of money. They need something new. Sound can be the answer, but"—he pointed his finger like a pistol at each of his brothers—"*only* as musical accompaniment."

Seeing Sam's disgruntled look, Harry smiled and added, "We could ultimately develop sound to the point where people *ask* for talking pictures."

Sam brightened. "*Now* you're talking." The three brothers laughed and Abe lifted his huge frame from the chair. "Okay," Abe said. "Let's do it."

Sam remained slouched in his seat, already turning over in his mind how they could make the thing work, how the synchronization operated, what kind of camera was needed, sound studios . . . He had to bank on the ability of the Western Electric Company and their Bell Laboratories to take care of the proper working of the instrument itself. It was up to them to make it right, as they would have just as much at stake as he and his brothers. Finally he said, "I'll go along with the orchestra idea on feature pictures, but that's a big project. We need to learn the process with shorter stuff. I don't want to get stuck just synchronizing movies with music. If people can be made to play instruments on the screen, they can be made to sing, too. "

Harry sucked in on the cigar and swirls of gray-white smoke swirled upward, creating ghostly images. "Okay, let's get the greatest artists and the best orchestras in the country. Let's have confidence in this and put all our muscle behind it. We'll know the result after we've opened the first show."

"What are we going to call this sound thing?" Abe asked. "What'd you call it, Sam? An apparatus?"

Sam said, "We just closed the deal with Vitagraph. Maybe we can use something like that. Vita . . . something."

The brothers decided to call the new sound system Vitaphone. The word meant "living voice." But there was a stumbling block: a man named Walter J. Rich. He had already obtained an option from Western Electric to see whether he could commercially develop the talking-picture apparatus.

Harry immediately got together with Rich, who had already spent $36,000 developing the device. Rich said he would sign a contract to share and share alike if the Warners would contribute twice his investment sum. Harry, having made the decision to go ahead with the project, and gaining the approval of Western Electric, agreed to Rich's proposals. Western Electric would provide the technical competence, Warners the artistic.

On April 26, 1925, the contract was signed. Harry issued this statement:

> Vitaphone will bring to the audiences in every corner of the world the music of the greatest symphony orchestras and vocal entertainment of the most popular stars of the operatic, vaudeville and theatrical fields.

Focusing on his own ideals of what the movies should do, Harry added that Vitaphone would also be used in the "educational, commercial, and religious fields." Then, referring to the movie producers who had not embraced the idea of sound, he went on to say:

> Do not forget that we are doing this single-handed today. We are doing it with our own money because we believe in it. We honestly believe that Vitaphone is going to do more good for humanity than anything else ever invented.
>
> We all know that if you and I can talk to one another, we can understand one another. If Lincoln's Gettysburg Address could be repeated all over the world, maybe the world at large would understand what America stands for. . . . If we have a message of friendship or enlightenment that can be broadcast throughout the world, maybe the nations will be led to understand one another better. Vitaphone can do all that.

The Warner brothers had suddenly thrust themselves into the sound picture business. Sam Warner was put in charge of the project. The first thing he did was to convert the old Vitagraph Studios in Brooklyn to making sound pictures.

The second thing he did was to marry an eighteen-year-old Ziegfeld Follies dancer named Lina Basquette.

In 1923, when she was only sixteen, Lina Basquette was billed as "America's Prima Ballerina" on the Ziegfeld Follies stage, where she danced and split laugh lines with Fanny Brice and W. C. Fields. Lina, under the watchful guidance of her mother, Gladys, had been a child star for Universal Studios, making thirty-nine featurettes as a dancer/actress. Lina later liked to say she was "the Shirley Temple of 1916."

Now, in 1925, Lina wore ermine, pearls, and peacock feathers. She later recalled: "I was a ballerina with the most beautiful legs in the world. Flo Ziegfeld loved long beautiful dolls with long legs, very slim, for his Follies. I was a little more buxom than the average, so he had to have me flattened down."

When Sam came to New York to witness Bell's sound presentation, he also chanced to see Lina perform in a musical comedy, *Louie the Fourteenth,* in which she was billed as the *première danseuse.* Sam, a thirty-seven-year-old bachelor who had known a lot of good-looking girls in his life, took one look at Lina and knew he was in love.

Other Voices

Lina Basquette

Sam fell in love with me when he saw me over the footlights in *Louie the Fourteenth* and sent backstage an orchid corsage with an invitation to dinner for both me and my mother, who was guarding me like the Holy Grail. Sam was bargaining with the one person who held the key to the merchandise, Mama. I could never figure out whether he was courting me or my mother, as she was closer to him in age and still very beautiful. I always suspected that Sam had an affair with my mother during the two months he took us out to "dinner."

Everybody liked Sam. They thought he was a big Irishman. He loved to go out with the boys, watch baseball and basketball. He didn't drink, but smoked a cigar now and then because everyone else did. He had sandy hair with flecks of gray and was tall. He always dressed in a three-piece suit and didn't like to wear "soup and fish," the white tie and tails that other stage-door Johnnies wore. I couldn't help liking Sam. He had an amiable, outgoing personality, and while he seemed "old," he was not unattractive. One of my showgirl friends let the cat out of the bag, informing me that he'd been quite a guy with the gals.

One day, my mother said to me, "Lina, you are going to marry Sam Warner."

My first thought was that Sam wasn't a Catholic, and I reminded Mama of this. She said, "Well, he doesn't look Jewish!" To my mother, Sam Warner had one major virtue—he was a motion picture producer—and it didn't matter about his religion, or age. He was a good catch.

The wedding was set for July 4, 1925, as the new Follies (which I was to star in) was to open two days later. Sam never gave me an intimate kiss until the night before we were married. It scared the hell out of me. I didn't know anything about men . . . and I felt his erection against me. I went to my mother after he left the apartment and said, "Please, I don't want to get married. Let's just postpone the wedding."

My wedding dress was chartreuse chiffon trimmed with lace and black velvet bows. I had a large picture hat to match. Sam had a rabbi perform the rites, at the rabbi's home, and Sam's best man was the banker Motley Flint. None of his brothers or parents were there.

Sam gave me a big wedding reception at the Biltmore Hotel in New York. Because of Prohibition, whiskey was camouflaged in coffee pots and drunk out of demitasse cups. Sam invited people from his New York office and some of my friends from the Ziegfeld Follies and I danced until 2 A.M.

The wedding night was spent on the fourteenth floor of the Congress Hotel. The apartment belonged to Sam's brother, Abe, and his wife, Bess, who were spending the summer in Europe. Sam put me on the couch in the living room and took off my shoes. I was wearing high heels, which I was not accustomed to as a ballerina, and my feet were covered with blood. I went into the bathroom and washed my feet, then went into the bedroom—and locked the door.

Sam knocked, twice, on the bedroom door . . . I panicked. I opened the window and straddled the sill with my long legs, looking down fourteen floors at the lights of cars going by. I sat there thinking about jumping. Then I had second thoughts: I had no real feeling for Sam, but there was nothing wrong with him. This was a new experi-

ence in my life. Besides, I wouldn't have to deal with my mother anymore. I walked across the room and turned the key.

Lina quickly learned that, as she said, "Marriage to Sam was not a bad deal after all. He was very sweet to me, and it was so good to get away from my mother. I discovered I had a pliant body, responsive to the series of exciting adventures that continued throughout the nuptial night." After a quick trip to Hollywood to meet Sam's parents (Ben and Pearl had moved into a bungalow next to the Sunset Boulevard studio in early 1925), the newlyweds returned to New York and Sam began the grueling task of making Vitaphone work.

There is an interesting footnote to the Western Electric contract that was signed a little over a week before Sam and Lina were married. My great-aunt Lina told me, "There was a lot of anti-Semitism going on in the days when Sam first signed the contract with Bell Telephone. The brothers never would have gotten the rights to the sound equipment if Bell had known Sam was a Jew." She claims it was her wearing of her little gold crucifix to a formal dinner with the directors of Western Electric, AT&T, and Bell Labs that put the deal over.

Sam began the renovation of the old Vitagraph studios in Brooklyn. He tried to insulate the thirty-by-fifty-foot space by hanging rugs over the walls and draperies from the glass skylight that ran the length of the room. Stanley Watkins, a Bell engineer, went around the set clapping his hands, listening for echoes. Sam, a big grin on his face, would follow him, hiding behind drapes and clapping his hands with perfect timing, convincing Watkins that he was making an echo. Watkins hung three times as many drapes as they needed.

Sam had already figured it out: the old studio wouldn't do for sound recording. There was no way they could blot out the rumble of the subway, which ran directly underneath the stage. The ground tremors, plus the pigeons scratching on the glass roof, drove him and his crew to take a year's lease on the Manhattan Opera House, where they hoped the working conditions would be more favorable. The semicircular orchestra pit was boarded to allow space for sets to be built and cameras to be set up.

Cameraman Ed DuPar tried a couple of test shots with the synchronization of the wax discs. He and Sam were dismayed to hear all sorts of crazy noises being caught by the delicate ear of the recording mechanism. "Surface noise," as Sam called it, was made up of a variety of microscopic sounds, partly electrical and partly mechanical. There was also "spot noise": footsteps, the shuffling of property men, sneezes, coughs, and other accidental sounds made by the crew. Everyone was told to move around in stocking feet. Even worse, Sam discovered the camera itself had to be squelched. Its gentle whir was now like the roar of a waterfall.

To correct the problem, cameras were set up in insulated boxes big enough to hold the cameraman and director. The overhead stage lights turned the boxes into ovens, practically broiling the men inside. The old arc lights made so much noise that Sam had to take time out for one of his technicians to develop a high-powered incandescent light. Sam commented: "The lights were hot enough to fry people. The opera singers opened their mouths and the heat reached their tonsils."

One day the camera booth's door came unlatched during the recording and the sound of the camera spoiled the take. After that, the poor cameraman was locked in. Also after that, Sam, who was directing the early tests, decided to remain outside the camera box. Harry and Abe would drop by occasionally to see how the tests were progressing, but stayed out of Sam's way. Jack remained in Hollywood, cranking out silent movies.

At the Manhattan Opera House, two sound experts, Stanley Watkins and George Groves, both under contract to Bell Laboratories, worked separate sound boxes, learning to skillfully vary the density and volume of the sound that came from the microphones placed high above the cameras. But that wasn't their major problem: There was no way of editing sound. The wax recording discs were sixteen inches in diameter, two inches thick, and highly polished. After the recording was completed, the disc was sent to the lab to be copper-plated. Because the disc could not be stopped once it started, each recording sequence had to be shot with a full ten-minute roll of film. Any error made by the crew or performer meant the whole sequence had to be filmed and recorded again.

It was exasperating work, and even Sam admitted to almost losing his nerve. To become proficient at using the sound system, the crew made several experimental films. Thomas Watson, the man to whom Alexander

Graham Bell had said "Mr. Watson, come here, I want you" on the first telephone transmission, delivered a short address in one of these first Vitaphone films. Distinguished and white-haired, Watson spoke directly to the audience from behind a desk laden with telegraph and telephone instruments, all of them marvels of the age of sound.

Harry told him to synchronize film with music, so Sam hired eight Russian singers to make a short feature called *The Song of the Volga Boatmen.* Much to Sam's chagrin, it turned out to be a comedy of errors.

A set was constructed in the studio depicting a riverbank; in the background on the opposite shore was a silhouette of onion-shaped domes and houses. The Russians were to drag a boat along the shore while singing the rousing "Volga Boat Song." Since only the bow of the boat would show in the scene, just the front quarter of the hull was built. A huge hawser was attached to the canvas hull. When the boatmen pretended to heave the boat upriver, it looked like they were tugging on the leash of a toy poodle. One of the Russians, a giant of a man, was enlisted to perform a tug-of-war offscreen with his compatriots.

Another unseen problem was getting the performers to stand in the exact spot to be recorded. The actors and singers complained that they were "slaves of the microphone." Sam had red spots painted on the floor so a performer would know just where to speak. No longer could an actor rush into the scene while exclaiming, "Darling, I'm here at last!" He had to hit his mark first.

As Sam became more familiar with the sound equipment, he began to make featurettes, borrowing talent from the Victor Talking Machine Company. Famous names like Enrico Caruso, who sang "Vesti la Giubba" from *Pagliacci,* and Eddie Cantor, who clapped his gloved hands and rolled his banjo eyes, appeared before the camera and microphone. George Jessel did one of the first talking shorts when he spoke lovingly to his mother on the telephone. But the one-reeler that hinted at what the future of sound movies might hold was a sequence showing a performer in black face and overalls walking out the door of a log cabin singing "April Showers." The singer's name was Al Jolson.

The purpose of the shorts was to be sandwiched between the silent feature productions that Jack (with the able assistance of Darryl Zanuck, who had been promoted to producer) was churning out at the Warner studio in Hollywood. Warners had only two theaters, the ones in New York and Hollywood, wired for sound. Theatergoers chuckled at the novelty of the featurettes and considered them good gags, but no one was really taking sound seriously. Audiences were hard to sell on the idea of sound, because they expected perfection from the start. Some people actually threw tomatoes at the singers on the screen.

Harry, with Sam's urging, realized it was time to take the next step: film a major production in which music was synchronized to the screen action. The brothers got together and decided to feature their superstar, John Barrymore, in the first full-length Vitaphone excursion into sound.

The property they finally chose was the swashbuckling adventure *Don Juan.*

9

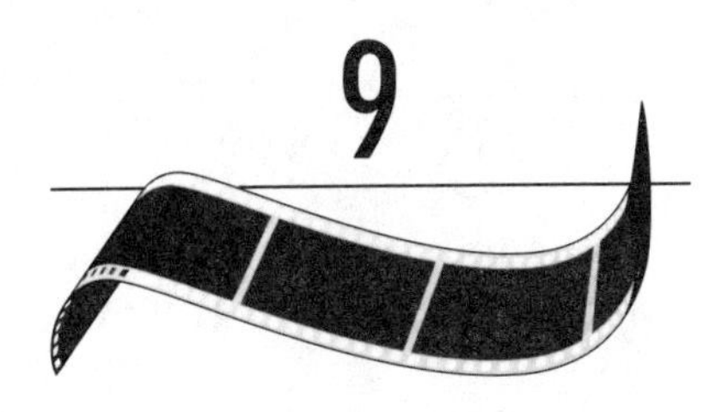

The 191 Kisses

Spring 1925

Jack Warner welcomed John Barrymore into his office and told him, "We're going to add sound to *Don Juan.*"

Barrymore looked down his aristocratic nose and turned slightly so the light from Jack's office window illuminated his profile.

"I see," Barrymore said.

"We're putting an additional two hundred thousand dollars in it," Jack added. "That cash goes for the orchestration and dubbing in some sound effects, swords clashing . . ."

"I see," Barrymore said, as he visualized the climactic scene where he heroically fights his way down the grand staircase in the palace of the Queen of Spain, crossing swords with . . . Ah, yes, and not only see it, but *hear* it. Well! Barrymore loosened the knotted belt of his full-length cashmere coat, took off his hat, and with a sweeping gesture spun it to the couch.

Jack went on, "We'll still film *Don Juan* like a silent here in Hollywood and mix the sound at the New York studio. It'll be the first

Vitaphone movie, so there's no way we can make a dime off it. So far we've only got the New York and the new Hollywood theater wired for sound."

Barrymore settled languorously onto the stuffed couch opposite Jack's desk. "I see." He saw a lot. *Don Juan* . . . Vitaphone . . . the first sound movie . . . This would qualify him for immortality in moviemaking history. Him and his bedmate of the moment, Dolores Costello.

Barrymore said, "Dolores will join the cast."

"That's impossible," Jack said flatly.

Barrymore rose swiftly from the couch, his eyes shooting daggers across the room. "I demand it!"

Jack stood up and pounded his fist on the desk. "I said goddamn well that it was impossible! I've signed Mary Astor to star with you."

Barrymore fumed but did not respond. He was a bit taken aback by Jack's outburst: The man was beginning to act like a movie mogul. Besides, Barrymore wanted to do the movie—to be a star in the first major motion picture with sound.

My great-aunt Lina Basquette had an interesting comment about Barrymore's relationship with Jack. She said Jack had sent Dolores Costello a radio-phonograph and an electrician to install it. The electrician wired it to a Dictaphone so it would transmit everything that went on in the actress's bedroom. A stenographer in Jack's office took it all down in shorthand. Lina said this was a way of getting Costello and Barrymore "over a barrel."

The script of *Don Juan,* the most ambitious movie the Warners had ever made, was concocted by Bess Meredyth from a pile of books about the great Renaissance lover. Alan Crosland, one of the studio's most successful directors and a drinking buddy of Barrymore's, was assigned to direct. Barrymore was the logical choice for the part of Don Juan, as he didn't need to act—he was playing himself.

Barrymore had decided to make this film his greatest achievement. He would eclipse Douglas Fairbanks as the screen's foremost acrobat, both in boudoir scenes and balcony swordplay. Barrymore would swashbuckle his way across the screen, scale castle walls, thrust his body down staircases, pausing for an occasional amorous interlude. The actresses who were the objects of his on-screen passions were Mary Astor (as

Adriana Della Varnese), Estelle Taylor (Lucrezia Borgia), a new actress, Myrna Loy, plus Phyllis Haver and June Marlowe.

Besides ending up with bruised lips, Mary Astor reported she was bent backwards over the wheel in the torture scenes so often and for such long periods that she suffered severe pains in her neck and back.

Barrymore's drunkenness was also another concern of Astor's. In one scene he was supposed to rush into the room, unravel his cape in a dramatic gesture, take her into his arms, and kiss her passionately. The kiss was no problem; the cape was. As the cameras whirred, Barrymore climbed a trellis into the room, twirled the cloak from his shoulders, and fell flat on his face before the astonished actress.

Barrymore had enough presence of mind to sober up for the final scene, in which he had to make a gigantic leap from the top of a stone staircase onto the body of Donati, the snarling villain.

Once the filming of *Don Juan* was complete, Sam began the tedious task of recording the musical accompaniment that would be mixed with the final cut of the movie.

Harry had decided to get the best orchestra in America. "It was one of the hardest jobs I ever had, to get a great orchestra for that picture," Harry remembered. "I had to do a lot of talking to musicians, convincing them the music would be heard throughout the world. Finally we got the New York Philharmonic. It cost us exactly one hundred and ten thousand dollars to add music to *Don Juan.*"

Harry engaged two noted musicians, Dr. William Axt and David Mendoza, to write the score. For the themes, they dipped into the works of classical composers, finally completing a score that was designed to fit perfectly (if the recording disc played at the same speed as the reel) with the movie.

Dr. Henry Hadley was assigned to conduct the orchestra. He studied the picture, rehearsed the score, one hand on his baton, the other on a stopwatch, as he noted how some parts would have to be speeded up, some played more slowly, some bars cut from the music, some added. Each member of the orchestra was tense. No one dared make a mistake or the whole take would be ruined. They realized that what they were doing had never been done before. If they did it poorly it might never be done again.

Harry and Sam decided to film eight Vitaphone shorts, featuring several prominent classical musicians, that would serve as a prelude to the showing of *Don Juan. The Song of the Volga Boatmen,* the short that had been recorded earlier, was added, and, for a change of pace, comedian Roy Smeck was filmed doing a vaudeville sketch called "His Pastimes."

This Vitaphone prelude upped the cost: The Metropolitan Opera Company charged the Warners $1,000 a week rental for use of their theater. Giovanni Martinelli sang twice for the camera and received $25,000. The Warners also had to pay $104,000 for permission to use the music the stars would sing. The brothers were investing several million in a gigantic gamble that the American people would like moving pictures with sound.

The Vitaphone prelude would be introduced on film by Will Hays, president of the Motion Picture Producers and Distributors of America, the organization that had been founded in 1922 to ensure that movies were dutifully self-censored. Hays would welcome the new medium of sound into the world of cinema.

Sam was in charge of filming these shorts in New York. He and Lina had leased a house in Great Neck, Long Island. Lina recalled that her "crucifix and Sam's 'gentile' impersonation and Anglo-Saxon name got us past the anti-Semitic restrictions." Lina loved the house. It came staffed with four servants: a butler, maid, cook, and a chauffeur who drove Sam to work at the Metropolitan Opera House in a Pierce Arrow. Lina would go along sometimes and they would stay in Abe and Bess's empty apartment in the city.

The long hours and constant strain were taking their toll on Sam's physical well-being. Lina, who had been aware of his sinus headaches when they first married, noted that "he ate aspirin tablets by the dozen and after every meal gulped milk of magnesia for a chaser."

In February 1926, Lina happily discovered she was pregnant. When she told her mother, Gladys scowled. "Rank carelessness."

Lina grinned wickedly at her mother. "But Mama, I did everything, including arabesques in bed, to get with child."

Lina wasn't sure how Sam would take the news. "He was proud of my beautiful body and didn't want me to ruin it with a baby," Lina recalled. To her surprise, Sam was jubilant. He practically did a pirouette around her, then grasped her in his big arms and kissed her long and lovingly.

Don Juan opened at the Warners Theatre on Broadway and 52nd Street in New York on August 6, 1926. A ticket cost $10 plus $1 tax, the record for the premiere of a motion picture, yet the 1,208-seat theater was sold out.

It was a hot evening and below the theater marquee a banner fringed with painted icicles proclaimed REFRIGERATED WASHED AIR. A woman stepped up to the box office and asked the cashier, "Is Vitaphone a new kind of refrigeration?"

Broadway from 51st to 52nd streets was jammed with curious onlookers, who had gathered to cheer as the first-nighters arrived. Estelle Taylor had part of her evening wrap torn by enthusiastic fans and husband Jack Dempsey had to show his prize-ring prowess to get her safely into the theater. The crowd gave a rousing ovation to musicians Efrem Zimbalist, Anna Chase, and Henry Hadley as they stepped from chauffeured limousines. (John Barrymore and Mary Astor had been held in reserve for the West Coast premiere.) Movie moguls Adolph Zukor, William Fox, Nick Schenck, and Lewis Selznick stalked into the theater convinced that what they were about to see was a crackpot idea, one that would break Warner Bros.

The lobby was decorated with eight illuminated paintings of Barrymore, Mary Astor, Estelle Taylor, and the rest of the cast. In one of the pictures, Barrymore held the hands of two of his female costars. Red lights from above flashed on and Barrymore appeared to be left with only one woman. Then blue lights flashed and he was back flirting with them both. In another painting, Barrymore was shown at Mary Astor's closed bedroom window. Red lights flashed—the window opened and Mary Astor opened her arms to him. The effect, dubbed "color absorption," was uncanny and fascinating—and a prelude to the uniqueness of the evening.

Harry looked down the row at his mother and father, Sam, and Abe (Jack had stayed in California to prepare for the Hollywood opening of *Don Juan*), then scanned the audience.

"We got the right crowd," Harry said confidently.

"What if the apparatus breaks down?" Abe nervously wiped his mouth with the back of his hand.

"It won't break down," Sam said through clenched teeth. His sinuses ached terribly and he reached into his pocket for aspirin.

Abe's huge hands gripped the wooden arms of the theater seat, practically tearing them off. He let go, flexed his fingers, and said, "It could maybe crack apart some way, spoil everything."

"Nothin's gonna happen to it." Sam stuffed several aspirin into his mouth and chewed.

"You sure?"

"I worked my ass off for six months to make sure."

Reassured, Abe relaxed a little. Very little. He still felt like he was in a crap game, but, he hoped, with loaded dice. "Then maybe we'll win."

"We'll win," Harry said. But even Harry wasn't sure. Whether they would be rich or bankrupt, wise men or fools, would soon be decided by the jury who sat around them.

At 8:30 the lights dimmed and the curtain slowly parted. Voices hushed. A white beam shot overhead, splashed on the screen, and Will Hays stepped toward the audience. Well, *he* didn't step, his *image* did—on the motion picture screen. He began to speak.

"My friends."

The hush deepened.

In his Hoosier twang, his voice synchronized with his lips, Hays said clearly:

"No story was ever written for the screen as dramatic as the story of the screen itself. Tonight we write another chapter in that story. For indeed we have advanced from that few seconds of shadow of a serpentine dancer thirty years ago when the motion picture was born—to this public demonstration of the Vitaphone synchronizing the reproduction of sound with the reproduction of action."

Sam watched Hays closely, relieved that the sound was in perfect synchronization with his lips.

"The future of motion pictures is as far flung as all the tomorrows . . ."

As Hays continued, Harry glanced at the audience in the flickering light. No one was moving; other than Hays's voice, the theater was in total silence.

"In the presentation of these pictures, music plays an invaluable part. The motion picture is a most potent factor in the development of a national appreciation of good music. That service will now be extended as the Vitaphone shall carry symphony orchestrations to the town halls of the hamlets.

"It has been said that the art of vocalist and instrumentalist is ephemeral, that he creates only for the moment. Now, neither the artist nor his art will wholly die.

"Long experimentation and research by the Western Electric Company and the Bell Telephone Laboratories, supplemented by the efforts of the Warner brothers . . . , have made this great new instrument possible, and to them and to all who have contributed to this achievement I offer my congratulations and best wishes."

Abe, a little embarrassed by all these congratulatory remarks, began to sink down in his seat as Hays concluded:

"To the Warner brothers, to whom is due credit for this, the beginning of a new era in music and motion pictures, I offer my felicitations and sincerest appreciation.

"It is an occasion with which the public and the motion picture industry are equally gratified.

"It is another great service—and service is the supreme commitment to life."

As Will Hays finished what newspapers were to call his "shadow speech," the audience applauded enthusiastically.

Abe took his handkerchief out of his breast pocket and mopped his forehead. "It's gotta work," he mumbled to himself.

The musical program started with the overture from Wagner's *Tannhäuser,* played by the New York Philharmonic Orchestra. Sam squinted at the screen, listening, watching as the camera moved in for a close-up of an instrument, back to a medium shot, then the complete orchestra. To Sam's relief, the synchronization was once again perfect.

The rest of the Vitaphone shorts followed: violinist Mischa Elman playing "Humoresque"; vaudevillian Roy Smeck doing jazz numbers on a guitar, ukulele, and banjo; Marion Talley of the Metropolitan Opera singing "Caro Nome" from *Rigoletto*; Efrem Zimbalist on violin; pianist Harold Bauer playing Chopin's Polonaise in A Flat; Giovanni Martinelli singing "Vesti la Giubba" from *Pagliacci*; soprano Anna Chase concluding the Vitaphone prelude with "La Fiesta," a lavishly costumed production number that served as musical prologue to *Don Juan.* The audience was fascinated by the intimacy with which each artist was revealed: Elman's delicate fingering of the violin; Martinelli's facial expressions. Vitaphone had brought the audience closer than ever before to accepting the illusion that the artists were present.

As the lights went up for the intermission, the audience applauded, cheered, even stamped their feet. Then the noise gave way to a concentrated buzz of excitement.

Harry, Sam, and Abe walked to the lobby, and like innocent bystanders eavesdropped on the excited first-nighters.

"Didn't believe it was possible," a portly man in a straw boater said.

"Never heard anything so marvelous," his companion sighed.

"Glorious."

"Beautiful."

"Divine!"

Harry passed out cigars to his brothers.

So far, the gamble was paying off. The real test would be the screening of *Don Juan.*

The audience filed back in and the feature began. The musical accompaniment was dramatic, but the audience also noted the sound effects: swords clashing, doors opening, the shattering of a glass. The closest any actor on the screen came to making a sound was during the 191 passionate kisses that Barrymore bestowed upon his dazzled costars. A woman reporter with a notebook and pencil tallied the kisses, then went on to write:

> These kisses were of all kinds, brands and varieties. Some were cool little pecks on the cheek, others the tender touch of lips to lips, but most of them were hot and long and drew envious murmurs from the females in the audience. In fact, most of the osculation was of such a tempestuous nature that Barrymore was seen wiping the perspiration from his brow after several of these ardent embraces.

Before the stirring finale the three brothers had received the audience's verdict. The enthusiasm of the crowd had said it: "You win!"

Several reporters buttonholed Harry, one asking, "Have you attempted to envision limits that Vitaphone or talking pictures may go to?" Harry paused, choosing his words carefully:

> I believe it would be far too wide to do more at this time than to state that we will be forced to encroach upon the domain of the legitimate stage. Just how far, however, I cannot state. For example, in the field of opera you heard tonight Martinelli sing that splendid Pagliacci number. And with the other Vitaphone numbers we barely scratched the surface.

Harry smiled and added, "Time alone will answer your question."

As Harry turned away, the reporters looked at each other.

"Did he mean talkies?"

Next day *Variety* devoted its entire issue to *Don Juan,* proclaiming

WARNERS' B'WAY SENSATION

Underneath was: REMARKABLE FIRST NIGHT CROWD ACCLAIMS VITAPHONE. One newspaper writer enthused:

> If an airship had landed from Mars there couldn't be more excitement . . . It was a glorious tribute to the vision of the men who had

backed it. It made the biggest producers in the movies sit up and rub their eyes. This thing that they had all condemned was rising like one of Aladdin's genies from the lamp of a single company and bewitching the public.

In Hollywood Jack Warner received a letter from his parents.

Dear Jack,

We never dreamed that we would live to see such a performance, and above all that we would be the parents of such wonderful boys. It is our fondest hope that we will live to a ripe old age so we can see you boys grow on and on. When four marvelous boys like you stick together through thick and thin, there is no question but that you will attain all the success you hope for. . . .

Lovingly,
Mother and Dad

Ben and Pearl had another reason to be happy. Three weeks after the opening of *Don Juan* they celebrated their golden wedding anniversary. The celebration was held August 26, 1926, in Youngstown, where Ben and Pearl could be near old friends. Annie, Sadie, and Rose were there with their husbands and children.

Annie and her husband, Dave Robbins, had remained in Youngstown, where he continued working in the Warner family theaters. Jack Jr. remembers Robbins as "a hulking person who seldom had anything to say. I see him plainly in my mind, but it is a silent picture which fades as fast as it appears. He seems to have done a good job, because I rarely heard my father say anything about him, which was not true when he spoke of his other brothers-in-law."

Louis Halper, who married the youngest daughter, Sadie, was a very bright and well-educated man. Because he was better educated than Jack, he became a public target of his ridicule. Jack Jr. remembers his father saying terrible things to "poor Lou," who had to sit in front of the family and take it. Jack would look at his brother-in-law, who ran a theater district for the company, and tell him that the theater business was being run by "schmucks." Sadie and Louis had two children, Sam and Evelyn.

Rose was known as a kind, sensitive, and warm woman, who subscribed to a long list of charities. Like both her sisters, Rose married a man who would work for Warner Bros. Harry Charnas was a likable man, yet Jack—when he was not picking on Lou Halper—would ridicule and demean Charnas. Rose and Charnas had one adopted child, Milton.

Harry and Rea arrived at the anniversary celebration with their son, Lewis, and daughters, Doris and Betty. Jack, Irma, and Jack Jr. attended, as did Abe and Bessie. Sam and Lina (who was seven months pregnant) were the newlyweds. Even David made the trip with his wife, Mary, from Boston, where he spent most of his time in a sanitarium because of his sleeping sickness.

The brothers had a special memorial brochure printed, and a poem written, which read:

Harry and Abe and Sam and Jack,
All of 'em hitting the home track,
All hell-bent to keep a date
With Dad and Mom in the Buckeye State.
And Dave and Sadie and Anna and Rose,
Hurrying swift as the wind that blows,
Hurrying back where they started from—
Back to the old town of Dad and Mom. . . .
All eager to give, in their own good way,
Their love—on the Golden Wedding day.

The brochure went on to say:

That friendly old hostelry, "The Ohio Hotel," was the place of fore-gathering. About the board were fifty guests, one for each year of the gallant pilgrimage. The Young Couple sat at the head of the table, all a-flutter with the magic of it all! In and out rushed diminutive bell-boys, bearing telegrams of congratulations from the Four Quarters of the Globe, and the Seven Seas, too, it seemed!

. . . At last the Bride and Groom, radiant, grateful, young-of-heart, arm in arm, climbing the stairs, together!

To such as these all days are Golden Wedding Days!

One wonders what studio staff writer was assigned to pen these ebullient words. The joyous occasion would be the last time the entire family was together.

In Hollywood, preparations for the opening of *Don Juan* at Grauman's Egyptian Theatre continued apace. The Vitaphone equipment had to be sent by rail from New York. Jack got a tip that some thugs were planning on wrecking it when it reached the freight yards in Hollywood.

He authorized his studio police chief and an officer from the Los Angeles sheriff's department to ride herd on the huge loudspeakers and electrical components as they were trucked to the theater. The tip proved to be false, and *Don Juan* made its California debut without any violence—but with plenty of fanfare.

The theater was packed with such stars as Charlie Chaplin, Buster Keaton, Harry Carey, Fay Wray, Pola Negri, Greta Garbo, and Wallace Beery. Samuel Goldwyn, King Vidor, Cecil B. De Mille, and other representatives of the movie studios also attended. If anything, the response outdid the one that greeted the New York opening. The headline in *Daily Variety* read VITAPHONE THRILLS L. A.

Convinced that they were on the right track with Vitaphone, Harry made a statement to the press:

> Vitaphone will thrill the world, for it will give to millions of people in the most remote localities the thrill that only music, the universal language, can engender. Shakespeare in his wisdom said, "The man that hath no music in himself, nor is not moved with concord of sweet sounds, is fit for treasons, stratagems, and spoils . . . Let no such man be trusted."

Harry, who had never read Shakespeare, had once again let studio writers unearth an appropriate quote.

Despite the rave reviews and the soaring price of Warner Bros. stock (which went from $14 to $54 a share within weeks of the New York premiere), the studio was in deep financial trouble. Filming *Don Juan* had been an enormously expensive undertaking. With only a few theaters equipped for sound, it was difficult to recoup the investment. The other movie producers hadn't jumped on the sound wagon and bought Vitaphone franchises. That's not to say producers weren't considering getting into sound, but buying a Vitaphone franchise from Warners rankled them. To install a projection and speaker system cost between $16,000 and $25,000 depending on the size of the theater.

My great-aunt Lina told me that Adolph Zukor, the head of Paramount, realized the future of the industry was in films that talked, and was ready to build sound studios. Unfortunately, Warners had the best technicians under contract. Zukor needed someone who knew sound equipment.

Someone like Sam Warner.

Other Voices

Lina Basquette

Sam came home one night and started pacing the floor. I could see that he was not himself and he finally said, "Baby, I've had an offer from Paramount. Zukor will make me an executive producer if I bring Vitaphone with me."

Paramount! I thought. Paramount was a class act, a prestige studio. Even with Warners' success, most people still considered them a third-rate Poverty Row studio.

I said, "Oh, darling, do it. Get away from your brothers. This is a real opportunity."

I talked and talked and fought and fought and even begged, and by midnight Sam said, "You're right, Baby. Tomorrow I'll tell my brothers I'm leaving the company." He gave me a big hug. "We're going to Paramount! Look, after I tell Harry and the guys, we'll celebrate. I'll rent the bridal suite at the Ambassador."

I said, "Me? In the bridal suite?" I was eight months pregnant!

I stayed home the next day waiting for Sam's call—but it never came. I remember sitting by the radio, hearing that Rudolph Valentino had died. At midnight I got undressed and went to bed. As I began to cry I heard Sam pull up in the gravel driveway. He tiptoed into the house and I could see the guilty look on his face as he came into the bedroom. I could smell the odor of stale liquor and cigar smoke. He never drank and rarely smoked, occasionally a cigar with the boys.

I started screaming at him that he'd promised. He began sobbing, saying that his brothers had got his mother and father and the sisters on the phone. "I just couldn't hurt them. Pops always talked about us boys sticking together."

It was over. I had to accept it. I guess the brothers had figured that if Paramount had wanted Vitaphone bad enough to try and hire Sam, then his "toy phonograph" should be taken more seriously.

A month after this confrontation, Lina felt her labor pains start. As Sam drove her to the hospital, Lina blurted, "I'm going to have the baby baptized a Roman Catholic."

Sam almost wrecked the car. "Don't *ever* tell that to my family."

Lina sat in sullen silence. Sam added, "If we have a boy, he's going to be circumcised."

"Nobody's going to put a knife to my baby!" Lina screamed.

That argument was settled when a daughter was born on October 6, 1926. She arrived, in Lina's words, "bruised, battered, and battle-scarred. She should have greeted me with, 'It was a tough fight, Mom, but I won!'"

Sam wanted to name the baby after Lina. She was set on naming her daughter after her patron saint, Bernadette. After another argument that sent Lina's temperature soaring ("and curdled my milk supply," she said), the couple compromised and named their baby Lita. In the movie gossip columns Lina had read about a girl named Lita Grey who was involved with Charlie Chaplin. Lina liked the name.

When the lease expired on the house in Great Neck, Sam and Lina took a seven-room apartment in Manhattan, not far from the opera house where he was still filming Vitaphone shorts. His work continued at a track-runner's pace, and his headaches grew worse. Lina begged him to take a vacation.

"I can't do that, Baby," Sam said. "We're way behind on the shorts Harry wanted. He's been putting more money into Vitaphone, converting theaters to sound, building four new sound stages at the Sunset lot . . . We have to make this thing work. We're spending a bundle."

Suddenly he squeezed his eyes shut as a spasm of pain shot through his head. Lina, seeing his reaction, cried, "You could be killing yourself. *Please* let someone else do it."

Sam tried to smile, but it came out a wince. "I can't do that, Baby. It would mess up two years of hard work. Besides, Harry's putting a lot of money into a new feature film. It's going to have singing . . ."

Lina pouted. "You've already done movies with singing."

"I told you, it's a *feature*."

Lina saw she couldn't change his mind, so she asked, "So what's the name of this singing epic?"

"*The Jazz Singer*," Sam answered.

10

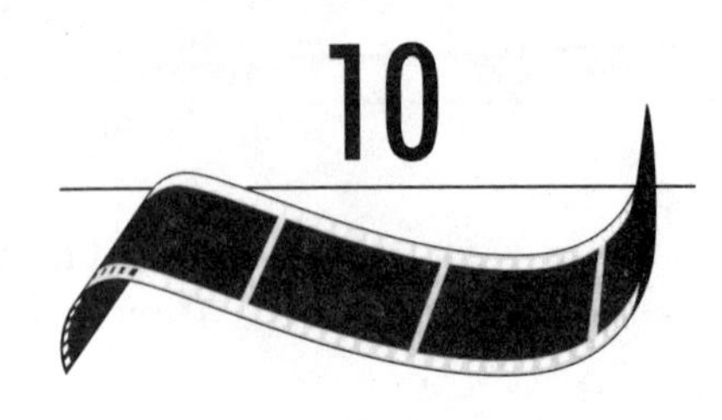

The Jazz Singer

"Everybody quiet!"

A sudden hush fell over the sound stage. Electricians froze next to their banks of lights, technicians took one last heavy breath. Actors, their mouths dry, swallowed.

Standing in stocking feet, the assistant director yelled, "No movement at all!"

A whistle blast cracked the air inside the studio. Then the shrill warning was repeated outside the building. The soundproof doors were sealed and locked.

The incandescent lights cast heat into the faces of everyone on the set. One of the electricians boasted: "Hell, you could light a cigar ten feet away off one of those damn things."

Then: "Turn 'em over."

A faint whirring came from the two glassed-in booths. The cameras were rolling. In another booth, sound technicians peered at the stylus riding on the wax recording disc. Microphones dangled over the orchestra. Conductor Lou Silvers looked over his shoulder at the director, Alan Crosland, and raised his baton . . .

Crosland made a circling motion over his head, a command to start the action.

In that dead stillness, a forty-year-old nightclub singer with the well-earned sobriquet "The World's Greatest Entertainer" opened his mouth to sing. But instead—as he had done a thousand times before on the stage of New York's Winter Garden theater—he began to talk.

"Wait a minute," he cried. "Wait a minute. You ain't heard nothin' yet. Wait a minute, I tell ya . . . You wanna hear 'Toot Toot Tootsie'? All right, hold on . . . Lou, listen, you play 'Toot Toot Tootsie,' three choruses, you understand, and in the third chorus I whistle. Now give it to 'em hard and heavy. Go right ahead . . ."

The baton whipped down, the orchestra began to play—and Al Jolson did what he was supposed to do at the beginning. He sang "Toot Toot Tootsie."

Alan Crosland, sealed in the camera booth, looked through the glass wall at Sam Warner and slashed his fingers across his throat. "Cut?"

Sam, whom Harry had put in charge of the overall sound production of the movie, smiled wearily, then shook his head and mouthed, "Leave it in."

Crosland shrugged, as if to say, "Okay, it's your baby." Warner feature films weren't supposed to talk. Not yet. Certainly not in this movie.

Sam knew he could have given talking pictures to the world almost two years earlier. But Harry had learned his lesson: Don't anticipate the public's taste. He was still convinced of the need to educate the public *up* to talking pictures one step at a time.

Harry, in an off-the-cuff speech he gave in Seattle six months after the opening of *Don Juan,* said, "The reason I'm working now instead of on a fishing trip is because I've resolved to stay in the picture business until I've driven out all the dummies and their directors."

In effect, Harry was ringing the death-knell for silent pictures, with what the Seattle newspaper reported as his "picturesque vehemence." The paper quoted him as saying:

> Vitaphone brings the talking movie within grasp and assures the revolutionization of the film industry. So far, Vitaphone has been utilized to augment the movies—in the way a musical prologue and orchestra do during a play. But we will soon develop the device to emphasize dramatic spots in the film itself. And then, the talking movie

> in which all the characters speak, just as they do on the legitimate stage, will ultimately and inevitably follow.
>
> But we are going a lot further than that before we are through. We are going to bring in a new generation of film stars. Al Jolson, Elise Janis—people like that. And the Fairbanks and the Pickfords will have to show a lot of stuff if they want to keep up with them. Fairbanks and Pickford and most of the other screen luminaries are dummies who have been collecting fabulous salaries, not for what they did themselves, but for what the camera did for them!

Strong words, and the meaning was clear: Sound movies were here to stay and actors had better learn how to talk.

Ever pragmatic (and constantly wary of force-feeding talking pictures to a reluctant public), Harry had authorized a second Vitaphone feature with synchronized orchestration, *The Better 'Ole,* which starred Charlie Chaplin's brother, Sydney. Harry then devised a three-step "sound" plan, which he outlined in a speech at Harvard University on March 30, 1927:

> This coming year we are making three pictures in which Vitaphone will play an important part. For instance, in a scene in which a stage forms an important part of the view, we intend to bring out the actors singing. Our first picture embodying that feature will be *The Jazz Singer,* with Al Jolson.
>
> Next, we are going to perform a wedding ceremony. As you see it today, a wedding ceremony performed on the screen takes place in complete silence. Nobody knows what the priest or the preacher or the rabbi has said. We will actually perform the wedding ceremony and try to make it as real as life.
>
> Third, we will have a world premiere December 9, 1927, at the Selwyn Theatre in New York in which John Barrymore and Dolores Costello star in *When a Man Loves,* a swashbuckling romance based on the story of *Manon Lescaut.* Actual parts have been written into the photo-drama which will involve speaking by the actors.

There was one major problem with Harry's plan: The brothers didn't know whether their studio could survive long enough to make it through 1927. *Don Juan* had all but drained the coffers. And *Don Juan* or no *Don Juan,* the studio had shown a major loss in 1926. The first half of 1927 looked even worse. Warners was so low on funds that New York bankers were turning a deaf ear to Harry's pleas for more loans. It was only through a major effort that allowed the studio to show a slight profit of $30,000 that money was advanced at all.

Even with bankruptcy staring them in the face, the Warners had sunk a staggering half-million dollars into the production of *The Jazz Singer.* The brothers had scheduled a program of forty releases for 1927, most of them standard silent features, including four Rin Tin Tin pictures, which were a virtual insurance policy. Darryl Zanuck, who had a knack for turning out silent film scenarios—Jack Warner said that the brash Zanuck "could write ten times faster than any ordinary man"—was supervising much of the studio's everyday production program. Jack still managed the studio, made important cast assignments, and oversaw the bigger movies.

The Warners believed—when almost everyone else was saying "talkies are the bunk!"—that the future of the motion picture industry was in sound. One other believer was William Fox, the head of Fox Studios, who was concentrating on making talking newsreels which he called Movietone. Fox's newsreels allowed the audience to hear as well as see famous people, such as Benito Mussolini, who, with his arms crossed stiffly over his chest, jaw jutting, greeted the American public with the words "I am very glad to be here speaking in front of the American nation . . ."

Calvin Coolidge presented a medal to Charles Lindbergh following his historic flight across the Atlantic: "As President of the United States I bestow the Distinguished Flying Cross—as a symbol of appreciation for what he is and what he has done—on Colonel Charles A. Lindbergh."

Even Lindbergh's takeoff from Roosevelt Field was filmed on Movietone, the drama of the moment mesmerizing audiences.

William Fox had signed a deal with the sound pioneer Theodore Chase to record sound directly on the margin of the film rather than on a wax disc. The technique, developed by General Electric Company, involved, as it was explained in 1926, "having a ray of light vibrate on the film just as a needle vibrates on a record. The wavy line that is made is transmitted back through an electrical sound projector on the same general principle by which a radio plays music."

Harry decided not to take a chance that the Fox sound system might turn out to be better. Borrowing even more money, he bought into Movietone.

"In order to protect ourselves in the future," Harry said, "we obtained all the rights and interests in the Movietone method. The arrangement was a mutual one. Mr. Fox has the right to manufacture his own pictures to be run on our machine. We have the right to make pictures with his machine."

It was a wise decision. By 1930, Warners would totally switch to the Movietone system. Vitaphone would prove too complex and delicate to be worthwhile. The fragile wax records wore out quickly, and four sets had to be provided in case of breakage. If the amplifying apparatus gave any trouble, it took a qualified electrician to repair it.

Throughout the showing of a Vitaphone film, the operator had to follow a cue sheet closely so he would know when to change the record, and that change had to be instantaneous. The cue sheets sent with the records had to be thoroughly studied in advance of a showing. Unless the system was monitored carefully, the audience was likely to hear squeaks and howls.

The big studios, MGM, Paramount, Universal, were playing wait and see. Together with First National and Producers Distributing Corporation, they concocted a moratorium called the "Big Five Agreement" whereby none of them would commit to any sound system until the winner had been decided, Warners or Fox. The studio bosses, committed to big silent productions, were hostile to innovation. Each found his own reason for not developing sound pictures.

Adolph Zukor: "The effect on the overseas market would be disastrous. Only a small part of the world speaks English."

Carl Laemmle: "We have too big an investment in the silent picture."

Louis B. Mayer: "Let them develop it if they can. Then we'll see about it."

Jesse Lasky: "Studios would have to be entirely rebuilt."

What kept Warners going was the firm belief in the future of Vitaphone and the good it could accomplish. Harry was now convinced it was more than just a money-making scheme or a fly-by-night entertainment gimmick. As he said:

> [Vitaphone] may even serve to eliminate war among the nations. We think of the film as the greatest of all the media for propaganda. We know that American movies are bearing a silent message of our progress to people inhabiting the globe. These same films may now go a step farther. They may even carry an actual message through speech spoken by some character, perhaps of America's doctrines for world peace.
>
> As for the educational possibilities of the instrument, I believe that not even the surface has yet been scratched, and that it will only be

> a question of time until most of the schools and colleges in the country will make use of it. Lectures by the greatest of all professors may now go the rounds instead of being merely presented in one classroom. Students will not have to travel far and wide to hear certain famous learned men. The mountain will move to Mohammed.

Yes, Harry, far beyond anyone else in the motion picture industry, had a vision of how this new medium of sound could be used. Whether his dream would become a reality depended on the audience. What Warners needed was a super success, a film that would prove to the rest of Hollywood, and the country's exhibitors, as well as the general public—as *Don Juan* had come within a whisper of doing—that talkies were more than just a footnote to the history of the cinema.

What the Warner brothers needed was a miracle.

And Sam Warner was watching it.

As Jolson finished singing "Toot Toot Tootsie," Sam looked over at Alan Crosland in the camera booth. The director, stooped over, hands thrust into his pockets, was dripping sweat in his airless cell, but he nodded vigorously, indicating they had it all on record—and film. No actor had coughed, no stagehand had dropped a hammer or stumbled over a prop, no truck had rattled by outside . . . they had it. Clean.

But should they take it again, this time without Jolson's ad-libbing? What was on film was the real Al Jolson at work. It was simply the way he performed, as if the audience were there cheering him on as they had done thousands of times before in nightclubs and theaters. To Jolson, the camera crew, the prop people, the extras, and the whirring camera were his audience. He had sung the only way he knew how, with every ounce of energy and charisma he possessed. And that, with his ad-libbing, would show on the screen.

As Crosland stepped out of the camera booth, mopping his face with a towel, Jack Warner breezed onto the sound stage, followed by one of his newly acquired yes-men. Jack was dressed as a dandy as usual, in a tailored sport coat, tie, and black-and-white spectator shoes. Lifting his white Panama and wiping a line of sweat from his forehead, he said, "Sam, how's it going? Jolie giving you any crap?"

Sam glanced at Crosland, who shrugged. "Naw," Sam said. "Jolson's okay. Got some good stuff on disc."

"Great. I'll see the rushes later." He shook his head. "You're lucky. I'm up to my elbows in banjo players across the street. Musical featurettes isn't my idea of moviemaking."

When Sam didn't respond, Jack looked at his brother closely, noting the dark rings under his eyes, the dog-tired sag of his shoulders. "Jeez, Sam, you look like somebody's been using you for batting practice."

"I'm okay—just abscessed teeth." Sam pulled a bottle of pills from his shirt pocket, shook two into his hand. "Gotta make an appointment to get them pulled. Sinus has been bothering . . ."

Jack elbowed his yes-man. "Teeth, hell. He's been shtuppin' that new bride of his till the wee hours."

Sam grinned good-naturedly. "Yeah, sure."

A long-legged extra walked by, swishing her rear provocatively. Jack didn't miss it; his eyes followed each exaggerated move. "Gotta run."

Sam watched as his brother caught up with the girl, then turned to Crosland, who stared at him for a moment before saying, "Well . . . do we keep Jolson's ad-libs?"

Sam pressed his fingers hard against his temple as if pushing the pain away. Finally he said, "The lines should stay."

"You sure?"

"Yeah, they stay."

"You're the boss."

"I've got another idea," Sam said, slipping back into his shoes and tying the laces, "but I've got to talk to Jolson about it."

Jolson stood before an electric fan, arms outstretched, as he listened intently to Sam. Sam wanted to film the singer in a new all-sound sequence. He even had a song in mind, Irving Berlin's "Blue Skies." He also had the scene, the one where Jolson's character, Jack Robin, comes home on the eve of the birthday of his father, Cantor Rabinowitz. The script called for Robin's mother to be there, alone, when he arrives, and he sits at the piano to sing "Blue Skies."

"I could get the script people to write in a few lines on title cards to bridge from your entry to the song," Sam said to Jolson.

Jolson responded, "Nah, no need to do that. I'll just kinda make it up as I go along . . . like I always do."

"Right." Sam nodded his head, feeling a little better. "You do it, Jolie, just like you always do."

"Yeah, yeah, sure." Jolson grinned, all eyes and teeth. "Leave it to me. I'll knock 'em on their asses. This ain't no Georgie Jessel playing the part."

The Jazz Singer, a Jewish drama starring an engaging young Jewish actor named George Jessel, had opened on Broadway September 4, 1925, to moderately enthusiastic reviews. The *New York Times* found its return-of-the-prodigal-son theme appealing and "so assuredly written that even the slowest of wits can understand it."

The play, which had started its life as a magazine story by Samson Raphaelson, was a tale of a cantor whose son leaves the synagogue to become a cabaret singer, then makes matters worse by falling in love with a gentile showgirl. Years later, the cantor becomes sick as Yom Kippur, the Jewish day of atonement, nears; there is no one to sing in his place. The mother seeks out her son, who must choose between his career and singing the Kol Nidre, the Jewish plea for forgiveness, in his father's place. Of course, after much deliberation, the son leaves his Broadway revue (on opening night!), returns home, and, dressed in the cantor's elaborate robes, sings the Kol Nidre.

The end of *The Jazz Singer* was innovative in that it resolved a conflict familiar to children of immigrants at that time. Everyone gets what they want: the traditional father reconciles with his son and hears him sing in the synagogue, but the son goes back to be a success on stage.

George Jessel considered *The Jazz Singer* to be totally within his bailiwick. When Harry bought the movie rights for $50,000, he let Jessel know the studio planned on letting him repeat his success on film. Harry told Jessel, "It will be a good picture to make for the sake of racial tolerance, if nothing else."

There was only one problem in hiring Jessel: money. When Jessel learned it was going to be a Vitaphone feature with singing sequences, he decided to up his price. It was the wrong move. In his 1965 autobiography, *My First Hundred Years in Hollywood,* Jack Warner remembered the hassle that followed:

> Along with the movie rights, Harry got a signed commitment from Jessel to play the lead for $30,000. . . .
>
> When everything was ready, I phoned Jessel and told him to start for Hollywood.
>
> "Now, Jack," he said, "when we set up the original contract there was nothing in it about all the extra singing and talking you now want me to do."
>
> "I guess I don't like your attitude," I said. "You want more money, right? How much?"
>
> "I want ten grand more."
>
> "You've got it," I said. I wasn't about to quibble over an extra ten Gs, because Jessel and everyone else in show business had been

breaking through doors to see the new films. "It's a deal, Georgie," I said. "Come on out and I'll give you a binding letter."

"The letter first, then I come out," he said stubbornly.

"Goddammit, Georgie," I snapped. "I give you my word."

"That's not enough. Your brother Harry will never go for the deal."

I didn't want to fight with Jessel over the phone. I had known him since he was a kid hoofer with Walter Winchell and Eddie Cantor in the Gus Edwards revue, and I resented the [implication] that I was some kind of nogoodnik.

"Don't talk [to me] like [I was] a child," I said. "I'm spending millions of dollars and I don't need Harry's okay for every single buck. If you want an extra ten grand you've got it."

"No," he said flatly.

I was suddenly fed up with this silly *schmendrick*. "Okay, Georgie," I said. "If you can't take my word, let's forget the whole thing."

He said: "All right, the deal's off," and he hung up.

Jessel gave his own version of the story in his 1976 autobiography, *The World I Lived In*:

When the original contract was signed, there had been no deal for sound pictures. I pleaded with H. M. [Harry Warner] that the least they, the Warners, could do was change my contract or give me a bonus. . . . As far as I was concerned, I wanted a new deal. He talked about "taking care" of me if the picture was a success. I did not feel that was enough. We had a very heated argument, and he closed it by saying that I would do no pictures for Warner Brothers or anyone else in Hollywood if I persisted. He took an oath on his family's life on this statement. Both Sam and Jack, however, felt this could be patched up, and at the designated time I went to Hollywood to start work. My first look at the scenario threw me into a state of shock, almost apoplexy.

Instead of the boy leaving the theater and following the traditions of his father by singing in the synagogue, as in the play, the scenario had him return to the Winter Garden as a black-faced comedian with his mother applauding wildly from a box seat. I raised hell. Money or no money, I would not do this version. . . .

Apart from the script, I was still insisting on being paid "up front" and in cash. I had been getting $2,500 for each picture from Warners.

> For the longer *Jazz Singer,* I asked for $5,000 and would have done it for that. But the same day I read the script all three checks the Warners had given me in New York for the three two-reelers bounced and caught up with me on the coast. This, I told Jack, was the reason I wanted my salary for *The Jazz Singer* up front.

Memories in the entertainment world can be extremely selective and few people let the truth interfere with a good story, especially if the facts take them out of the center spotlight. No doubt there are truths in both their tales. But as it turned out for Jessel, the story had an unhappy ending.

After his conversation with Jack Warner, Jessel looked up his old friend Al Jolson, who was headlining for a week at the New York Biltmore. Jolson consoled his pal, telling him he was right. Jack Warner was wrong. Two days later, Jessel picked up a newspaper and was shocked to read that Jolson had signed a contract to star in *The Jazz Singer.*

Jack Warner twisted the knife deeper when he wrote:

> . . . Jessel probably would have been great in the part. He would be remembered in history as the first actor to utter words on a movie screen. He knows it, too, and I am sure there have been times when the knowledge has curdled his heart. *Too bad, Georgie. You should have known my word was enough.*

Whether Jessel would have been known as the first actor to utter words on the screen is a moot point. Had Jolson not been hired for the part, that honor could have well gone (in accordance with Harry's plan to film *When a Man Loves* as the first talkie) to John Barrymore. Actually, Al Jolson was not even Jack's second choice for the lead in *The Jazz Singer.*

Eddie Cantor, the great blackface showman, was headlining at the Orpheum Theatre in Los Angeles. Jack made a call. Cantor worried about the effect of a risky sound picture on his stage career. "I won't do it," he told Jack. "Besides, it's Jessel's part."

"I just kissed Jessel off!" Jack stormed back.

The Warners put their heads together and decided on Jolson. He was expensive, it was true, but Harry argued they could offer him a lump sum and a percentage of the film's profits. Jolson's price was $75,000, one third down in cash, the rest at $6,250 a week until the balance was paid. The contract was signed.

Sam and Jack made every effort to make *The Jazz Singer* the best film they had produced. Warner Oland (an actor who would achieve fame as Charlie Chan) was chosen to play Cantor Rabinowitz. Eugenie Besserer, an excellent character actress, was cast as the mother.

For the scenes in the synagogue, set design teams observed services at the Orchard Street Synagogue and constructed an exact replica on the sound stage at the Sunset Boulevard studios. The same design principle was used to create the scenes at the Winter Garden. Although some sequences were actually shot at the Winter Garden, such as the final scene with Jolson on the runway singing, many interior portions of the theater were reproduced in the studio.

Sam preferred to film most of the scenes at the studio because of the soundproof stage and the fact that the recording equipment and camera booths were difficult to transport. Jack Warner Jr. recalled a day he spent on the set when he was eleven years old.

Other Voices

Jack Warner Jr.

My most vivid memory of those early years was when I spent what seemed an eternity cooped up inside one of the boxlike booths used to isolate the noise made by the rackety Mitchell cameras. Al Jolson was to sing Kol Nidre in place of his dying cantor father and as it was the climax to this crucial film, I wanted to be close to the action. I knew all of the crew, and one of the cameramen invited me to share the booth with him while the scene was shot and recorded.

The door was sealed shut on us to keep the camera noise inside and the air outside. After many minutes of Jolson cantilating, then several retakes, our meager supply of air was slowly being used up and we both were getting faint and dizzy. Finally, when the end of the longest scene ever shot up to then for a sound film came, the camera booth door was unlatched and we staggered out of the nearly airless box. It was a scene from a picture that was to give new life to Warner Bros.—and near-suffocation to me.

Filming *The Jazz Singer* was a technical nightmare, but it wasn't the only worry Sam Warner had. Director Alan Crosland quickly discovered that Al Jolson was not the world's greatest actor, even in this eye-rolling, chest-pounding, hair-pulling era. The singer could make love to a theater filled with cheering fans, but it wasn't easy doing the same for the cold eye of the camera.

Jolson realized he was having difficulty working in front of a camera. "Everything was new and strange to me," he recalled. "I would do one scene five times with tears in my eyes and Crosland would say, 'Do it again—and put some feeling into it!'"

Sam and Jack were hoping he could pull it off by just being himself, "The World's Greatest Entertainer."

The Warners kept their fingers crossed.

The last take on *The Jazz Singer* was made on August 27, 1927. The film was developed that night and viewed by Alan Crosland, Sam, and Jack. No one said much. Crosland, locked in his stooped-over stance, tugged at his tie and said, "Well, that's it."

"Yeah," said Jack, worry coloring his tone. "Now if the public will only go for it."

In a muffled voice Sam said, "They will." He sat listlessly in his seat. Getting the movie in the can had been the biggest battle of his life; he had put everything into it, carrying the burden of making it work on his shoulders.

Jack looked at his brother. In profile Sam looked like a pug fighter. Jack knew the reason: Two years earlier Sam had thrown a stuntman, Slim Cole, off the studio lot. Angered, Cole returned the next day and without warning swung his fist and landed a hard blow on Sam's nose, breaking it. It was not the first time it had been broken, but this time the splintered bone could not be set properly. Since then, the injury had been giving Sam painful headaches.

"You okay, Sam?" Jack asked.

"Sure . . . I'm . . . fine. Jaw aches from these damn abscessed teeth . . . I think there's something wrong with my sinuses."

"Maybe you better see a doc," Jack said.

"It's okay." Sam lifted himself wearily out of the chair. "It'll get better." He started for the screening room door. "I'd better get home to Lina."

Jack winked at him. "Yeah, she'll fix you up." But it didn't come out as light as he wanted. Jack was worried about his brother.

Other Voices

Lina Basquette

Sam was under a lot of pressure during the filming of *The Jazz Singer*. There were many problems with Vitaphone, problems with Al Jolson, who wasn't an actor, and problems with George Jessel, who was always crying the blues about not getting the part. Harry and Abe were spending more money on the movie than they had ever spent in their lives and they were crying about that.

To add to his worries, Sam made a couple Vitaphone shorts, one with Gus Arnhein and his band. They needed a ballet dancer, so Sam asked me to do it. Because I was family, I wasn't paid anything, but Sam saw that I got a beautiful saddle with silver inlay.

I wanted to go back to work, so I got an agent and was hired by FBO, a Poverty Row studio, to do a dog opera, *Ranger of the North*. Ranger was a Rin Tin Tin clone. The movie was filmed in the High Sierra near Lone Pine, a couple hundred miles north of Los Angeles, so I was gone for several weeks. When I got back, someone in Paramount's casting office had seen the rushes and wanted to sign me up for a feature film, *Serenade,* with Adolphe Menjou. I was to play a Russian ballerina who danced on men's hearts. Quite a switch from "Priscilla of the Pines."

I asked Sam, "Do you mind if I do this film at Paramount?"

"Paramount!" he cried. "Oh, jeez, what will the family think?"

"I want to be an actress."

"There's no actresses in the family!"

"Well, there is now!"

He finally came around to the idea, telling me, "You go over there and be a big success." I knew he was thinking how *he* could have gone to Paramount—and been a big success.

Sam started complaining more and more about his migraine headaches and sinuses. But, I mean, I didn't know what the heck sinuses were.

The Jazz Singer was scheduled to premiere at Warners' New York theater. The Warner brothers were predicting that the movie would be "without a doubt the biggest stride since the birth of the industry." At least that's what they were telling the public. If the gamble failed, the brothers would be out of the motion picture business.

October 6, 1927

Opening night.

The evening was cool and clear as New York society stepped out of taxis and entered the theater. A huge poster of Jolson, just his eyes, mouth, collar, and white gloves outlined on a black background, greeted the crowd. A line surged urgently about the box office. But there were no more seats.

Al Jolson was attending reluctantly. He thought the movie was going to be a flop. "I wanted to go away and hide," he recalled. "The brothers got me by the collar, threw me on a train and packed me off to the opening."

Inside, the lights dimmed, and a short Vitaphone feature, filmed on the streets of New York and accompanied by a chorus singing "East Side, West Side," was shown. The audience applauded perfunctorily. Nothing new. This kind of stuff they had seen before in other featurettes.

Bobby Gordon, who played the jazz singer as a young boy, was introduced on the screen by a brief title. In a clear voice he sang "My Gal Sal." Still the same old stuff, another vaudeville act.

Then: *The Jazz Singer.*

On the sound synchronization disc the 120-piece Vitaphone Orchestra played a jazz theme as the young boy, Jackie, unable to conform to Jewish tradition, leaves home to make his way as a singer. Then suddenly he is a man, sitting in a restaurant eating a breakfast of ham and eggs (a bit of heavy-handed breaking-with-the-past symbolism). The scene shifts to a nightclub, where Jackie is talking to a friend, played by William Demarest. The title card reads WONDERFUL PALS ARE HARD TO FIND.

Al Jolson as Jackie walks to the nightclub stage, opens his mouth and sings a forgettable song, "Dirty Hands, Dirty Face." Jolson finishes, bows to the audience, and says the first words ever uttered in an American feature motion picture:

"Wait a minute . . . wait a minute. You ain't heard nothin' yet!"

The audience acted as if a volcano had erupted. Jolson had spoken to *them.* Sure, there had been Vitaphone speech before in short pre-show films, but this was different. *This was a story.* A story that reached for the heart, a story that shook emotions. The audience felt it was living with the actors.

Then, shockingly, the music background came up, lips began to once again move silently, and title cards flashed on the screen: THERE ARE LOTS OF JAZZ SINGERS, BUT YOU HAVE A TEAR IN YOUR VOICE.

A great line by Jackie's girlfriend, but it wasn't spoken. The audience felt cheated. They wanted to hear *talk.*

They needn't have worried. Sam Warner had seen to that when he asked the entertainer to "bridge a few lines between the choruses of 'Blue Skies.' "

And now, on the screen, Jackie sings a chorus of "Blue Skies," then stops, and in that wide-eyed Jolson style turns to his mother and says, "You like that, Mama?"

"Yes," the actress Eugenie Besserer replies.

"I'm glad of it. I'd rather please you than anybody I know. You know, darlin' . . . will you give me something?"

"What?"

"You'll never guess. Shut your eyes, Mama, shut 'em for little Jackie. I'm going to steal something." (He kisses her on the cheek.) "Mama, darling, if I'm a success in this show, we're going to move from here."

Besserer shows surprise. She has no idea what to say or do.

"Oh, yeah," Jolson continues excitedly, "we're going to move to the Bronx. Lotta nice green grass up there . . . whole lotta people you know, the Ginsbergs, the Guttenbergs, the Goldbergs . . . a whole lotta bergs."

By this time the audience was in shocked silence, afraid to make a sound.

Jackie holds his mother's hand. "I'm gonna buy you a silk dress, a nice pink dress . . . take you to Coney Island . . . I'll kiss you and hug you, you see if I don't."

As he goes back to singing, his father, the cantor, walks in and cries, "Stop!" Once again the synchronized background sound comes up, the actors' lips move in silence and the title cards flash on. The audience groans at this silent intrusion.

Minutes later the movie ends in a theater scene with Jolson doing his rousing blackface number, "Mammy."

As the final title, THE END, came on the screen, the audience stood and cheered hysterically, refusing to sit down until Jolson, who was weeping unashamedly in his box, walked onstage to take his bow.

Actor William Demarest said he felt cold shivers going down his spine, the same feeling he'd had upon hearing the *Titanic* had struck an iceberg.

Irving Berlin, composer of "Blue Skies," was unable to hold back his tears.

Walter Wanger, a young Paramount executive, raced into the lobby and phoned his boss, Jesse Lasky. "Jesse, this is a revolution!"

Director Frank Capra recalled, "It was a shock to the audience to see Jolson open his mouth and hear words come out of it. It was one of those once-in-a-lifetime experiences to see it happen on screen. A vision . . . a voice that came out of a shadow."

The *New York Times* called it the "biggest ovation in a theater since the introduction of Vitaphone."

It was the greatest night in the history of the Warner family. Yet not one of the four brothers was there to witness the moment of triumph.

Twenty-four hours earlier, Sam Warner had died.

11

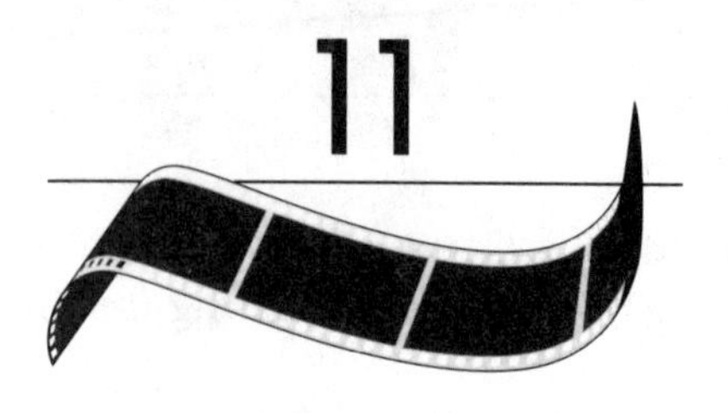

Sam

Sam Warner had planned on taking the train from Los Angeles to New York for the October 6 opening of *The Jazz Singer*. Jack, who was going to travel with him, noticed that his brother looked even sicker than he had during production.

"I knew he had worked day and night for weeks," Jack recalled, "losing weight and deepening the hollows under his eyes, but I had seen him hit bottom many times before and then climb up out of the dark with new strength."

A week before the trip, Jack told Sam, "Maybe I better cancel the trip to New York. I'll call Harry and Abe, tell 'em . . ."

"Nah, I'm okay, just need a little rest." Sam took out a handkerchief and blew his nose. A blotch of blood appeared on the cloth. Jack looked away as Sam tried to hide it.

"I don't think I should go," Jack said over his shoulder as Sam wadded the handkerchief and stuffed it back into his pocket.

"Go ahead. I've just got a hell of a headache."

It was more than a headache. By the end of September, Sam was so unsteady on his feet he had to be led to a chair. He was in great pain.

Lina had just started work on *Serenade* for Paramount when Sam was taken to the doctor. He complained that he felt as if his head was about to explode. The painkillers weren't having any effect. The doctor advised him to have several teeth that were badly abscessed removed immediately. Sam didn't see any sense in having the extractions done in a hospital and said he'd go to a dentist. The doctor strongly recommended the procedure be done in the hospital under anesthesia. Sam reluctantly agreed and was admitted to the California Lutheran Hospital in Los Angeles for the treatment. Lina, shocked that her husband was going to a hospital, asked to take a room next to his.

Sam told her it was nothing serious. "Hell, Baby, it could have been done in a dentist's chair." Then, trying to allay her fears, he added, "I just need a checkup and some treatment on my sinuses. I'll be out of here in a couple days. It's okay. I could use the rest." He told her to go back to Paramount and "knock 'em dead. Give Cuddle-Me-Close [Sam's nickname for baby Lita] an extra kiss for me."

Relieved by Sam's sprightly manner, Lina returned to their luxury bungalow at the Ambassador. Taking Lita from Emily, the full-time nanny, she hugged her daughter tightly and cooed, "Your Da Da is fine, he's going to be fine . . ." Early the next morning Lina called her husband's hospital room.

"Sure, Baby," Sam said, "I'm feeling okay. The head still feels like a balloon about to burst, but they're giving me something for it."

That afternoon, still in her makeup and dancing costume from a scene in *Serenade,* Lina went to Sam's bedside. Pearl was there, leaning over Sam's bed, patting his hand, dabbing her eyes on the bedsheet. Ben Warner stood like a statue next to her. Four physicians who had been called in to diagnose Sam's problem hovered nearby. Into this grim tableau pranced Lina in her Mazurka costume.

Sam, shifting his eyes to Lina's ruffled outfit, told her she looked great. Lina kissed his cheek and felt the heat radiating from his skin. His hands, holding tightly to hers, were as dry as parchment.

Sam wearily tried to ask her about her day at the studio but she interrupted. "You're on fire with fever." She looked at the four doctors who had broken from their huddle and were watching her. "What are you keeping from me?"

Sam said, "These guys, the doctors, think the whole problem is with my sinuses . . ."

"An acute mastoid infection," one of the doctors said.

Confused, Lina gripped Sam's hand harder.

"They're going to operate tomorrow," Sam told her.

Lina felt her muscles tighten with fear. Operate? How could they operate on Sam? He had always been so strong. What would she do if . . . She was only twenty, she had a baby daughter . . . She looked into his greenish-blue eyes, which were now dull and lifeless, and told him she wanted to stay with him.

"No, no, Baby. You've got a movie to make. Just stick with it. This will all be over and I'll be riding my horse in a couple days."

The next day during lunch break at Paramount, Lina slipped away from the set and hurried back to the hospital. The operation for a frontal sinus infection had been performed and Sam was under the effects of the anesthetic. He was completely swathed in bandages. To Lina he looked like he had been mummified. She sat by his bed—and prayed, and waited.

By evening the effects of the anesthesia had worn off and Sam felt a little better. Lina stayed by his bedside until after midnight, stroking his hand. He felt so hot! She didn't know what to think. Her once vital husband looked so small and helpless.

The next morning, Lina called Paramount and canceled her shooting schedule. The doctors went into a white-coated conference and agreed that a serious infection had localized somewhere in their patient's head. Another operation was planned.

The following day, Lina brought Lita to the hospital. Sam was awake but groggy after the operation. Lita, just a few days from her first birthday, was terrified when she saw her "Da Da" under all the bandages. She screamed. Sam started talking to her in a soothing voice, calling her "Princess Cuddles" until the chubby-cheeked little girl went to his arms.

Lina, trying to hold back the tears, told Sam, "Remember Lita's birthday party is only a week away . . . you must be there."

Sam squeezed Lita and nodded his head.

That evening Sam had a serious relapse. The doctors prepared for a third operation. Lina called Harry in New York. In tears she explained that Sam's condition had worsened. She told him she didn't know what to do. She was terrified. Harry tried to calm her, saying that his father, Ben, had talked to him each day about Sam's condition. What Harry didn't say was that he had also talked to Dr. George McCoy, the hospital's chief physician, who had told him that a brain abscess had developed and the team of doctors were deeply concerned about the outcome of the operation. Harry advised Lina that Abe had left New York two days earlier and would arrive in Los Angeles the following day. Harry said he would catch a train the next morning, Saturday. Jack would follow as soon as he could break away from last-minute preparations for the opening of *The Jazz Singer.*

That night Lina stayed in a room across from Sam's, sleeping fitfully.

On Saturday, Harry grabbed two New York specialists and took the first train he could get to Los Angeles, a 2,500-mile trip. A few hours out of New York, Harry received a telegram. The operation had not been successful; Sam's condition was "serious."

Frantically, Harry counted the minutes as the train chugged through depots and took on other passengers. It was moving so slowly! He was less than halfway to Los Angeles. The train seemed to be making an agonizingly slow crawl across the Arizona desert when Harry got another telegraphic report that Sam was dying. Harry tried to rent an airplane (a hazardous move in itself), but the only cross-country plane was laid up for repairs. He chartered a special train to make the last dash to Los Angeles. It was now Tuesday.

Dr. McCoy issued a bulletin that Sam Warner's condition was "very low." Jack immediately left New York by train.

The fourth operation was imminent when Abe arrived. Pneumonia had now set in. Abe and Lina clasped hands and stood by Sam's bed. In a weak voice Sam told Abe and his wife that he didn't think he was going to pull through the next operation.

Lina kept saying to him, "You're not going to die." She later remembered thinking, "I couldn't believe it. I was a good Catholic. There must be a miracle, a miracle, miracle . . ."

Sam muttered for Lina to come closer. "I love you, Lina." His right hand slipped from her grasp and his breathing came in hollow rasps from his chest.

Sam was taken to surgery, where a delicate operation was performed at the base of his skull. The family was given no assurance that his life could be saved. Throughout the night Abe and Pearl and Lina remained at Sam's bedside.

Lina, clinging to the faith of her youth, kept calling out, "Dear God, *do something!* Please let Sam live." Over and over through the night she prayed, fingers moving endlessly over her rosary. At 3:22 A.M. a small, strange sound came from Sam's throat and his hand became rigid in Lina's. Her fingers stopped their movements on the beads. . . .

Sam Warner died from cerebral hemorrhage on Wednesday October 5, 1927. He was thirty-nine years old.

Harry Warner arrived in Los Angeles three hours after his brother's death. Jack Warner got the tragic news as his train passed through Chicago.

Sam Warner's estate was estimated at over $1 million. Lina had no idea of this until she read the newspaper headline:

WARNER RICHES GO TO BROTHERS
WIDOW GETS LIFE INTEREST IN $100,000
OF FILM MAN'S $1,000,000 PROPERTY

Lina was to receive the dividend income from the $100,000 for the rest of her life and at her death the principle would revert to the estate. A trust fund of $100,000 was set aside for Lita, from which the child would receive dividend income until she was twenty-one years old, when the principle would be paid to her. A paragraph in Sam's will leaving $50,000 cash to Lina had been stricken, as had a clause that voided her dividend income should she remarry. Lina was also awarded $40,000 from an insurance policy of Sam's, plus her personal wearing apparel, the household goods, and automobile.

The major clause in Sam's will stated that his 62,500 shares of Warner Bros. stock (worth $1 million in 1927) would go to his three brothers.

In 1927, to a twenty-year-old girl with a budding film career, it seemed a fair settlement. She had just signed to do a movie, *The Noose,* with Richard Barthelmess. For that film she would be making $750 a week.

Sam's funeral.

Crowds of fans pushed their way into the Bresee Brothers funeral parlor in Hollywood, not to pay homage to Sam Warner, but to gawk at the movie stars in attendance. A "mystery mourner" draped in a long black veil (a type who often appeared at movieland funerals) flung a spray of flowers across Sam's casket. She was forcibly removed by studio security people.

Lina numbly carried Lita in her arms and walked up to Sam's casket. She leaned over and let the child place a little hair ribbon next to Sam's hand. Lina kissed a photograph of herself and the baby and laid it on his chest. One shoe of Lita's, the first pair Sam had bought his daughter, had also been put in the casket.

As the brothers and family sat somberly in their seats, a eulogy was read:

> O friend. We've shared the fortunes of the road,
> On many a long hard climb, in many a weather,
> But now we've reached the parting ways—
> no more, old comrade, may we rove together.
>
> It will be strange! You had such zest for life.
> Such mirth, such golden castles in your dreaming—
> You might have been a knight of Camelot,
> With nodding plume and lance and armor gleaming.

A memorial service was held at the Warner studio, where Sam's favorite horse was saddled with boots that were reversed in the stirrups. "Taps" was played on a bugle. Sam was interred in the family mausoleum at the Home of Peace Cemetery in Los Angeles on October 9, 1927.

Three days earlier *The Jazz Singer* had opened in New York. Sam Warner had dreamed that motion pictures would one day talk, and now that vision was a reality.

And now Sam was gone. He was loved by all for his charming, self-sacrificing, and fun-loving nature. What no one realized until his death was that he was also a stabilizing influence in his brothers' relationships. He had managed to keep the peace between Jack and Harry. That peace would eventually erupt into a war between the Warners.

There is a curious footnote to Sam's death.

In 1977 I was attending a ceremony at the Hollywood Palladium honoring the fiftieth anniversary of the "talkies." A commemorative stamp had been issued for the occasion and Jack Warner, the only surviving brother, was there to accept the honors. I was seated in the bleachers and after a while I introduced myself to the elderly man who was sitting in front of me. It was William Demarest, the actor who had played Al Jolson's friend in The Jazz Singer. *Learning I was Harry's granddaughter, he seemed delighted to see me. We became immediate friends.*

After the pomp and circumstance was over, and Jack had offered his usual corny gags and thank-yous, Demarest pulled me out of earshot from the rest of the people in the room.

"They killed him!" he whispered urgently. "They killed Sam."

Before I could say anything, he continued, "I worked with Sam. He was a good man. He didn't agree with what they wanted him to use the medium of film for, so they got rid of him."

I felt privy to a very deep secret and urged him to continue. I sensed his urgency to share this with a relative of Sam's he could trust.

He continued: "Sam didn't die of natural causes. He went in for a very simple operation. The doctor was paid to make it look like his death was unavoidable. Sam wouldn't agree to what they wanted, so they got rid of him. They killed Sam."

I asked who "they" was. He said, "They wanted to use the film for propaganda purposes and Sam would have nothing to do with it, so they got rid of him."

I felt the weight of carrying this untold story lift from him. He was relieved of the guilt from not saying something sooner. Later, I felt like I was part of a mystery plot—a true-life mystery—an angle that no one yet knew about. The burden he had carried for fifty years was now transferred to me.

We talked for a few more minutes and I noted that he had not been drinking. He invited me to come to his home and have tea and meet his wife sometime in the near future. I gladly took his number. Unfortunately, I never got around to meeting with him again; he died in 1983.

I had decided to begin my research for this book prior to meeting Demarest. I had known there was a reason I had to be at this event. This was more than I had expected.

My great-aunt Lina said she couldn't conceive of anyone doing something "so diabolical." She added, "William Demarest must have been reading too much about the Borgias."

Jack Warner Jr. said, "Bill Demarest was a wonderful guy who could not hold his liquor and when loaded indulged in flights of fancy. This murder plot is based on an oversoaked imagination. Some pretty dreadful things happened at Warner Brothers, but this was not one of them."

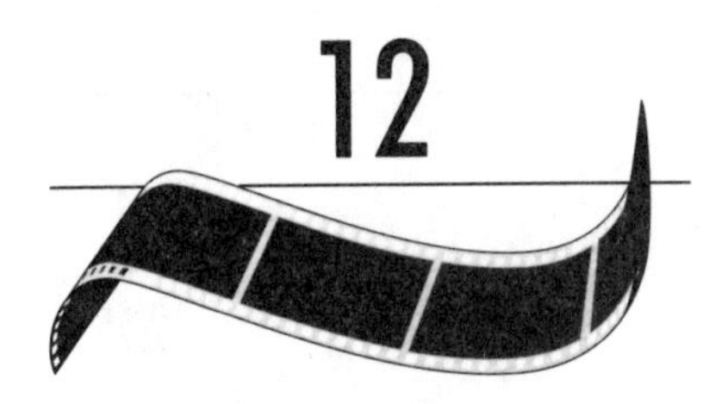

From Poverty Row to Moguldom

Shortly after Sam's death, Abe was quoted in a *Screenland* magazine interview: "The three of us who are left will carry on, and I believe we will accomplish more work in one day than any other trio of men will in three, not because we are smarter, but because we trust each other implicitly and don't have to waste time with petty executive jealousies."

In December 1928 the *Cleveland Plain Dealer Magazine* reported: "There has never been another so closely knit family in the film industry. These brothers, so tightly bound by family ties and genuine love for each other, formed a terrific flying wedge in the battle for a new kind of film, the Talkie."

"The dicks will be here at ten o'clock, but they must not find . . . Eddie . . ."

There was a long pause as the two hoods (who looked like stooges from a Keystone Kops flick) pondered what the gangster boss had said. The boss, razor thin, hair slicked back from his forehead, leaned forward and added:

"Don't you understand?"

"You mean . . ."

In a sinister voice, "Take him . . . for . . . a ride."

"Oh . . ."

This dialogue, spoken into a microphone hidden in a flower vase on the boss's desk, may have been trite—but it was dialogue. The movie, *Lights of New York*, was the first all-talking feature film. Strangely, neither Harry or Jack had scheduled it to be anything more than a Vitaphone short. They had not made the decision to launch a season of full-length talking features.

In early 1928, to satisfy the sudden, insatiable demand for sound pictures, Warners quickly made several movies with talking inserts (called "goat glands" in the trade): *Tenderloin,* a murky melodrama starring Dolores Costello that had started as a silent and had fifteen minutes of dialogue added; *Glorious Betsy,* again with Dolores Costello; and *The Lion and the Mouse,* starring Lionel Barrymore. The Warners had yet to produce a feature-length sound movie.

In February 1928, still stunned by Sam's death, Harry and Jack traveled to Europe to oversee the London, Paris, and Berlin premieres of *The Jazz Singer.* Before leaving, Jack told Bryan Foy (one of the Seven Little Foys of vaudeville fame), who was in charge of the Vitaphone shorts program, to make a $12,000 two-reel gangster film titled *Lights of New York.*

Foy wrote the script with Murray Roth and Hugh Herbert, a comic actor and sometime screenwriter. Then Foy decided to make it into a full-length all-talking picture. Actually, he didn't decide, it just grew. The crew, seeing that Jack Warner was gone, and convinced that they had this sound thing conquered, kept elaborating on the plot as they shot the film. Jack had left sister Sadie's husband, Louis Halper, in charge, figuring he wouldn't do anything unexpected. But Halper, seeing that Foy and his crew were running out of money to finish the film, cabled Jack in Europe and asked for an additional $15,000 to make the two extra reels. Although Jack wondered why the film had more than doubled its original budget, he agreed to the extra expense. When he returned and saw the daily production reports he was about to fire Foy—until he screened the film and saw the potential of a few million bucks at the box office.

"We previewed it in Pasadena," Foy recalled.

> There was a line around the block and the theater had a huge sign announcing THE FIRST ALL-TALKING PICTURE. We showed it in New York at the Strand, and theater owners Charles and George Skouras and several other exhibitors began to bid for it by the third reel. Jack stopped the

> film and made a deal right then. In the first week at the Strand it did forty-seven thousand dollars.

When the fifty-seven-minute feature was released nationwide, it turned into a $2 million hit.

Jack went one step further. He quickly produced *The Terror,* advertised as a "Titleless Talking Film." There weren't any visual credits at the start; the cast of players, and names of the film's staff, were announced from the screen by the shadow of a masked man (actor Conrad Nagel). The public rushed to see it.

Not everybody thought talkies would last. Mary Astor, one of Warners' leading ladies, said, "I thought pretty soon that we'd get back to silents. It was only a gadget to play with. I didn't take it seriously."

Douglas Fairbanks visited a sound studio with English art director Lawrence Irving and said, "This is of no real concern to us," and immediately started work on a $2 million silent about the Three Musketeers, convinced silent movies would never completely disappear.

Two years later, after the audience rejected this last silent epic, Fairbanks realized it was all over and said, "The romance of motion-picture making dies here."

Charlie Chaplin continued to stick his toothbrush mustache to his upper lip, swing his little cane, and make silents such as *The Circus.* He told reporters he'd never do a talkie. Even Al Jolson, after viewing *The Circus,* said, "It's the greatest thing I've seen. Dey's my sentiments."

Jolson was still the hottest property in Hollywood, so Jack cast him in a second film, *The Singing Fool,* offering him a staggering $150,000, which Jolson wisely took in Warner Bros. stock. The movie had a few more lines than *The Jazz Singer* but was mostly songs by Jolson. The singer knew he needed a big number, so he called the songwriting team of De Sylva, Brown, and Henderson, who had composed many of his old hits and who had been assigned to create several songs for the movie.

Jolson told the songwriters he wanted a tune about "a little kid who's dyin' and his daddy's cryin' his heart out." The songwriters thought the idea was silly, but they obliged with what they considered a schmaltzy tune. The new song, "Sonny Boy," had audiences weeping into their handkerchiefs.

"Sonny Boy" became the first tune to sell a million copies of sheet music. *The Singing Fool* was an out-and-out smash and brought in $5.5 million, a record that stood until *Gone With the Wind.* By the end of the year, the Warners were making nothing but all-talking pictures and cashing in. The million-dollar theater they built on Hollywood Boulevard, and which almost cost them eating money, began to pack them in with the new fad.

In the spring of 1928, a reporter from the *Sunday News* interviewed Jack and noted:

> I talked to Jack Warner in his Hollywood office, and he was almost hysterical and raving about the way money was rolling in. He could think of nothing else. He told me so often that the dough was rolling in faster than they could count it that I threatened to ask him for a loan if he didn't talk about something else.

Up to this time, the brothers had agreed not to accumulate individual fortunes. Everything had gone back into the studio. In 1926 they had started a family holding company, Renraw (Warner spelled backwards), which served as a symbol of family solidarity, embodying the Three Musketeers' maxim, "All for one, one for all." Renraw also served as a joint bank account. Until 1929 the brothers' total salaries were only $100,000 a year. Thereafter it jumped to $500,000. This sum was paid by Renraw and each brother took whatever additional money he needed to live comfortably on, no more. Abe said, "If I wanted a sum of money, no matter how large it was, I would take it. And neither Harry or Jack would ask me what I wanted with it. We love and trust each other."

Renraw, as a private company, acted as a 6 percent Santa Claus whenever Warner Bros. Pictures Inc. was hard-pressed to raise cash. During the late 1920s, because of the upward trend of the stock market, Renraw was well-heeled. When Harry needed extra money for development or to buy companies and theaters, Renraw periodically sold some of the Warner Bros. stock (the brothers controlled 300,000 of 350,000 shares) on the curb for cash. Banker Waddill Catchings called this "turning the firehose on the market." Renraw would then lend its money to Warner Bros. Pictures Inc. without security at current banking rates. This was tricky. By 1928, the brothers were down to their last 60,000 shares of stock. It would take them three more years to regain their old stock position of 300,000 shares.

The success of *Lights of New York*, *The Terror*, and *The Singing Fool* drove even the nonbelievers in the major studios out of the silent trenches to join the mad scramble to reorganize and meet the demand for talking pictures. By the spring of 1928, silent stages were being torn down and sound put in. Engineers who knew anything about sound could command whatever salary they liked. Costs soared and the old relaxed atmosphere in Hollywood disappeared. Actors were terrified. Would their voices be compatible with sound? An army of extras became a howling mob of hopefuls. It was no longer how good-looking you were, but whether you could talk.

In response, elocution training schools popped up in the Yellow Pages: "School of Voice Culture," "We Teach Dramatic Speaking." Their numbers increased so rapidly that they had to be given a separate section in the Hollywood telephone book. Studios required that stars have their voices recorded. Heartthrob John Gilbert's voice tested rather high. Foreign actors who couldn't speak English had to return to their homelands. Pola Negri, filmdom's Polish vamp, saw her contract go unrenewed. Vilma Banky was hampered by a Hungarian accent. Of the greatest concern to MGM was their superstar, Greta Garbo. How could they expose her thick accent to the public? They waited over two years while she underwent elocution lessons, then in 1930 decided to take the plunge and cast her in *Anna Christie.*

In the movie Garbo saunters into a roadside bar and says, "Give me a whiskey . . . ginger ale on the side. And don't be stingy, baby."

Garbo talked!

Harry knew he couldn't hold his lead for long. He knew that as soon as the big-timers could make enough talkies to supply their theater chains, Warner Bros. would once more be just another independent producer. To forestall this, in one big burst of family pride and financial sorcery, he went on a spending spree.

Harry called Jack to New York for a special meeting with Abe in their Manhattan office. Jack twisted in his seat and looked around at the office's oriental throw rug, the heavy oak furniture, the plain walls, and the high windows that overlooked the stores on Eighth Avenue. Abe's office was through a connecting door, furnished in the same spartan manner. A second door led to an apartment in which Jack had an office he used on trips to New York. At least it wasn't plain: pictures of movie stars and movie posters covered the walls.

Harry immediately stated his intentions: "I'm going to put in a bid for First National Pictures."

"You're gonna *what*?" Jack looked at Abe, who had practically slid out of his swivel chair and was choking down a lungful of cigar smoke. "You're gonna buy . . ."

"First National," Harry repeated with a sly smile, relishing his brothers' shock. "The organization is floundering and so it's ripe for a takeover."

"It would make more sense if First National bought *us,*" Jack said. "*We're* going to grab up one of the biggest studios in the industry?" He rolled his eyes and snapped his fingers. "Huh, nothing to it." Actually, Jack was relieved by the announcement. He figured Harry had called him to New York to chew him out for some recent amorous escapade. There had been several dalliances over the last few years. Pushing those thoughts aside, Jack concentrated on Harry's decision—buy First National?

The hot-shot First National organization had been formed in 1917 by a group of independent producers to break Adolph Zukor's domination of the industry. First National had grown rapidly. It now had the largest theater chain in the country and had also emerged into an important production company, with a studio in New York and a sprawling lot in Burbank. Jack couldn't wait to get his hands on *that* studio, as well as National's dazzling list of reigning stars, big names like Kay Francis, Constance Bennett, Basil Rathbone, and a new girl, Loretta Young. He'd love to try her out, but he had a rule: Studio stars were taboo. An occasional starlet, okay; a secretary, sure. What Harry didn't know . . .

Jack had been faithful to Irma until around 1923. Then he began to tell her he had to work late editing a picture, attending a conference—he could always come up with a good excuse. The marriage went on fairly well because Irma was ignorant of Jack's expeditions into the bedrooms, parlors, and baths of the women he came into contact with in the business of freezing glamour on film.

Yet Jack would maintain that he was truly in love with Irma. In early 1926, he had honestly told a reporter that his wife was his "inspiration." He was quoted as saying:

> All that I am, I owe to my wife. It sounds like regular movie "blah" to hear me say it, but this is the exception that proves the rule. Wherever I go my wife goes with me. I couldn't do my work without her near to encourage me.

"We'll also buy the Stanley Corporation," Harry was saying.

That woke Jack up; Abe choked again on his cigar. The Stanley Corporation was one of the country's biggest chains, with 250 theaters in 75 cities and every state, including the luxurious Strand on Broadway, one of the elegant picture palaces from the silent days.

Harry went on: "We know most exhibitors are . . ."

"A bunch of stubborn tightwads," Abe finished.

"Exactly. When it comes to spending their own money. Their refusal to put in sound equipment means we're being locked out. Since we have so few theaters equipped, we have to look to the future. With the Stanley theaters we can ensure that all our future films have nationwide distribution."

Abe, the cautious one, said, "What would Sam think about all this?"

The mention of Sam's name chilled the room and no one spoke. Finally Jack intoned, "Sam dreamed about sound theaters all over the world. He'd agree."

Abe slammed his huge fists down on the glass-topped desk. "Let's grab them up!"

"Uh, wait a minute. Harry, let's get back to First National." Jack flicked the ash of his cigar on the standing brass ashtray at the side of Harry's desk. "How are we going to get control of the company?"

"I propose to buy forty-two thousand of the company's seventy-five thousand shares."

"Hell, Harry, why not buy the whole damn mess?" Jack said.

Harry looked at his brother sharply. "I *intend* to buy the 'whole damn mess.' I found that Fox West Coast Theaters is going to take over the remaining third of the stock. No matter; we will still have controlling interest. Fox can be bought out later."

"How much is this little shopping spree going to cost us?" Jack asked.

Harry enunciated the words slowly: "One . . . hundred . . . million . . . dollars."

Jack whistled. "That's a lot of moola." Then, worried, added, "We're taking a big chance."

Harry smiled, in his element now. "That's the name of the game. We have to take big risks to make big money."

"Yeah, and we could be out on the street again." Jack was thinking of the three acres of rolling farmland in the foothills just north of Beverly Hills where he had started construction on a fifteen-room Spanish-style house. Hell, it was a mansion. And he could lose it all.

Harry said, "If we hadn't taken risks twenty years ago, Abe would still be selling soap. And Jack, you'd be crooning ballads in some speakeasy. We *need* new studios to equip for sound, we *need* theaters to show our films . . ."

Jack raised his hands as in surrender. "Okay, okay, I agree. But it would be nice to know *what* films we're going to be showing next season."

"I plan on coming to Hollywood in October. We'll map out next season's shows then—after I have this business settled."

"*If* we're still in business," Jack mumbled, still not sure this was the right thing to do. He could lose his plush office on Sunset, but most of all the power that he, Jack L. Warner, studio chief, had so recently attained.

Harry had heard Jack's remark, but didn't respond. He looked at Abe. "Do we *all* agree on the First National and Stanley buyouts?" He slapped the flat of his hand down on the big desk. Jack got up and put his hand over Harry's; Abe's big paw covered theirs. In unison they said, "Agreed."

Jack started to walk out of the room when he heard Harry say in a flat tone, "I need to talk to you."

Jack felt like an icicle had been jabbed in his back. He turned. Harry had his hands crossed over his chest, fingers tightly meshed. He looked like a school principal about to confront a wayward student. "The family's upset."

At that, Abe spun on his heel and left the room. He was not about to be the mediator in this one.

Harry worked his entwined fingers against one another, leaned back in his chair, and fixed his eyes on Jack. "Now . . . it embarrasses me to bring this up, but as your older brother, I must talk to you about your . . . behavior."

Jack let his features melt into a rubbery mask as he prepared himself for what he knew was coming—the old family lecture about morality.

Harry continued: "We are stepping into the limelight more and more each day. These acquisitions I have planned will force us further into the public's view. Our personal lives cannot be, well, tasteless. Any cheap, immoral behavior by one member of the family would reflect on all of us."

Jack sighed. "Okay, Harry, what are you trying to say?"

"I'm talking about your relations with . . . with people other than your wife . . ."

"It's none of your business."

Harry controlled his temper. "I know all of the temptations you are exposed to at the studio . . ."

"I said, it's none of your damn business!"

Harry's face was a tight mask. "It is as long as you're part of this family—and a growing studio that has *all* our names on it."

Jack leaned toward his brother, knuckles on the desk. "Talk to me all you want about holding down production costs, but my personal life is damn well my own."

Above: The Warners posing for a family photo shortly after arriving in America. Back row, left to right: Abe, Harry, Ben, Pearl, Annie. Front row: Sam, Fannie, Rose.

Left: Harry, age 15.

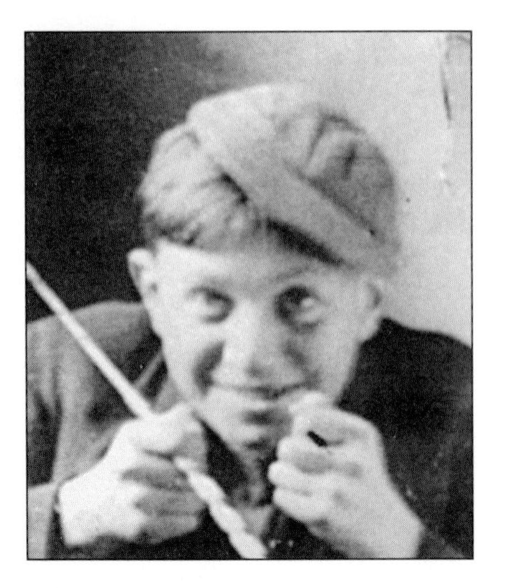

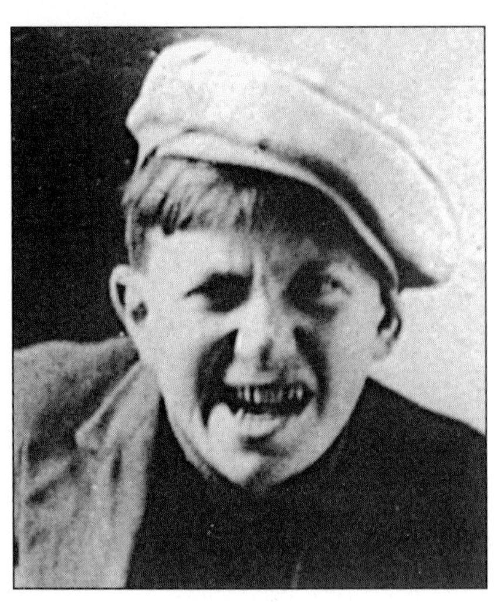
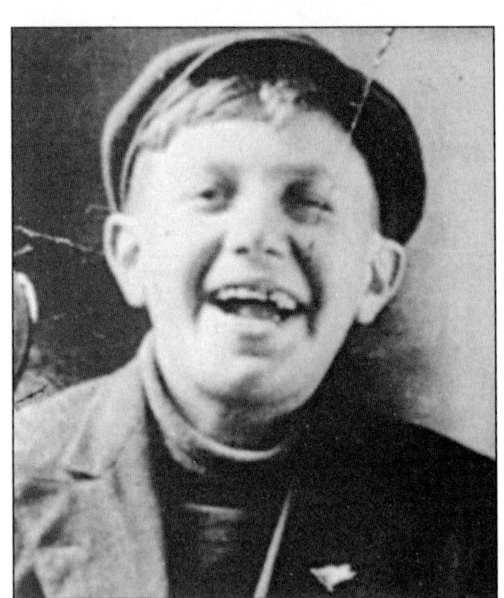

Photo strip taken at Banner Photo shop, where Jack was a barker, in Youngstown, Ohio, 1900.

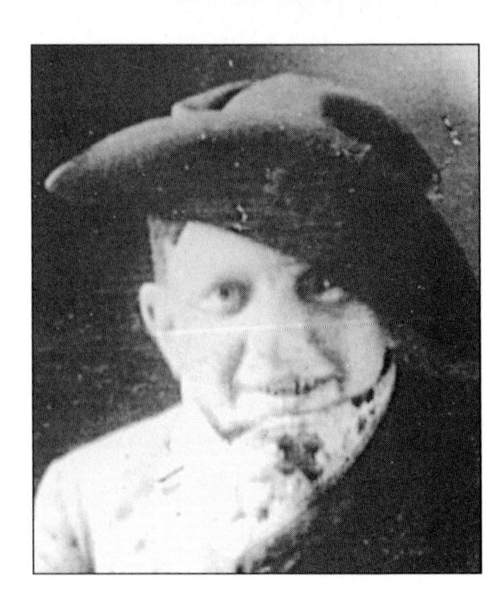

The Duquesne Film Noise

PUBLISHED MONTHLY BY

DUQUESNE AMUSEMENT SUPPLY CO., Inc.

Licensed by Moving Picture Patents Co.

103-105 Bakewell Building, Pittsburg, Pa. 235-237 Monticello Arcade, Norfolk, Va.

SAM WARNER, Editor Price—100 Kazamblos—American Money Free: JACK WARNER, Asst. Editor.

Entered in Every Moving Picture Theatre as First-Class Matter.

Volume I APRIL, 1909 Number 3

WE ARE THE ONLY FILM EXCHANGE ISSUING ITS OWN MAGAZINE—WATCH OTHERS FOLLOW

S. L. WARNER
Editor and Manager Norfolk Office of Duquesne Amusement Supply Co.

J. J. WARNER
Asst. Editor and Manager Film and Supply Dept. Norfolk Office Duquesne Amusement Supply Co.

JUST RECEIVED PATHE'S PASSION PLAY.

Three thousand and two hundred feet in length and highly hand-colored. Write today for rates. We also have a great quantity of advertising matter at a very low price, consisting of dodgers, hand bills, half-sheet, three-sheet, [illegible] all other advertising matter. The management of the [illegible] to capacity, and this shows you that this great play is still in demand.

We have three copies of Pathe's Earthquake and three copies of Vitagraph we are renting at a reasonable rate, also lithographs. First in first served. Write either office, as we have both in each office.

When in need of good films and would like to arrange communicate at once with the Dequense File Exchange. Quantity and quality are always in demand, and if you are a licensed member this firm will stand by you. Give us an order and you will get the best. The firm with the goods will stand a test.

NO EXHIBITORS' CONTRACTS.

The Patents Company people declare that they have never required a contract signed by exhibitors, and that, therefore, the story that they had withdrawn their demand is a misapprehension.

They say they merely required exhibitors to sign an application for a license for their theatres and a payment of $2 weekly for that license. The instructions telegraphed to Chicago last week were designed to clear up a misunderstanding among the renters. The application, they say, is not in any way to be construed as a contract, since it does not bind the exhibitor to any terms.

NOTICE.

Vaudeville Department.

Mr. Harry Mitchell, formerly with Harry Von Tilzer, Music Publishing Co., is now in charge of our large Vaudeville Booking Department, and will be glad to meet [illegible]

Sam and Jack's film exchange newsletter, 1909.

Top: Irma and Jack when they first met in 1914 in San Francisco. Furnished courtesy of Jack Warner Jr.

Right: Baseball star Milton Warner, age 18, in Rayen High uniform in 1914.

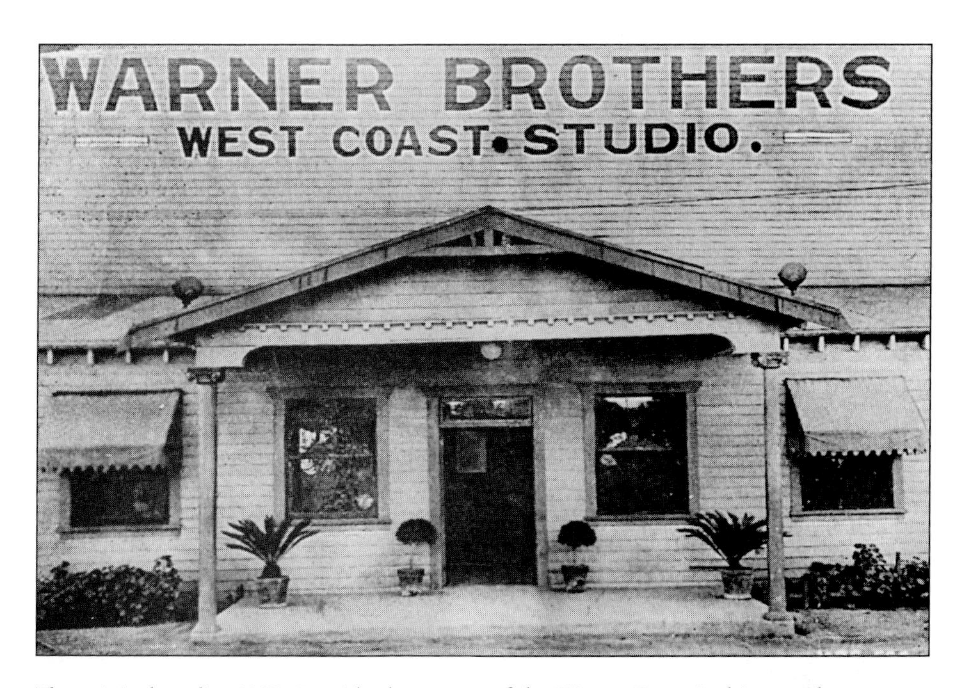

The original studio, 1918. Furnished courtesy of the Warner Bros. Archives at the University of Southern California.

Ben is cameraman, as Jack plays director in 1920.

Above: Pearl and Ben with their grandchildren in 1921. Left to right: Pearl, Lewis (standing), Jack Jr., Doris, Betty, Ben.

Below: The brothers in their New York office, early 1920s. Left to right: Sam, Harry, Jack, and Abe.

Above: Abe, Sam, and Jack planning the Sunset Boulevard studio in 1922. Furnished courtesy of Jack Warner Jr.

Left: Harry checking the books in 1923.

Left: Glamorous Lina Basquette in the mid-1920s.

Below: Sam (left) admiring the celluloid with an unidentified man and William Koenig (right), general studio manager, in 1925. Furnished courtesy of the Warner Bros. Archives at the University of Southern California.

Harry's home in Hancock Park, Los Angeles, 1925–1929.

Right to left: Doris, Betty, and friend in 1925.

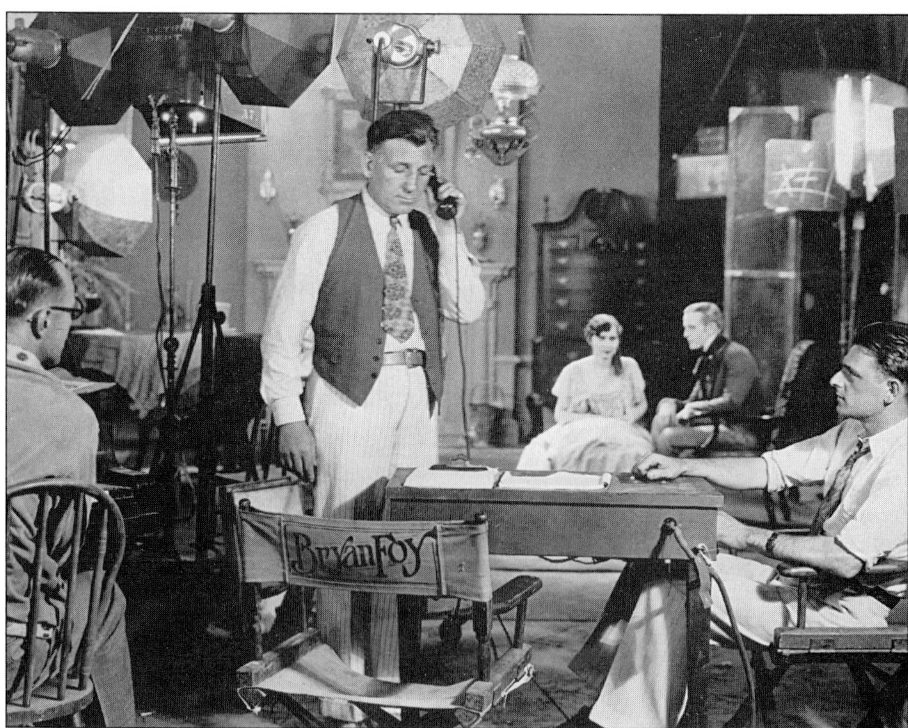
BryanFoy

Facing page, top : Harry, Rea, Betty, Doris, and Lewis in front of their Hancock Park home in 1925.

Facing page, bottom: Sam and Bryan (Brynie) Foy on the set in 1926. Furnished courtesy of the Warner Bros. Archives at the University of Southern California.

Above, top: Standing tall: Harry, Jack, Sam, and Abe in 1926.

Bottom right Jack showing off the 33⅓ rpm sound disc for *Don Juan* in 1926. Furnished courtesy of the Warner Bros. Archives at the University of Southern California.

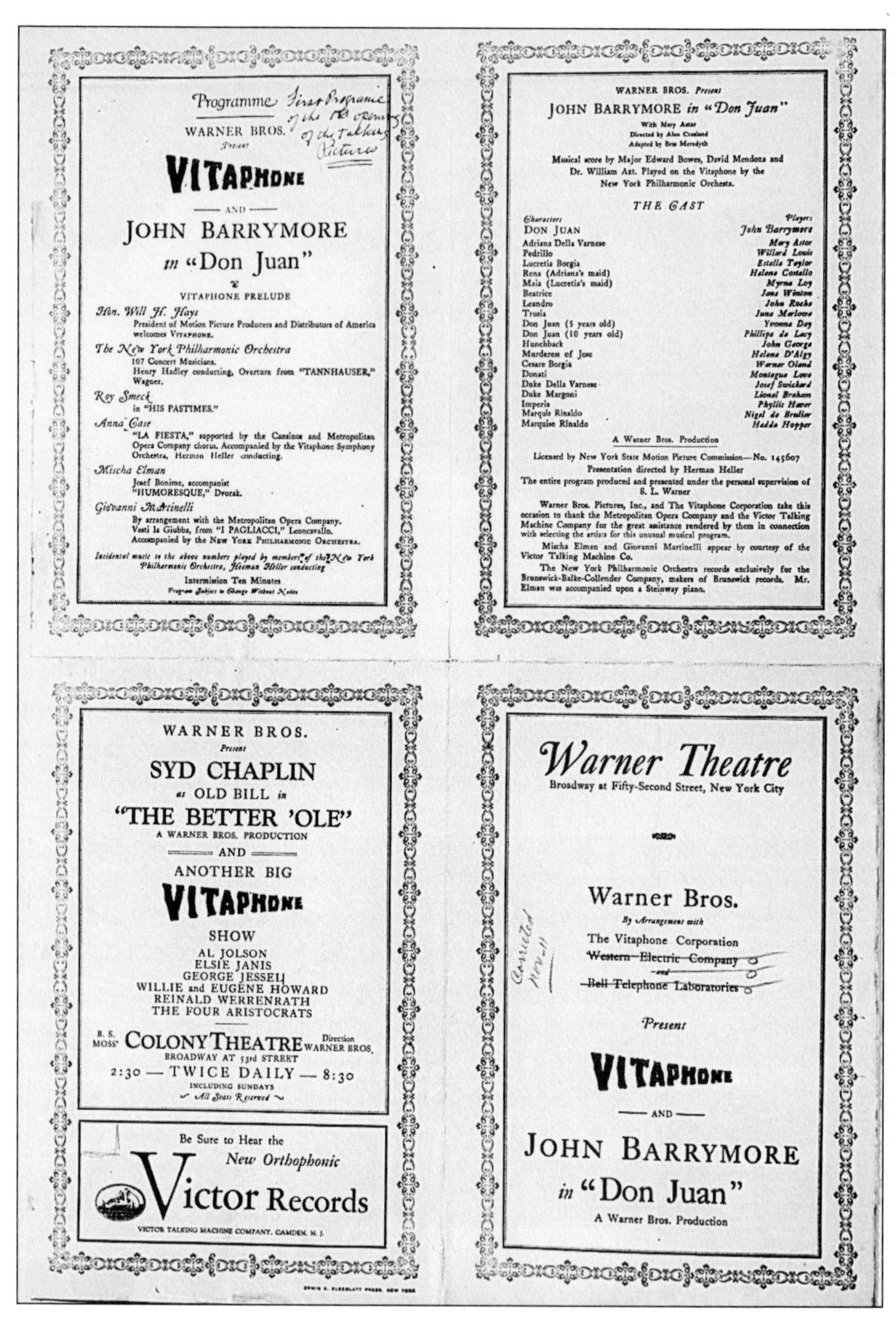

Programme

WARNER BROS.
Present

VITAPHONE

— AND —

JOHN BARRYMORE
in "Don Juan"

VITAPHONE PRELUDE

Hon. Will H. Hays
President of Motion Picture Producers and Distributors of America welcomes VITAPHONE.

The New York Philharmonic Orchestra
107 Concert Musicians.
Henry Hadley conducting, Overture from "TANNHAUSER," Wagner.

Roy Smeck
in "HIS PASTIMES."

Anna Case
"LA FIESTA," supported by the Cansinos and Metropolitan Opera Company chorus. Accompanied by the Vitaphone Symphony Orchestra, Herman Heller conducting.

Mischa Elman
Josef Bonime, accompanist
"HUMORESQUE," Dvorak.

Giovanni Martinelli
By arrangement with the Metropolitan Opera Company.
Vesti la Giubba, from "I PAGLIACCI," Leoncavallo.
Accompanied by the NEW YORK PHILHARMONIC ORCHESTRA.

Incidental music to the above numbers played by members of the New York Philharmonic Orchestra, Herman Heller conducting

Intermission Ten Minutes

Program Subject to Change Without Notice

WARNER BROS. *Present*

JOHN BARRYMORE *in "Don Juan"*

With Mary Astor
Directed by Alan Crosland
Adapted by Bess Meredyth

Musical score by Major Edward Bowes, David Mendoza and Dr. William Axt. Played on the Vitaphone by the New York Philharmonic Orchesta.

THE CAST

Characters	*Players*
DON JUAN	*John Barrymore*
Adriana Della Varnese	*Mary Astor*
Pedrillo	*Willard Louis*
Lucretia Borgia	*Estelle Taylor*
Rena (Adriana's maid)	*Helene Costello*
Maia (Lucretia's maid)	*Myrna Loy*
Beatrice	*Jane Winton*
Leandro	*John Roche*
Trusia	*June Marlowe*
Don Juan (5 years old)	*Yvonne Day*
Don Juan (10 years old)	*Phillipe de Lacy*
Hunchback	*John George*
Murderess of Jose	*Helene D'Algy*
Cesare Borgia	*Warner Oland*
Donati	*Montague Love*
Duke Della Varnese	*Josef Swickard*
Duke Margoni	*Lionel Braham*
Imperia	*Phyllis Haver*
Marquis Rinaldo	*Nigel de Brulier*
Marquise Rinaldo	*Hedda Hopper*

A Warner Bros. Production

Licensed by New York State Motion Picture Commission—No. 145607

Presentation directed by Herman Heller

The entire program produced and presented under the personal supervision of S. L. Warner

Warner Bros. Pictures, Inc., and The Vitaphone Corporation take this occasion to thank the Metropolitan Opera Company and the Victor Talking Machine Company for the great assistance rendered by them in connection with selecting the artists for this unusual musical program.

Mischa Elman and Giovanni Martinelli appear by courtesy of the Victor Talking Machine Co.

The New York Philharmonic Orchestra records exclusively for the Brunswick-Balke-Collender Company, makers of Brunswick records. Mr. Elman was accompanied upon a Steinway piano.

WARNER BROS.
Present

SYD CHAPLIN
as OLD BILL *in*

"THE BETTER 'OLE"
A WARNER BROS. PRODUCTION

AND

ANOTHER BIG

VITAPHONE

SHOW

AL JOLSON
ELSIE JANIS
GEORGE JESSEL
WILLIE and EUGENE HOWARD
REINALD WERRENRATH
THE FOUR ARISTOCRATS

B. S. MOSS' COLONY THEATRE Direction WARNER BROS.
BROADWAY AT 53rd STREET
2:30 — TWICE DAILY — 8:30
INCLUDING SUNDAYS
All Seats Reserved

Warner Theatre
Broadway at Fifty-Second Street, New York City

Warner Bros.
By Arrangement with
The Vitaphone Corporation
~~Western Electric Company~~
~~and~~
~~Bell Telephone Laboratories~~

Present

VITAPHONE

— AND —

JOHN BARRYMORE
in "Don Juan"

A Warner Bros. Production

Program for the opening of *Don Juan*, 1926. Furnished courtesy of the Warner Bros. Archives at the University of Southern California.

"DAD" AND "MA" WARNER CELEBRATE GOLDEN ANNIVERSARY

THE GOLDEN WEDDING DAY

August 4th, 1876—August 24th, 1926.

Harry and Abe and Sam and Jack,
All of 'em hitting the home track,
All hell-bent to keep a date
With Dad and Mom in the Buckeye State—
And Dave and Sadye and Anna and Rose,
Hurrying swift as the wind that blows,
Hurrying back where they started from—
Back to the old town and Dad and Mom.

Dad and Mom for these fifty years,
Have shared life's trials, hopes and fears
And now the children who've learned to roam,
Are hurrying, scurrying, flurrying home.
How Dad and Mom will gloat to see
All of 'em home, like they used to be.

For there's one gift that's far above
All other gifts; we call it—Love;
It's like the answer to every prayer,
To know, when old, that the children care.
Rose, Anna and Sadye—Dave and Jack—
Abe and Harry—are taking Love back;
All eager to give, in their own good way,
Their Love—on the Golden Wedding Day.

MA WARNER

DAD WARNER

HARRY

SADYE

ROSE

ANNA

ALBERT

JACK

DAVE

SAM

The Golden Wedding of Mr. and Mrs. Benjamin Warner was celebrated in the old home town, Youngstown, Ohio, on the evening of August 25th, 1926. All the boys were there, Harry, Abe, Sam, Jack! All the wives were there and the children! All the best of the old friends and the best of the new! The hardships of the half-century had faded out! Happiness had faded in! It was a glorious scene! Sun up! Not like afternoon at all! Like morning! New fires lit! Old fires rekindled! A re-awakening of everything! A re-blossoming! A scene that might have been aptly titled: "Came the Dawn!"

That friendly old hostelry, "The Ohio Hotel," was the place of fore-gathering. About the board were fifty guests, one for each year of the gallant pilgrimage. The Young Couple sat at the head of the table, all a-flutter with the magic of it all! In and out rushed diminutive bell-boys, bearing telegrams of congratulation from the Four Quarters of the Globe, and the Seven Seas, too, it seemed!

Then came the tender ebb and flow of the old-time music, the dash and crash of the new! And the dancing! Then the well-wishing! And the good nights. And at last the Bride and Groom, radiant, grateful, young-of-heart, arm in arm, climbing the stairs, together!

To such as these all days are Golden Wedding Days!

Pearl and Ben's Golden Wedding Day celebration with their children in 1926.

A family gathering in the mid-1920s, possibly Ben and Pearl's golden wedding anniversary celebration in 1926. Back row, left to right: Sam, Dave Robbins, Jack, Irma, Abe, Bessie, Lewis, Doris, David Warner, Harry, Rea, Harry Charnas. Middle row: Lina, Sadie, Annie, Ben, Pearl, Rose, Louis Halper. Front row: Milton Charnas, Shirley Warner, Evelyn Halper, Betty Warner, Jack Warner Jr.

Lina, baby Lita, and Sam in 1926.

Left to right: William Koenig, general studio manager; Darryl Zanuck, Jack Warner, Al Jolson, Sam Warner in 1927.

Irma, Jack Jr., and Jack Sr. at a premiere in 1928. Furnished courtesy of Jack Warner Jr.

Right: Jack posing with special award for *The Jazz Singer* in 1928. Furnished courtesy of Barbara Warner Howard.

Bottom: Harry and Rea with the banker Motley Flint in the late 1920s.

Top: Lewis Warner in 1930.

Bottom: On a Warner's set in the late 1920s left to right: Harry Charnas, Pearl Warner, Sadie Halper, chimp, Rose Charnas, Milton Charnas, and unknown lady. Furnished courtesy of the Warner Bros. Archives at the University of Southern California.

Above: Ben and Pearl on the set around 1930.

Right: Jack chats with Dr. Albert Einstein at a studio luncheon in his honor in 1931.

Left: Ben Warner dressed for an opening in 1932.

Below: Jack with Leon Schlesinger, head of Merrie Melodies and Looney Tunes, clowning around in the mid-'30s.

Right: Abe with his second wife, Bessie, in the 1930s.

Below: Doris and Mervyn LeRoy's wedding in 1934. Left to right: Betty, Harry, Rea, young Lita, Doris, Mervyn, and Mervyn's parents.

Above: Actress Marilyn Miller and Jack in the early 1930s. Furnished courtesy of the Warner Bros. Archives at the University of Southern California.

Right: A power handshake between Harry and Ben in 1934.

Above: Jack and Abdul, his masseur and "best friend and pal always" in 1935. Furnished courtesy of Barbara Warner Howard.

Right: Ann and Jack saying their marriage vows in 1936.

Above: Ann and Jack Warner playing around in 1936. Furnished courtesy of Barbara Warner Howard.

Left: Barbara, age 4, and Jack skipping together in 1938. Furnished courtesy of Barbara Warner Howard.

Harry giving orders at his daughter Betty's wedding to Milton Sperling in 1939.

Salvador Dali's painting of Ann Warner, 1942. Furnished courtesy of Barbara Warner Howard.

Harry introducing Bette Davis to a national commander of the American Legion in 1942. Furnished courtesy of the Warner Bros. Archives at the University of Southern California.

Hal Wallis dining with Ann and Jack in the early 1940s. Furnished courtesy of Barbara Warner Howard.

Top: Ann Warner's daughter, Joy Paige, playing Annina Brandel in *Casablanca* in 1943. Furnished courtesy of Barbara Warner Howard.

Bottom: Harry showing off his new colts to Olivia de Havilland in the late '40s. Jockey to his left.

Top: The three Warner sisters with Harry at a 1940s gathering. Left to right: Rose, Annie, Sadie.

Bottom: Jack and daughter Barbara in front of their Beverly Hills home in 1950. Furnished courtesy of Barbara Warner Howard.

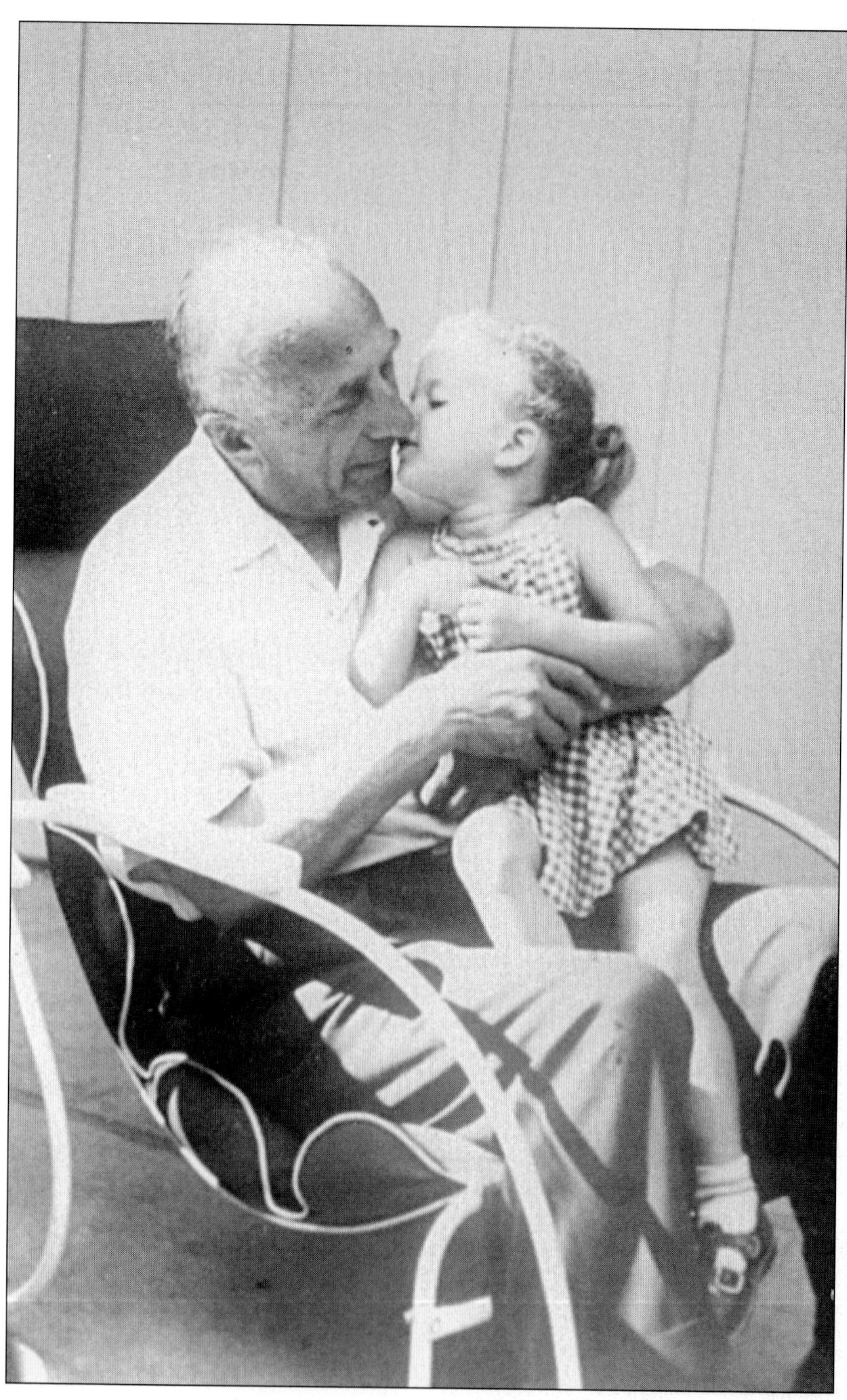

Young Cass giving her Grandpa Harry a smooch in the early 1950s.

Top: Jack Warner Jr. in the early 1950s. Furnished courtesy of Jack Warner Jr.

Right: Harry at the the racetrack in 1953.

Bottom: Jack in the trophy room at the studio in the mid-'50s.

Above: Harry joking with the Lone Ranger and Tonto on the studio lot in 1955.

Left: Harry with a Wild West life-size cutout of Jack, 1955. Furnished courtesy of the Warner Bros. Archives at the University of Southern California.

Above: The Warners studio executive dining room, 1956. Seated, left to right: Ben Kalmenson, head of distribution in New York; Jack; Mervyn LeRoy; Steve Trilling. Standing: Bill Orr, Jack Jr. Furnished courtesy of the Warner Bros. Archives at the University of Southern California.

Below: Rea and Harry's 49th wedding anniversary, 1957. Left to right: Lita, Doris, Rea, Harry, Betty.

Jack Warner on his way to the South of France in the '60s. Furnished courtesy of Barbara Warner Howard.

Ronald Reagan paying his respects to Jack at a tribute to Jack upon his retirement from Warner Bros. studio in 1969. Furnished courtesy of the Warner Bros. Archives at the University of Southern California.

Harry rose, his face red. "What you are doing is . . ."

"No! What *you* are doing is bullshit, family bullshit!"

The two brothers stood glaring at each other for a moment before Jack spat out the words *"Just keep the hell out of my life!"*, then stalked out of the room.

Harry sat alone for a long time after Jack left.

September 13, 1928

In a bold move, Harry borrowed $100 million from Goldman, Sachs and Hayden, and Stone and Company, and bought First National Pictures. As a result, Warner Bros. stock rose from $28 to $130 a share. With the purchase of First National came the one thing that Warners did not have: prestige. Even the success of their talking pictures hadn't changed their image as a penny-pinching studio specializing in schlock.

Harry was out to improve that image and continued his buying spree. He bought Witmark, Remick, Harms, and other music publishers. As a first step in his grooming of his twenty-year-old son, Lewis, to inherit the leadership of the fast-developing Warner empire, Harry put the young man in charge of Warner Bros. Music.

Abe Warner became the head of the company's theater division. Jack still ran the studios on Sunset Boulevard, but placed Darryl Zanuck in charge of the First National studios. Harry stayed in New York, manipulating the purse strings. He bought the radio, record, and phonograph divisions of Brunswick-Balke-Collender Company for the company's patents, its record factory, and its stillborn 16mm home talkie projector. In particular, Harry wanted the company for its record presses needed to make Vitaphone discs. Harry had the presses, giant affairs weighing tons each, moved across the country to the Sunset studio, then sold off the rest of Brunswick-Balke-Collender, which he and his brothers agreed was a loser because all it made now were pool tables, bowling alleys, and bowling balls.

The record presses were soon scrapped when sound-on-film became the favored sound projection method, while pool tables and bowling balls made Brunswick a highly profitable company. Harry later said, "When I make a mistake, I make big ones!"

Although Harry lost $8 million on the Brunswick deal, it didn't slow him down: He bought radio companies, foreign sound patents, and a lithograph company. He even produced a Cole Porter musical on

Broadway, *Fifty Million Frenchmen.* His financial wizardry became the talk of Wall Street. "He has the buzzing persistence of a mosquito on a hot night," one banker said. "And the toughness of a brothel madam," replied another.

May 16, 1929

The line of chauffeur-driven automobiles inched their way along Hollywood Boulevard until one by one they reached the entrance of the two-year-old Roosevelt Hotel. A uniformed hotel captain opened the first car door and out stepped Douglas Fairbanks, resplendent in white tie and tails. The rows of exuberant film fans screamed their approval, the men doffing straw boaters to wave, the women waiting expectantly for—*her.*

Fairbanks reached inside the open car door and grasped the fingers of his movie star wife. The roar of the crowd became hysterical as the first shapely ankle reached for the red carpet. Mary Pickford emerged, gowned in shimmering blue and gold.

Arm in arm, Hollywood's reigning king and queen waved to the crowd. Fairbanks mumbled under his breath, "This is the *last* time I'm *first* at anything."

Pickford, her perfect teeth sparkling in the floodlights, whispered under her breath, "Darling, you *are* the president."

As the president of the Academy of Motion Picture Arts and Sciences, Douglas Fairbanks was there to preside over the first annual awards dinner. The winners had been announced to the press three months earlier, so there would be no surprises this night. Because *The Jazz Singer* had not been nominated for Best Picture (the Academy governors thought it unfair for a talkie to compete against silents; the movie was nominated for Best Writing and Best Engineering Effects), Jack declined to attend. He sent Darryl Zanuck in his place to accept a special award for "the outstanding pioneer talking picture, which has revolutionized the industry."

As Fairbanks and his wife entered the hotel and walked toward the Blossom Room, the huge ballroom that would seat the 250 award recipients and special guests, the couple passed the table holding the twelve golden statuettes (not yet nicknamed "Oscars") being presented to the winners, and the twenty additional certificates for honorable mention.

Stars began to file into the Blossom Room: Gloria Swanson, Warner Baxter, Joan Bennett, Richard Dix, as well as studio heads Harry Cohn and Louis B. Mayer. The assembled stars dined on jumbo squab perigeaux, lobster Eugénie, Los Angeles salad, terrapin, and fruit supreme.

After dinner, Louis B. Mayer, the founding father of the organization, spoke of the "closer relations between the artistic and business sides of making pictures."

Everyone there knew this was an evening of show business, a chance to demonstrate to the public that moviemaking had grown up and was now a prosperous, respectable, and creative industry. Fairbanks was quoted as saying that the Academy had been founded two years earlier to ensure that the image of the movies would never be tarnished and stressed how strong an influence motion pictures had been on the world—and would continue to be.

Mary Pickford passed out the awards: Paramount's thrilling war movie *Wings* won the Best Picture award; Janet Gaynor stepped forward to receive the Best Actress award for her role in *Seventh Heaven*; Emil Jannings, star of *The Last Command,* had picked up his statuette for Best Actor a month earlier.

The last award was for *The Jazz Singer*'s special contribution to the sound revolution. When Zanuck accepted it, he said, "This award is dedicated to Sam Warner, the man responsible for the successful usage of the medium."

Al Jolson sang for the dinner guests, then said, "I notice they gave *The Jazz Singer* a statuette, but they didn't give me one." He rolled his eyes. "I could use one; they look heavy, and I could use a paperweight. For the life of me, I can't see what Jack Warner will do with one of them. It can't say yes."

As Zanuck left with the award for Warners' efforts in sound synchronization, no one realized that this first Academy Award year would be the last in which a silent film took home the prize for Best Picture.

Most movie moguls now recognized Warner Bros. as a major force and Harry was acknowledged as the No. 2 man in the industry, second only to Nicholas Schenck of the Loew's theater chain, which owned MGM. Harry lived up to the billing when in late autumn he took the final step in acquiring First National Pictures, buying Fox West Coast Theaters' one-third interest. With the new studio under Warners' wing, Jack decided to move all their production facilities to the First National plant in Burbank.

On first touring the studio, he was stunned by the huge staff and the lavish production budgets. "Harry will never believe this," he told his new production manager, Hal Wallis. "He squeezes nickels until they come out dimes."

First National had been spending $300,000 apiece on its class-A films, while Warners averaged $100,000. Although he would have loved that kind of budget to work with, Jack knew Harry would never agree. Cost per picture had to be lowered to Warner level.

In 1929 Warner Bros. produced eighty-six feature films, forty-five of them under the First National banner. The most ambitious and also the longest (at 135 minutes), was *Noah's Ark,* a retelling of the biblical story in which all but the first 35 minutes was done in sound. Its notoriety came not from its epic style, but the cataclysmic flood scene in which no shots were faked. The floodgate of water and the collapse of the temple descended not on stuntmen, but on unwary extras from Central Casting, several of whom drowned.

As the filming continued, Harry became concerned about the movie's soaring production costs. He later said:

> *Noah's Ark* was budgeted at $600,000. The cost rose and when it reached $940,000, I told the producer: "Never mind the final three weeks of shooting. I don't want a million-dollar picture. This film stops Saturday. Take a death scene." On Saturday the shooting was stopped, by filming a fight-to-the-death scene. And this movie, from which was lopped the final three weeks' work, made a fortune for the studio.

Other Voices

Jack Warner Jr.

Harry's fascination with the Old Testament suggested the story for *Noah's Ark.* Zanuck was assigned as the screenwriter and another amazing character became the director. He was one of Harry's more spectacular discoveries, a talented and imaginative Hungarian named Michael Curtiz, who took to the Old Testament with the same gusto earlier displayed by another film genius, Cecil B. De Mille.

The great scene from *Noah's Ark* I best remember watching shot on the Vitagraph back lot involved a soaring temple gate at the top of a very steep flight of steps reaching high into the pre-smog Hollywood skies. I took off from school to be present at the re-creation of this important piece of biblical history, the launching of the world's first cruise ship. The two-by-two animals were being rounded up at the various zoos and pet shops, and the rain spouts and wind machines were turned on full force, quickly converting the lot into a quagmire and running up the company water bill to make Harry holler all the way from New York.

It was the big confrontation scene just before the flood where one of the good guys races up the temple steps, sword drawn, and cuts down the guards at the gate, one of whom was to roll all the way down to the ground. It was a scene that screamed for Errol Flynn, but as he had not been discovered yet, a stuntman was hired to do the job. The stuntman, who was no fool, rolled up in a tight ball all the way down the stairs to protect himself from damage.

Director Curtiz jumped up from his chair beside the camera and ran over to the rolled-up man, screaming in his trademark scrambled English that the "domfool don't know a stairway how to fall down and should turn in his stunting license!"

Curtiz fixed the poor recumbent wretch with a middle-European glare and went on, "This, Mister Stunt-faller, is how I vant!"

He then raced all the way up the stairs to the temple gates, collapsed dramatically and pirouetted down the steep flight, arms and legs flailing, the perfect picture of a heroic death. Curtiz lay at the foot of the stairs, breathing heavily, then shrieked loudly in Hungarian until his remarks were translated and an ambulance summoned.

Fortunately, only his left leg had been broken. He was rushed off to the hospital while the company called lunch. They waited an hour until he returned in a wheelchair and in a leg cast to resume his place beside the camera. He cleared his throat and yelled at the stuntman, "Now, you dummy—get your ass up those stairs and fall down the vay I showed you."

One of 1929's successes was a musical, *On With the Show,* billed as "The first 100 percent natural color all-singing production." The term "natural color" was all hype, as the film was made in a two-color technique that gave the illusion of different colors. Three-color movies were several years in the future. That year Warners also made *Gold Diggers of Broadway,* a film that was memorable only because it used the same two-color technique.

Color, as well as other industry innovations, had gripped Harry's imagination. The thought of how much Sam would have loved to tinker with the new inventions kept the window to the future open for Harry. In a speech to the League of American Penwomen in Washington, D.C., in 1929 he said:

> The next five years will make the motion picture miracles of this decade, vivid as they have been, seem only a preparatory period. The absolute perfection of synchronization is here. The mechanical brains of the producing companies are now concentrating on such problems as full, natural color and three-dimensional film. The latter will make the screen seem not a flat surface, but a complete room or a countryside, with a perspective in true values. A sculptural quality will be added to movement, sound, and color. The stereopticon of childhood days is undergoing a magical transformation.

Harry was not idly boasting. A year earlier, a Frenchman, Henri Chretien, had invented a system that used an anamorphic lens to compress a very wide image onto a standard 35mm frame. In Harry's mind this was a prelude to three-dimensional movies. (It wasn't until 1952, though, that the system was used for CinemaScope.)

Journalists began to tag Harry with such titles as "the uncrowned monarch of the 'Talkie' kingdom" and "the godfather of the talking screen." Because of his philanthropic gifts, he was also called "the man who brought charity to Hollywood."

Harry was pleased with the recognition, not for egotistical reasons, but because of the credibility it gave the studio. In Harry's mind, the future of Warner Bros. was assured, not just creatively, but financially: In 1929 the company's net profit was over $14 million.

In late 1929, Harry bought a twenty-two-acre farm in Mount Vernon, New York, and moved there with Rea, their three children, Lewis, twenty-one years old, Doris, sixteen, and Betty, nine. Harry and the family had lived in Hollywood from 1923 to 1926, as he wanted be closer to the studio—and Jack. The Hollywood home, as Betty Warner remembers it, had "huge grounds, a pool, a tennis court and play yard. It was the biggest place we had ever lived. Rea had Lewis, Doris, and myself painted on the formal living room ceiling as cherubs, complete with ribbons tied around our middles." Harry soon tired of the Hollywood atmosphere, though, and returned to New York, taking an apartment near the Hudson River, where he and his family stayed for a year and a half.

Other Voices

Betty Warner

The New York apartment was small, with a very typical middle-class atmosphere. The smell of food got into the furniture and seeped into the halls. When you came to your floor on the elevator, you knew who had been cooking—and what. Mother loved living in New York, but my father wanted a farm with horses and cows. He told her he'd build an elegant home in the country.

Mother said she wouldn't live in the Mount Vernon house unless it was made to her specifications, and insisted on an imposing marble staircase where she would be able to make a grand entrance at parties. My father built the staircase and a big beautiful foyer.

She went back to the city, where they were dismantling famous homes, homes that had belonged to the original 400 families, like the Astors, and bought whole rooms—walls, ceilings, and floors—and transplanted them to Mount Vernon.

The walls of the living room were of an elaborate material, damask in gold and green. There was also a portrait of my mother in a velvet dress and pearls. She got a grand piano. Nobody in the family could play it, but no matter where we moved after that, the grand piano went with us.

The house also had a billiard room, a sun room, and downstairs a projection room, complete with a small stage for party entertaining. The space could also be used as a bowling alley. We had a huge swimming pool, tennis court, and a pond that we skated on in the winter, and rowed on in the summer.

I remember my father brought in these enormous jars, six feet tall, which he got from an old theater that had been torn down. I would get in the jars and hide. They were wonderful toys for a nine-year-old.

My father loved the farm. He had all kinds of animals, pigeons, sheep, cows, dogs, and rabbits. He continued to practice and be proud of his butchering and breeding of prize chickens. He thought of himself as a first-generation Jewish cobbler and butcher turned movie mogul. It was a miracle to him that he had accomplished what he had with so little education, and that he was able to talk to people like bankers and lawyers. He didn't enjoy books because reading was not easy for him. But believe me, he could read an agreement or contract.

He was a workaholic tempered by his life on the farm and his outside enthusiasms for golf, horseback riding, the races, and traveling. He arrived at the New York office every morning by eight o'clock, but he'd be home for dinner every night. He was always busy. He didn't sit quietly for long periods of time. He could sit in a chair, take a twenty-minute nap, and be up and going again. In the evenings, he'd play bridge with Mother's friends, or shoot pool, or bowl. There was no business at home.

He was loved by his employees. If someone on his staff had the sniffles, he'd prescribe something for a head cold; if someone was having trouble making house payments, he'd suggest a way of reducing the payments. He spent a good deal of his time taking phone messages for his secretaries. He was a combination of father confessor and one-man Salvation Army.

Also in 1929 Abe, at his wife Bessie's insistence, bought a large estate in Rye, Westchester County, known as Caradel Hall. Designed in the Italian Renaissance style, it was situated on one of the finest waterfront sites of Long Island Sound. Abe paid $2 million for the house and grounds, which consisted of seven acres. The estate had a pebble-strewn beach for bathing, and a pier for Abe's motorboat. (Jack Jr. remembers the boat blowing up one day when several members of the family were aboard; although the incident created a lot of excitement, and an unscheduled swim for shore, no one was injured.)

The main residence came complete with rare tapestries, paintings, and furniture. In addition to a large living room and library, there was a wide hall that led to a massive ballroom two stories high, fitted with an organ suitable for elaborate entertainment and use as a private theater.

Yes, Abe had come a long way from selling soap.

Jack had built his fifteen-room Spanish colonial-style mansion in Beverly Hills and in early 1929 had moved with Irma and Jack Jr. into the new residence. But with the success of Warners' talkies, and with the money pouring in, the house quickly seemed inadequate. Jack bought an adjacent piece of land and put in a pitch-and-putt golf course with two ponds. He also acquired three mansions on adjoining properties, demolished them, and added the lots to his estate.

While the brothers were enjoying the good life, the country was undergoing desperate changes. The year 1929 was a year of activity in America: Herbert Hoover had succeeded "Silent Cal" Coolidge as president, construction had begun on the Empire State Building, and a coach from Notre Dame named Knute Rockne had lifted his football team to a national championship.

Then came Black Monday.

On October 23, 1929, the stock market crashed.

Part III

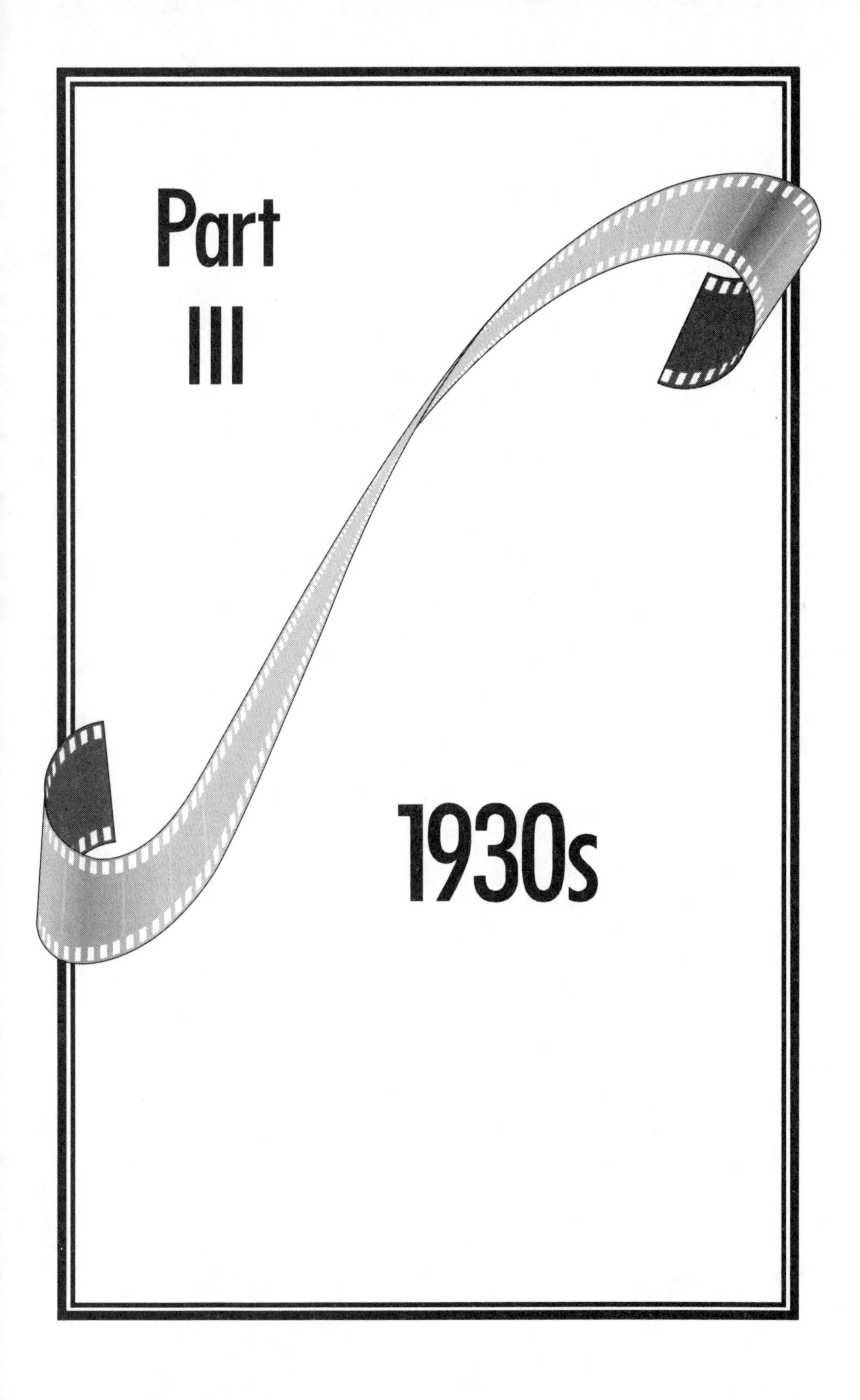

1930s

13

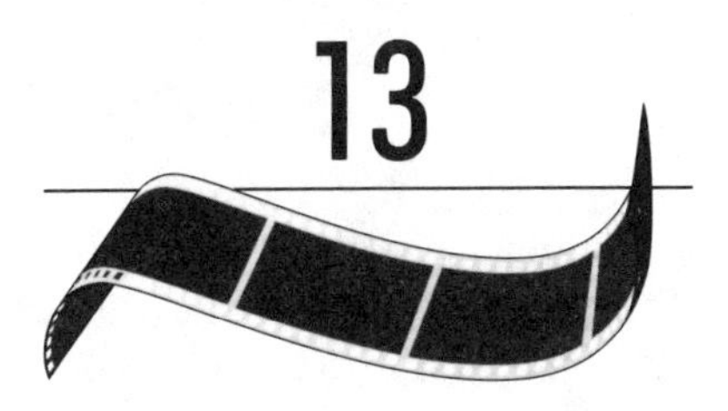

Triumph and Tragedy

By early 1930 the Depression was well under way. But as businesses across the nation were closing, Warner Bros. kept making films. Sound movies were enough of a new attraction to keep hard times at bay for a while. The public was using the movies as a release from daily woes. Men who would normally leave home for work in the morning now lined up in front of cinemas, buying tickets to sit in the comforting dark until it was time to go home. The only problem was scraping up the 35 cents admission.

Bill Schaefer, Jack Warner's personal secretary for forty-five years, recalled:

> You had to sell a lot of tickets at that price in order to make a profit. The theaters were changing programs in the middle of the week and there were always double bills, so that meant four pictures would be used a week. They seldom ran longer than one week. So we had to make a lot of pictures during the year. In fact, we were making about seventy a year.

> Some of them had ten-day or two-week shooting schedules. These were the so-called "B" pictures. Shortly after I started in 1933, they put Bryan Foy in charge of the "B" movies. A lot of times, Foy would see a good set on the lot for a picture he had in mind, and he would go in and use it. Other producers complained because Foy's picture would be out before the "A" picture that the set had been designed for.
>
> Foy was also known to look at the total number of pages in a script, find a section in the middle, tear it out, hand it back to the writer and say "Bridge it."

Harry, who firmly believed that if Warner Bros. stood still they were walking backwards, continued to buy properties. By the end of 1930, Warners owned 51 subsidiary companies, including 93 film exchanges, 525 theaters in 188 American cities, and the huge studios in Burbank and on Sunset Boulevard. This diversification had paid off. While other stocks vanished from the Wall Street exchange, Warner Bros. stock was valued at over $200 million and shareholders reaped a dividend of $12 million. The company employed 18,500 people and provided them with salaries that totaled a $36 million annual payroll.

Then things began to get worse, even for Warners. Although President Hoover kept saying "Prosperity is just around the corner," the American public was standing in bread lines—not movie lines. Attendance at theaters dropped drastically. People could no longer afford the 35 cents for two hours of escapism. Ticket prices were lowered, but attendance continued to fall off. The year 1931 ended with net losses for Warner Bros. of almost $8 million (which would grow to $14 million in 1932).

Harry sent out the alarm from New York: Costs must be cut in half. If the show business crowd had forgotten that Warners could be tight with money, they were quickly reminded. Harry ordered the expenses of running theaters slashed $123,000 a week. Marquee barkers were made to buy their own Listerine. Everyone from executives to errand boys took pay cuts. High-priced contract stars took a 50 percent pay cut for eight weeks. Actor Joe E. Brown's name was sold to Quaker Oats. Reluctantly, Harry authorized the dismissal of 900 employees.

In an interview for *American Photo Play,* Harry remarked, "I don't care so much for myself, but it is those who are dependent on us who have been caught in the whirlpool. A constant stream of people have been in and out of the studio in the last few days, telling tales of losses that are almost unbelievable. They come to me with tears in their eyes. I am trying to do all I can to help them."

But there were signs the Depression was abating. Franklin Delano Roosevelt had assumed the presidency in 1933 and the nation's mood had changed. The Warner brothers, to everyone's surprise, had backed Roosevelt's candidacy. Harry hadn't been happy with a Senate investigation of a stock sale of his during the Hoover administration and felt it was time for new leadership.

The brothers organized a special train to take a bevy of stars to Washington for the inauguration, including Ginger Rogers, James Cagney, and Bette Davis. In return for Warners' support, Jack bragged, Roosevelt had privately offered him an ambassadorship. He told everyone he declined, saying, "I think I can do better for your foreign relations with a good picture about America now and then."

But Jack Warner Jr. recalls, "FDR never offered an ambassadorship to my father. I heard him say that he wished it was true, but his old presidential pal never came across. My father was jealous of Louis B. Mayer, who, rumor had it, was offered the ambassadorship to Turkey by President Hoover. J. L. wished Mayer had taken it, just to get him out of Hollywood."

Roosevelt's campaign song, "Happy Days Are Here Again," had fit right in with the Warner brothers' mood. Harry was particularly jubilant, as he had added a new child to his family—not a son or daughter of his own, but Sam and Lina's little girl, Lita.

In early 1930, Harry had decided it was time to become the legal guardian of Sam's three-year-old daughter. Her mother, Lina Basquette, was—in Harry's opinion—a tramp, unworthy to bring up a child with the Warner name. He would raise Lita with his three children.

Since Sam's death, Lina had met—as she said—"a fair share of wolves who did not bother to don sheep's clothing." Harry Cohn, boss of Columbia Pictures (then a cracker-box outfit on Poverty Row), was one of the first to knock on her bedroom door, but she threw him out after he delivered the all-time cliché line, "I can make you a big star!" She didn't need Harry Cohn. She had an audition with Cecil B. De Mille.

De Mille had been searching for someone to star in his latest project, *The Godless Girl* (1929). Stalking around his office in his riding boots and jodhpurs and swinging a riding crop, he auditioned Lina—by thrusting his hand down the opening of her blouse and cupping her breast. Lina said that when she wrenched free, De Mille intoned, "My godless girl must have a beautiful body, exquisite legs, and a form so perfect that the old masters in art and sculpture would worship it . . . Now, take off your blouse!"

Lina declined, advising the director that she had "all the essentials for a godless girl."

The film caught De Mille in the changeover to talkies, so he tacked on a few scenes with dialogue, grumbling that his star's late husband was the idiot who'd started the whole damn sound mess. De Mille knew nothing about the new medium, and the result was an amateurish hodgepodge of sound and silent scenes. Still, Lina got a lot of notoriety from working with De Mille. She also became enamored of De Mille's chief cameraman, J. Peverell Marley, with whom she began a torrid affair.

With her future in motion pictures looking bright, Lina grew reckless with her money. She blew Sam's $40,000 insurance settlement on stocks and bonds that were duds, and the interest from her and Lita's trust funds just covered daily expenses. She hung around with Jean Harlow, Clara Bow, and Carole Lombard, a threesome known as "The Bad Girls." "Pev" Marley proposed marriage and Lina accepted, even though he said he wouldn't live with "that kid" in the house. Lita was given to Lina's mother and stepfather, Gladys and Ernest Belcher, to raise.

That was enough for Harry. He decided to force Lina to give up her daughter. He would buy Lita. His lawyers contacted Marley and made an offer: a $100,000 cash settlement to Lina and a $300,000 trust fund for Lita. Marley told Lina to accept it. She refused, as she was having second thoughts about the poor treatment she felt she had received when Sam died.

In her 1991 autobiography, *Lina—De Mille's Godless Girl,* she wrote that on his deathbed, Sam had Abe and Jack clasp their hands over his chest and promise they would do nothing to hurt her. "My wife gets everything that's mine," Lina reported Sam as saying. "Jack, Abe, you swear to me you will be decent to my wife . . . My share is Lina and Lita's. Do you promise?" As she remembers it, Abe and Jack, like "mechanical men," muttered the words, "We promise, Sam, we promise." A poignant scene, but Jack did not arrive in Hollywood until two days after Sam died. Jack Warner Jr. later suggested that much of what Lina wrote in her autobiography was the concoction of a very active, self-serving imagination.

Marley urged Lina to take the Warners' offer. He wanted the $100,000. He kept hammering away at her, telling her that Lita would thank her someday, that the little girl would be better off with cold cash than she would with warm motherly love. At the time, Lina, a twenty-two-year-old, fun-loving rising star who preferred partying to baby-sitting, agreed, rationalizing that the Warners would provide a better family life for Lita.

On March 19, 1930, Harry Warner and his wife, Rea, became legal guardians of Lita Basquette Warner and took her to their new estate in Mount Vernon, New York. The newspapers reported that the Warners "settled $300,000 upon their niece in a trust fund."

The tabloids had a field day with the news, painting Lina as a heartless hussy who had abandoned her child for "thirty pieces of silver." The reporters implied that the Godless Girl's sins had rubbed off on the actress who played her.

A month later, the newspapers reported that Lina was "determined to have her child with her this summer." If necessary, she would take the matter to court for a legal ruling. She said she wanted the world to know that she did not "sell her baby":

> I am still her mother, and the Warners are only her aunt and uncle. Although she makes a home with them, I can see my baby at any time and have her with me at any time. She is coming out from New York in June to be with me for three months. It is cruel for people to say I gave her away for adoption or "sold" her.

A year later, Lina, now divorced from Marley, tried to commit suicide by taking poison. She told the reporters that she wanted to kill herself out of loneliness for her baby:

> They took Lita away to New York and promised they would send her back this summer. They have put me off a week at a time, promising, but they do not send my baby. I wanted her to be with me. I wanted her to dance, to ride, to swim, to play. I wanted her to have a golden life. That's why I drank poison. I felt so depressed about the futility of fighting them and their immense wealth. I just suddenly became afraid I would never see my baby again.

Lina Basquette was almost right—she would see her daughter only twice in the next twenty years.

It was explained to Harry's daughter Betty that Lita had to be brought into the family because her father had died and her mother couldn't take care of her. Harry said he wanted to raise Lita because she was the daughter of his "beloved brother." Betty remembers Lita as being "adorable and cute. Life didn't change. I never considered her as anything but my sister." Lita remembers her early days with Harry and Rea fondly.

Other Voices

Lita Warner

Mount Vernon was the best place I ever lived. I loved it. I called Rea and Harry "Mother" and "Dad." I knew that they weren't my real mother and father. I *always* knew they were my aunt and uncle, although Harry and Rea *never* referred to themselves that way. When someone would say something to me about my "aunt and uncle," it would become very confusing. Rea would get very angry at anyone who'd say, "Oh, Lita's your niece." They told everyone I was their daughter, yet I was never legally adopted.

I believe the main reason Harry wanted me with his family was so I would be brought up as a Warner. He didn't want my mother to get my father's money, but neither did he want to deprive me. Harry wasn't a tyrant, he was just a very strict man. He was basically very sentimental, but didn't know how to express it.

I don't remember seeing Lina at Mount Vernon, although I was told she did come to the house when I was very young. I was nine when we moved to Los Angeles and I saw her there. The only other time was years later when I was married. I have no real animosity toward my mother. I don't feel she abandoned me; I don't think she really did. I believe pressure was put on her. She's a character and I can see why the family didn't want me to live with her. With Lina I would've had a wild life. It wouldn't have been the greatest thing to have been brought up by her.

Lina and Lita were not Harry's only concern. Jack Warner, who had professed to Darryl Zanuck that he felt like killing himself for "being unfaithful to that dear, precious wife of mine," continued to bed down an occasional starlet or secretary. And then he even broke his own rule—he started playing around with a cute honey blonde who was more than a starlet or a secretary. She was a star.

Marilyn Miller's singing and dancing had an infectious sparkle that had made her Florenz Ziegfeld's brightest stage personality. In the late 1920s the great showman starred her in his musicals *Sally, Sunny,* and the Follies. Jack, who had watched Miller perform on Broadway, was captivated by her enchanting beauty, vivacious personality, and petite but well-rounded body. (Actually, Miller was similar in coloring and size to Irma.) He brought Miller to his Hollywood sound studios to re-create her roles in *Sally* and *Sunny.*

Marilyn Miller knew how big a star she was and made demands that Jack readily acquiesced to: a Rolls-Royce to take her to the studio, and a star's dressing room, remodeled the way she wanted it, with French antiques and a Roman marble bath. Studio workers, who knew how frugal Jack was, were amazed at these extravagances. They soon realized that Jack wasn't visiting the *Sally* set just to oversee the production. He would show up at all hours of the day—and evening. When a break was called, Jack would disappear into Miller's dressing room, holding up shooting, until the diminutive star, a bright smile outlined in fresh red lipstick, would emerge to continue the day's work.

It didn't take long for the studio spies to report to Harry in New York that Jack was having an affair with the studio's latest Broadway import. Harry grabbed the phone and read his brother the Old Testament version of the riot act. *"I want you to end this affair,"* he summed up.

"Harry, all I can tell you is . . . boys will be boys." And Jack hung up the phone. The days when his older brother could lecture him about personal behavior were over. At age thirty-eight, as the studio boss of the fastest-growing company in the movie industry, Jack had achieved the status of a newly crowned monarch. He now ruled over his own principality, a little kingdom of serfs, subjects, and stars. He wasn't a shadow of Harry Warner. He was Jack L. Warner—executive producer in charge of production.

Jack had become a flamboyant personality, part clown, part demigod, a character that would have been impossible for his staff writers to create. He dressed like the song-and-dance man he would have liked to have been in his youth, in white pants and a blazer, and he cracked borrowed jokes that were funny only to his yes-men. His forays into original humor were more embarrassing than amusing.

Seeing the French word for fish, *poisson,* on a menu, he'd ask the waiter, "Oh, you serve poison here?"

When an interviewer stated, "I understand you're a raconteur," Jack replied, "That's right, I play a hell of a game of tennis."

When feeling overburdened by the demands of his position, he'd sigh and utter his favorite expression, "Uneasy lies the head that wears the toilet seat."

In executive meetings, when he grew weary and would lose track of what was going on, he'd say, "I'm from the Bon Ton Woolen Underwear Company and I haven't got any idea what you guys are talking about." Jack Jr. secretly printed and distributed cards that read:

THE BON TON WOOLEN UNDERWEAR COMPANY.
JACK WARNER, PRESIDENT.

Jack Warner had become a clown with clout.

He loved having his name in the news and his face in the limelight. He never missed an opportunity to get his picture taken with famous folks, be they actors, politicians, or even scientists like Albert Einstein. When the theory of relativity came up, Jack responded, "Yeah, I have a theory about relatives, too—don't hire them." Then he flashed his ever-widening grin. Einstein didn't get the joke.

For photographs, Jack positioned himself at the left-hand side of the lineup ("captions always read left to right," he'd say), his shoulder and body twisted in a one-quarter turn toward the camera, one hand stuffed in his pants pocket, head cocked at a jaunty angle, a big toothy grin on his face.

Although a clown on the outside, Jack Warner had an uncanny ability to select the right motion pictures for production. He also surrounded himself with the best men available to take a lot of the workload off his shoulders.

Darryl Zanuck had become a filmmaking machine, running the studio like an assembly line. He had a huge chart the size of his desk divided up into the fifty-two weeks of the year. In the left-hand column of the chart he'd list fifty-two stories and on the right side the major stars at the studio. Then he'd get together all his writers and say: "Okay, this year gimme three Bogarts, four Bette Davises, four, no, make that five Cagneys . . . I need a George Raft by Christmas . . ." Then the writers would go back to the Writers' Roost and create original stories around the actors' names.

One of the writers Zanuck hired was a nineteen-year-old who had become obsessed with films, Milton Sperling. Sperling had taken his first studio job as a messenger boy and shipping clerk at Paramount's Astoria studios in Long Island City, then went to work at United Artists, whose studio boss was Joe Schenck. When the person who hired him quit, Sperling (perhaps out of a misplaced sense of loyalty) walked into Schenck's office and said, "I quit, too!" As the young man walked out the door, Schenck yelled, "Who the hell are you?" He kept yelling as Sperling disappeared down the hall, "Who the hell are you to quit!"

Other Voices

Milton Sperling

I was sent to see Zanuck because he was looking for a personal secretary who knew how to take shorthand. I said I could—but didn't know how to take down one word.

At his office I was told I could enter when a buzzer sounded. I waited half an hour and when the buzzer went off I opened the door—and ran nose first into another door. Rubbing my bent beak, I realized the double doors were used to seal off sound. I opened the second door and entered a vast room, one larger than I had ever been in before. At the far end was a desk, and seated against the walls like in a ballroom were Jimmy Cagney, producer Hal Wallis, and four other guys I surmised to be writers. They were staring at an empty desk. Then they looked at me, then back at the desk.

Suddenly, Darryl Zanuck popped up from behind the desk, aimed a polo mallet at me, and went, "Rat-tat-tat-tat-tat-tat . . ." like the stick was a Tommy gun. I stood there bewildered.

"Fall down, you son-of-a-bitch, you're dead!" Zanuck screamed.

He started shooting at me again with the polo stick, annoyed that his first bullet hadn't hit me and even more annoyed that I wouldn't fall down dead. I figured that he was acting out the way he wanted Cagney to assassinate his rival in a movie, so I dropped in a heap to the floor. (I was soon to discover the film was *The Public Enemy,* Cagney's first feature for Warners.) The next thing I saw was a pair of polished shoes near my face and heard Zanuck's voice say, "Good work, kid, you can get up now."

Sperling became Zanuck's conference secretary and sat in on all the writers' meetings, making notes of the ideas that popped into Zanuck's head. Only Zanuck's ideas, no one else's. Sperling would then type them up and send them around as "The Sermon of the Day."

Sperling, still a teenager when he began working for Zanuck, loved his job, feeling like he had been "sworn into a sort of citizenship." Although he considered Zanuck and Jack Warner "insecure men," they created around themselves a "sense of permanence, as well as a sense of security and loyalty."

"You even had T-shirts with the studio's name on it," Sperling said. "It was just like being a subject in a small kingdom." Sperling did notice, though, that the behavior of his feudal lord was sometimes bizarre.

Other Voices

Milton Sperling

One day I was walking with Zanuck after typing his conference notes, and he had his usual court of studio stooges hanging with him, when he stopped, zipped open his fly, and started pissing against the wall of one of the sound stages. There he stood like a street urchin, talking over his shoulder while girls in costume walked by him.

He had this long electric cattle prod and if people turned away from him at his desk, he'd jab them with it. When they jumped in shock, he'd howl with laughter. Zanuck especially liked this one guy better than anyone else, as he leapt like a ballet dancer when he was goosed with the prod.

One day Zanuck called me saying he urgently needed some notes I had been typing, so I rushed over to his office, only to find him gone. His and Jack's offices adjoined each other and I heard voices in the executive bathroom they shared. I walked in and there was Jack, hunkered down on the toilet, taking a shit. Zanuck was standing next to him, talking, his hand on the toilet chain, ready to flush. Although my face turned red upon seeing this intimate scene, neither one of them paid any attention to me. Zanuck, who was accustomed to pursuing Jack to the bathroom, just grabbed the pages from my hand and went on talking. I heard the toilet flush as I left the room.

But I didn't consider Zanuck's behavior too unusual. Shortly after I arrived in Hollywood, Sam Levine, a B-movie producer, promised me $500 to write a script and when I delivered it to his office on Poverty

Row a sheriff was standing outside the door. "We're closing the joint down," he growled.

"But he owes me money."

"Good luck!"

I knocked on the door. "Mr. Levine, it's your writer, Milton Sperling."

After a moment I heard a voice: "The cop still there?"

"Yeah."

"Ask him if you can come in."

The sheriff said, "Sure, you gotta get your money."

I opened the door a crack and I sneaked in. In the semi-darkness, I could see Levine with his pants down around his knees, hobbling around the room, jerking open drawers. "Wait a minute," he said, "I'll be with you in a second." I noticed a woman spread out on the couch, half naked, looking at him.

"Christ," Levine said, popping open a desk drawer. "I know I have a fuckin' condom around here someplace." He snapped his head at me. "How's the script?"

"Great."

"No shit? Well, just wait a minute until I can find that goddamn condom!"

I was waiting to get paid, the girl was waiting to get laid, and the sheriff was waiting outside to raid the joint. That was Hollywood at the beginning of the '30s!

I soon became embroiled in Jack's nocturnal maneuvering as one of his "beards." He always had to have a little cutie or two to dally around with, and to do this, he needed some excuse to tell Irma so he could stay out late at night. Friday night became the big night when Jack sneaked out with some girl. Jack's father and mother, who were Orthodox Jews, had a Friday-night supper which Jack told Irma that, as a dutiful son, he couldn't miss. The reason Ben wanted his son there was not for the traditional Friday-night Sabbath ceremony but because the old man loved to play pinochle after dinner. Jack told Irma, who was suspicious about her husband, that he and his father would play pinochle until midnight, then he'd come home. Jack began to take me along as his Friday-night excuse, mainly because I was Jewish and had some Orthodox background. My task was to accompany Jack to the dinner, then stay with the old man and play pinochle with him, while Jack left the house and got laid.

While Jack carefully cooked up alibis to cover his amorous adventures, Zanuck blatantly advertised his conquests. If they could be called conquests. Each day promptly at 4 P.M. Zanuck had a "sex siesta." A pretty little thing from the starlet pool would slip into his office and close and lock the door. Everyone in the office would stop working for half an hour while Zanuck dallied. No one dared even call in because they knew Zanuck was "occupied."

Groucho Marx once remarked about his family, "It has always been a matter of wonder to me how one set of parents could spawn so many different kinds of children." The same could be said of Harry, Abe, and Jack. Each had his own personality, dreams, and desires. While Jack preferred the company of show business people, the other brothers were living soberly while also enjoying the riches of their rise to the top.

Abe, massive, genial, down-to-earth, had no enemies; everyone liked him. Harry once said of his brother, "He's known as 'Honest Abe' and is the most popular man in the movie industry." It was very close to the truth.

He was happy with his sophisticated second wife, Bessie. Because of his lack of education, Abe felt the need to have, as he said, "a classy-type woman" as his mate.

Arthur Steel, Bessie's son by her first marriage to Jonas Siegal (who had changed the family name from Siegal to Steel), remembered Abe this way.

Other Voices

Arthur Steel

Everyone called him "Maj." He was awarded the honorary Army rank of major in World War I for the work Warners had done in making wartime propaganda films. Harry, as president of the company, was supposed to receive the award but was out of town when it was presented. Abe stood in for his brother. He loved the title and from that time on was addressed at work as "Major Warner," never Albert or Abe. His immediate family called him Abe, except his wife, my mother, Bessie, to whom he was always Major.

To me he was Albert.

Albert never tried to act refined, like a Wall Street tycoon or something like that. He was not a man who would hold back from belching at the dinner table. He didn't need to be known as a man of high taste. He'd look at valuable paintings, but they didn't mean a thing to him. Bessie bought expensive furniture and antiques and had a jade collection, but he never paid any attention to any of it.

Albert was not a guy who spent a lot of money on himself. In fact, if Bessie had not spent the money, it never would have been spent. He eventually moved into a house in Florida and it was grand, but that was because Bessie wanted it. He liked having a chauffeured Cadillac, but even that was an embarrassment. He rode up front most of the time with the chauffeur.

When Albert was ready for dinner, he'd sit down at the table and Bessie would say, "Major, why don't you wait?" When guests were invited, he'd sit in his own little den with the newspaper or his racing form and when he was through reading, he'd come out, shake everybody's hand, and say, "Hello, how are you?" Then he'd ask Bessie, "Dinner ready?" She'd say, "No." He'd grumble, "Yes, it is," and go into the kitchen with a plate from the dinner table and serve himself out of the steaming pots and pans, then sit down at the table and eat. There could be ten guests for dinner, but he'd begin eating. He never had any qualms about doing it.

When he was done eating, he'd light up a cigar. Bessie would warn, "Maj . . ." but he'd keep on puffing. Sometimes she'd say, "Dear . . ." or, "Major, you can't smoke now." This was a continuing ritual.

He rarely spoke at the dinner table, except to growl, "Bunch of bums." He liked to use the word "bums" when referring to anything from politicians to businessmen: "They're just a bunch of bums." If a member of the family borrowed money from him, he would no longer respect them. After that, if they'd offer an opinion on any subject, he'd say, "Whatta you know? You're just a bum."

After dinner, all the men would go into Albert's den, close the door, and play gin all night while fogging up the place with cigar smoke.

When the *Daily Racing Form* would arrive about eight in the evening, he'd take it and go to the den. He loved the horses. He loved being in Florida in the winter. He loved television, which he'd turn on anytime he wanted, even at dinner, and watch the news or Westerns. Many times I'd trail into the den and sit and watch with him. Finally, he'd say, "How's it going?"

"Fine."

"What are ya doin' now?"

"I'm in college."

"How do ya like that." And he'd study the racing form.

He was concerned about his lack of education and in 1928 told a reporter, "As I look back, if I have any regrets, it is that we brothers didn't get more education. Most of us stopped in grammar school. I had one year of high school because I played football. But often when I have to get up to address conferences, employees' organizations and the likes, I wish I had the gift of eloquence and the training that comes from a university education. Perhaps I could reach my own men better."

He had a huge heart and a very gruff manner, and the points of contact between his interior and exterior were very hard to see, but you knew they were there. He liked what he and his brothers had done for their parents and once said, "We built them a fine home in Hollywood where they can lean back and take it easy. But they've worked so long that they can't get out of the habit and every day they drive over to the studio and keep an eye on things."

Albert was sweet to Bessie in a very quiet way. He didn't show a lot of affection in public, but it was there. They slept in the same bed, although he had a separate room. She was very frail, not weighing more than ninety pounds. He was such a massive man, with a big head and huge hands.

Albert went to work every day when in New York. He'd go in to check his investments (he had a little typewritten list of his stocks in his pocket), make some phone calls, order cigars from Dunhill. He never smoked before lunch, but he couldn't wait to eat, then have his cigars, which he smoked constantly thereafter.

Every day at 1 P.M. he'd go to the racetrack, where he had a box. He loved steady betting, placing wagers of $50 to $100 on each race. He'd say, "I lost a couple hundred" or "I had a big day, won a couple hundred." At the end of the year he knew he would have lost a few thousand dollars. That was preordained.

Albert didn't like people to make a fuss over him just because he was a Warner brother. He liked to walk into a party, sit down, and not be noticed or recognized. He was never one to put his head one inch above the level of observation; in that he was far different from his two brothers, especially Jack.

In the early 1930s Paul Lazarus, a young man just out of college, went to work for Warners in New York in the company's press department. He was told by the personnel manager, "The bottom rung is open. Want the job?" so he took it.

Other Voices

Paul Lazarus

Harry and the Major's offices were on the top floor and I worked on the ground floor. The Warner building had department-store show windows at street level which were filled with billboards advertising Warner movies. Above the top floor was a free-standing building that was the projection room.

The Major was in charge of distribution and was also the treasurer. Whenever my boss had a stiff problem in distributing a film, he had to get the blessing of the Major before he could release the movie. He'd explain the situation and the Major would look at him, then lift his huge hands, clap them together, and say, "Do the best you can."

One time I was in an elevator along with Harry and the Major when one of the publicity staff said sotto voce, "Boy, was that movie a stinker." From the back of the elevator, Harry's finger shot forward, waggled, and like a teacher scolding a child for a wrong answer he said, "We make no stinkers. Every one of Warner Brothers' pictures is a good one!" The brothers never deviated from the party line.

Harry Warner never succumbed to the temptations that were certainly available to him as president of Warner Bros. He didn't even like to be in the company of stars, and was ill at ease around the studio's sensual leading ladies. He was content to sit at his dinner table with people whose names were not on the *Social Register* or in *Who's Who.* Good humor was written across his features and in a crowd he could pass as a businessman living on a modest income. Studio employees thought he bore an uncanny resemblance to one of Warners' stars, George Arliss.

An article in the *New York Morning Telegraph* in 1933 described Harry thus:

> Whenever adversity is in the ascent he destroys hysteria by his own imperturbable calm. In his business conferences his geniality blows rancor out of the window. An inoffensive sense of humor finds profound support in Talmudic proverbs, which he continually quotes. In the heat of a discussion, he introduces a light irrelevance by interrupting with, "Are you finished? Shah—now let me tell you something."
>
> Inquisitively thorough—his persuasive powers result from complete knowledge of his subject.
>
> Eats lightly—chiefly cereals. Used to be a heavy smoker—but abstains completely now. Does not drink. He has golf ambitions and fifteen sets of golf clubs. Hopes to lower 80's into 70's. Never goes horseback riding after having been thrown and sustained a broken leg.
>
> Tireless worker. Possesses the highest form of determination—that which expresses itself through concentration. Loved by his employees and associates. Respected by his competitors as a brilliant executive. Admired by all for the goodness which overflows into many charitable channels.

The one love of Harry's life, other than the studio, was his wife, Rea. She in turn loved Harry, and she also enjoyed the status offered by the company's success.

Other Voices

Betty Warner

My mother had her clothes made by well-known New York designers. My father was very generous and loved to see her dressed up. She had a fur coat for every occasion, in every style and color imaginable, and an enormous amount of jewelry to go with her outfits. Every hair was always in place, and she never went out without a hat and a pair of gloves and shoes that matched. My dad was different. Once a year Mother would buy him a suit, several shirts, and underwear, because he never went shopping.

My mother was a beautiful woman. She was five-foot-two and weighed 105 pounds. Her hair was black and abundant, her brown eyes warm and expressive. I loved her skin—alabaster, soft and white. She never went in the sun without an umbrella or hat to shade her face. Around the house she wore specially designed lingerie and nightgowns. Her handkerchiefs were embroidered with her monogram and were trimmed in lace. She never accepted Kleenex as a substitute.

Rea was very vain about her figure and wore tight clothes. In those days when everyone had wasp waists, she had the smallest waist that anyone could have. She always wore a boned corset, even after they had gone out of style. To keep her tiny waistline, she would even stop eating. Even when she was in her seventies, she could proudly declare that she weighed the same as the day she got married. Between the cinched-in corset and the dieting, her health was ruined. All her life she had a lot of stomach trouble.

Even though she watched her diet, she loved pastries. She didn't cook, but she hired wonderful chefs. A cook had to know how to make *schnecken,* a German coffee cake with raisins, cinnamon, and a honey glaze. She'd have company to the house every Saturday and

Sunday, a sit-down lunch or dinner for fifteen to twenty people, and serve these wonderful pastries. She wasn't trying to impress people, she just loved to have them over, mostly close friends and family.

She comported herself with dignity and expected people to respect her and treat her as someone important. She took her place in society very seriously and always introduced herself as "Mrs. Harry Warner," never "Rea." When she went to temple, she sat in the first row, which was a sign of being the most important person in the congregation.

Yet people, especially young people, came to her for advice. She was very warm to them; she really liked people. It wasn't evident all the time, just in special one-on-one situations.

Rea never learned to drive, nor do I remember Harry ever driving. He traveled to and from the office in Manhattan by chauffeured limousine, which would come back home to take Rea to lunch, or in the evening to drive them to the theater or opera. Rea loved the opera. Harry slept through the performances.

We had that limousine and chauffeur all through the Depression. I hated it. By then I was well aware of the difference between us and the people on the street, the people in soup lines, the people selling apples. My dad was very socially conscious and caring about the homeless and helped them get jobs. If anyone came up to him on the street and asked for money, he'd give them some from his pocket.

Every Chanukah he would take Doris and me to an orphan asylum or a hospital. It was a required activity. For me it was terrible because I was known as Harry Warner's daughter, not Betty Warner. I never had a name of my own. I was Little Miss Benefactor. "Here's this rich kid, here to give you a doll." I didn't feel good about it: I just felt lucky.

My parents were religious but never forced it on us kids. But I was Jewish. I never felt for a moment that I was not Jewish. We had a menorah, the special candleholder for Chanukah, but we also celebrated Christmas. Not with Grandpa Benjamin or Pearl! Heavens no! They never would have allowed that. There was always a Christmas tree and presents. We all believed in Santa Claus, but the word "Jesus" was never spoken in our house. Passover was the big holiday. Dad would start from page one of the Seder, the special ceremonial meal, by holding up a tray with three matzos on it and saying, "This is the bread of affliction that our fathers ate in the land of Egypt . . ." By the time he got to page twelve, my mother would say, "That's enough, Harry, let's eat," or, "Harry, it's time for another glass of wine." At that time, my mother and father wanted to forget the old country. My dad never talked about Russia.

Never.

My father loved my mother; he always thought she was adorable, even though they picked on each other constantly. Anytime he said something to displease her, she'd say, "Oh, Harry." She said, "Oh, Harry" a lot. He would pretend he had a hearing aid and that it had a dial so he could turn the volume up or down. When she started to chastise him, he'd pretend to twist the dial and say, "I'm turning you off."

Mealtime could be pretty tense. It was an opportunity for Harry and Rea to yell at Lewis and Doris (and sometimes me) because of something we had done wrong, usually at Lewis, who played music on his phonograph day and night—loudly. His suite was next to my parents, and the noise drove Harry crazy. Because of these squabbles, it was hard to enjoy our chef's wonderfully prepared meals.

The most important thing in both their lives was Lewis. He was six-foot-one, very handsome, dark hair and eyes, long dark lashes, a strong nose (broken later playing polo), smart, full of fun, and extremely sexy. He was, of course, my hero and protector.

Lewis always invited his friends, young high school kids from Worcester Academy, and later college friends from Columbia University, to the house. They took me out to football and hockey games, to proms: I always had these gorgeous guys, these beautiful hunks of men around me. I was their mascot. Two of them became my special friends, and one gave me his fraternity pin when I was fourteen!

With Lewis around, the house was always a hubbub, music on the phonograph, someone playing the piano, a movie in the projection room . . . I remember Dick Powell, Joan Blondell, Ruby Keeler, and Al Jolson coming over to the house for such evenings. The place was a party from morning until night, all week long. It just never stopped. Harry didn't care for the noise, but Mother loved it. She loved the kids, the fun and the life they brought to the family.

Father was more concerned about what Lewis would be to the company. Jewish tradition dictated that the father hand the business to his son, so Lewis was groomed to be head of the studio. That was okay with Lewis. He loved the picture business. He had executive ability. Despite a lack of training, Lewis had taken over the music publishing companies Dad had bought and made them into a profitable part of Warner Bros.

A trade publication reported in 1929:

> Lewis Warner and Edwin Morris, Sam Morris's son [Sam was general manager at Warner Bros.], have startled and stunned the old birds in the music world by their clever manipulations and by their youthful up-to-date ideas and methods of conducting their business. . . . They are wonderful fellows. Great boys. Business boys. Boys with vim and vigor . . . likeable boys. In fact, lovable boys. Boys who have not been spoiled by money. Boys who are not conceited and boys who are democratic. The wonderful expressions regarding Lewis and Edwin are duplicates of the same expressions that are always said about their fathers.

My father preened when he read the story. Lewis would someday be president of the company. Harry and Rea had everything they wanted from life.

I don't know what life would have been like had Lewis lived.

One of Lewis's new friends was the actress Joan Blondell, who had made her film debut in the 1930 Warners movie *Sinner's Holiday.* (She was paired with James Cagney; both were playing roles they had created on the Broadway stage.) When Harry heard about his son's infatuation with an actress, he had the same feeling he had upon learning of Jack's affair with Marilyn Miller. Sex and business didn't mix. Lewis had just graduated from college and would soon be going to Hollywood to work at the studio, but Harry decided it would be a good idea if he took a vacation, a kind of cooling-off period. Lewis decided to go to Cuba.

Lewis had had an infected wisdom tooth extracted in New York, and although his doctor told him to stay in the city until it healed, he took off for Havana with his friend Herbie Copeland. While there the gum became infected and the twenty-two-year-old Lewis fell ill. A Cuban dentist tried to treat him but only made things worse. Harry was informed of Lewis's serious condition and chartered a plane to fly his son to Miami, where he was put on a train and brought to a New York hospital. There, surgeons immediately operated to isolate the infection. But the infection had gone into Lewis's bloodstream, perhaps the result, the doctors said, of flying in an unpressurized airplane.

For Harry, standing at his son's bedside, it was like those terrible, agonizing days of Sam's fatal illness. This time he was right there with his son, instead of rushing to arrive too late, but it was no different. He could only wait by his bedside—and pray.

To Harry's relief, his son's condition slowly improved. The doctors felt they had the infection under control. Then, five weeks after being hospitalized, the young man was stricken with pneumonia.

Betty Warner went to see her brother in the hospital. "He was all bandaged up," she remembered,

> and it terrified me. I couldn't stand the idea of my beloved brother, my special, dynamic brother, swathed in bandages. I didn't want to see him this way, but my mother pushed me to his bed. Lewis talked to me and he was so adorable. Even with all the bandages he joked with me, tried to make me feel better. I left in tears.
>
> I was supposed to go back to him the next day, but I couldn't. I just couldn't. I was sitting in my dressing room at home when someone, I can't remember who, came in and said, "I'm sorry, your brother is dead."

It was Saturday, the fifth day of April, 1931.

14

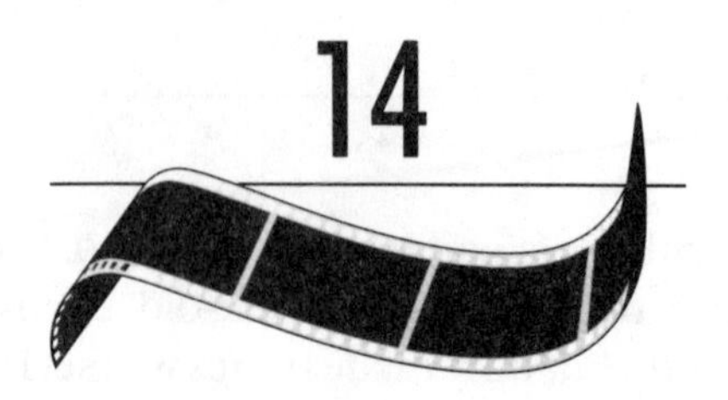

Gangsters, Good Guys, and Dancing Feet

After Lewis Warner's funeral, Betty remembers, she found her mother lying on the couch in the dim light of their living room:

> She had a damp washcloth over her face and she was crying. I sat down next to her and held her hand, telling her that nothing had changed between us, even though she'd had this great loss, and even though she missed Lewis terribly, I was still there. She grasped my hand tightly, but she never stopped crying. I don't think, ever. She visibly cried for the next three years and remained in her darkened bedroom most of the day. She wouldn't allow music in the house because when Lewis was alive each room was filled with laughter, friends, and music. I was still allowed to have friends over, but her grief permeated the house.
>
> She wore black for a full year, then purple because that was the next stage in mourning. She removed herself from all of us. She was alone.
>
> Dad had to keep going, to run the business. And now he had to do it without his son. His dream of Lewis leading the next generation of Warners into a new era of movie making was shattered.
>
> That's when he thought of Doris.

Doris, also devastated by Lewis's death, became more rebellious and dated boys Harry didn't approve of. She spent a lot of time away from the house with Lewis's two best girlfriends, Gwen, a dark beauty, and Flora, a lovely blonde.

It was the middle of the night when Harry came for his oldest daughter. He stood over her bed for a moment, then clicked on the bedside lamp and shook her arm.

"Come. Get dressed," he said.

"What . . . ?"

"We're going out."

Doris started to protest, then in the glow from the small lamp she saw the distraught look in his eyes and rose from the bed, hurriedly put on some clothes, and followed her father out the door. Without saying another word, he, who rarely drove anywhere, drove them to the Warner office in Manhattan. She followed him through the dark corridors and into his office.

"Sit down in my chair." His voice was without emotion, without tone. To Doris it was as if he were speaking from the grave.

Then he said, "You're going to take Lewis's place."

The confused girl started to say, "I don't under . . ." when Harry cried, *"You will learn the business."*

For the next several hours until the gray light of morning began to invade the office, Harry bombarded his seventeen-year-old daughter with statistics, thrusting ledgers, books, charts, and graphs in front of her bleary eyes. She tried to follow what he was saying, realizing the depth of his pain, but it was too much for her to comprehend. Doris's life had been one of collecting movie star autographs, not managing the business that made the stars famous.

As the first rays of sunlight bore through the window in slanted shafts, Harry looked at his sleepy, bewildered daughter—and stopped. He looked at her with fear in his eyes, as if he had been awakened from a bad dream.

And Harry Warner started to cry.

He knew it was hopeless. No one was going to replace his son.

No one.

Nine days before Lewis's death, Jack Warner Jr. had celebrated his fifteenth birthday. He was in the second half of his freshman year at Beverly Hills High School and had confessed to a prying newspaper reporter that he would probably follow in the footsteps of his father in the film business. As summer vacation began, he asked Jack Sr. to let him work at First National Studios. Mervyn LeRoy, director of *Little Caesar,* learned of the request and asked the young man if he'd like to work as "fifteenth assistant director" on the film. Jack Jr. took the job—as assistant script clerk—and at the end of the first week stalked into his father's office and demanded to know why he wasn't getting a standard wage for the work he did, pointing out that messenger boys at MGM got $18 a week and his check amounted to only $15. Peace was restored when Jack Sr. told his son that he was getting a free ride to the studio every morning and taking his meals at home.

Then Lewis died and Jack Jr. felt as if he had been crushed with a massive weight. After all, he and Lewis were destined to be the next generation to mastermind Warner Bros.: Lewis in Harry's place as company president; Jack Jr. succeeding *his* father as production chief. But now . . . ? How could he do it alone? The thought lingered—without an answer.

Jack L. Warner was far from abdicating his throne, a fact that Darryl Zanuck was just beginning to realize.

Milton Sperling recalled:

> In the early 1930s I knew that Darryl would never be given the studio, although, at the time, he was confident he was next in line. He was the wonder boy, the dynamic executive with the energy and bubbling ideas that had carried Warners into the new era of talking pictures. He felt he *was* Warner Brothers. Then he began to realize that Jack Warner had no intention of making him the heir. Jack would have cut off his right arm, even his head, rather than hand over the studio to anyone. That included his brothers.

And, as it would turn out, his son.

When, in the early part of the Depression, the heads of Paramount, RKO, Fox, and Warner Bros. decided on 50 percent pay cuts for their personnel, the cuts did not apply to top executives, which disturbed Zanuck enough to decrease his own salary from $5,000 to $2,500 and announce to the studio employees that there would be a quick end to the cuts. To Zanuck's chagrin, Jack and Harry prolonged the cuts for another nine weeks. That was too much for Zanuck. On April 15, 1933, he released a statement to the trade newspapers:

> On April 10, as Head of Production of Warner Brothers Studios, I announced that the salary cuts decided upon on March 15 last would

> be restored immediately. This promise has now been repudiated, and since a matter of principle is involved and I obviously no longer enjoy the confidence of my immediate superiors, I have sent my resignation to the Chairman of the Company, Mr. Jack Warner.

Zanuck quickly accepted an offer from Joe Schenck to join Twentieth Century (which would later merge with Fox). The following day, Jack summoned Hal Wallis, who was still head of the publicity department, to his office. Harry was there and they both shook Wallis's hand. Jack grinned and said, "Well, you're it."

Although he had a new production head, Jack continued performing the tasks he enjoyed: watching the dailies, editing films, supervising the selection of screen properties, and discovering new stars.

"You see those ears?" Jack Warner said the words out of the side of his mouth to Mervyn LeRoy as they watched the screen test of the young actor. "I mean, they stick out like a couple wind socks."

"Honest, Chief, he's sexy enough to have all the gals in the audience eating out of his hand. He could play the part of Masaro . . ."

"Mervyn, you don't hear me so good. The guy has ears like an elephant."

"He'd be just right in *Little Caesar* . . ."

Jack looked at LeRoy, whom he called "a great director—with a small 'g,'" and said, "You spent five hundred bucks on a test of this big ape? That's five hundred big ones down the toilet." Jack got to his feet and started for the door as the test film spun through the last few feet and the screening room lights came up. He turned to LeRoy. "What'd you say was the name of this guy?"

"Gable, Clark Gable."

Jack walked through the door, trailing the words: "He'll never make it."

Douglas Fairbanks Jr. played the part of Masaro in *Little Caesar* and Clark Gable was hired by MGM, where a clever makeup man pinned Gable's ears back with adhesive tape and the actor became one of the biggest box office stars of all time. For years afterward, Jack would say to his casting people, "Get me a guy with big ears. You hear me—*big ears*!"

Jack Warner Jr. took his share of ribbing from the old hands when he first reported to First National at 10 A.M., briefcase in hand, to start work on *Little Caesar,* the first of Warners' "social conscience" crime movies. He wrote of his experiences in a newspaper report.

Other Voices

Jack Warner Jr.

I entered stage six to find *The Gorilla* company at work. I nearly stepped on Bryan Foy, who was directing it. He told me that I was not wanted on his picture, but I could try stage eight, where *Kismet* was shooting, and ask where *Little Caesar* was hiding.

John Dillon, director here, told me *Sunny* was on the back lot and maybe the director there knew where I belonged. The director of that picture told me to try stage four. There was Mervyn LeRoy with *Little Caesar,* the machine guns and the rest.

My only squawk was when one of the cameramen sent me to the camera department for a bucketful of sprocket holes.

I hope that next year the studio will be so pleased with my work they will take me on with a big raise in wages and a long-term contract.

In *Little Caesar,* Edward G. Robinson, playing the role of Rico (who was clearly based on Al Capone), made such an impact on audiences by snarling lines like "No buzzard like you will ever put the cuffs on Rico" that he would always be identified with gangster parts. In reality the actor was afraid of guns and had to have his eyelids taped to keep them from blinking when he pulled the trigger.

Another crime movie, *The Public Enemy,* made a superstar out of the diminutive, fast-talking Jimmy Cagney. Audiences loved it when Cagney staggered down a dark street after the big shootout with rival gangsters at the end and said, "I ain't so tough," then fell facedown into the gutter. *The Public Enemy* also made Mae Clarke a minor star when Cagney, in the movie's most celebrated scene, smashed a grapefruit into her face. The actress had been assured by director William A. Wellman that the grapefruit would not actually touch her: the scene would be shot in such a way as to appear that it had. The shocked, sour look on Mae Clarke's face was genuine.

Warners kept making gangster movies, but not simply because stories of tough guys and machine-gun wars were popular. They felt an obligation to inform their audiences. Making socially conscious films—films that dealt with problems from a moralistic perspective, exposing injustices, and suggesting some type of action that would improve the social system—was their forte. Oftentimes their stories were documentarylike in their accuracy, which was one of the ways they assisted audiences in understanding characters like the criminal Cagney played in *The Public Enemy.*

Harry was quoted in an article in the *Boston Post* as saying:

> Gangster pictures are not responsible for the wildness of youth, nor are there too many gangster pictures . . . Gangster pictures properly presented should have a good effect. They are intended to point out the lesson that crime does not pay. With proper home training, they should assist in keeping kids from turning into delinquents.

A year earlier, Harry had begun a series of crusading biopics about distinguished historical figures. When he announced the first of these films, he said: "In *Disraeli* we will make a talking picture about the great British statesman, with an English cast, headed by the inimitable stage actor George Arliss, and in this production we will prove that it is possible to make an artistic as well as a commercial success."

Disraeli received an Academy Award nomination for Best Picture of 1930, but lost to Universal's *All Quiet on the Western Front.* Arliss, who had become a film star at age sixty-one, won the Oscar for Best Actor.

Arliss went on to make nine more movies for Warners, including the biographies *Alexander Hamilton* (1931, from a play Arliss co-wrote), and *Voltaire* (1933). For the 1932 film *The Man Who Played God,* Arliss personally chose a new studio actress to costar with him—a young blonde named Bette Davis. It was a lucky break for Davis, as she had decided to leave Hollywood after little success with Universal, where studio boss Junior Laemmle had remarked in her presence that she had "as much sex appeal as Slim Summerville." Bags packed, Davis received a phone call from someone who identified himself as George Arliss.

"Sure, and how are you, old boy," Davis replied in a fake English accent, believing the caller to be an impostor. Fortunately, she stayed on the line long enough for Arliss to convince her he was the genuine article.

In 1932 Harry and Jack decided to make *I Am a Fugitive from a Chain Gang* into a social drama. Jack Warner Jr. recalls:

> As a teenager, I was an avid reader of detective story magazines and books and came across one titled *I Am a Fugitive from a Chain Gang.* I passed it on to my father, who passed it to the story department, who had a reader make a half-page synopsis (the longest piece of reading my father would bother with) before committing to spending anything.

> From there the rights were bought from the author, Robert E. Burns, and the story went into production. I received a $25 bonus!

Burns's gripping autobiographical account of the savage cruelty he had suffered when wrongly convicted of a crime and sentenced to a Georgia chain gang was just the kind of morality piece Harry wanted for Warners. Burns, who had dictated the story to his brother, a priest, was still a fugitive (he was not pardoned until 1944) and had to secretly slip into Hollywood to advise director Mervyn LeRoy on the script. Paul Muni, who had received a Best Actor nomination in 1929 for *The Valiant,* was cast in the lead.

During the rehearsal of the last scene the studio lights inadvertently failed when the fugitive's girlfriend asked him, "How do you live?" Muni's final line came from the darkness: "I steal." Realizing the scene's dramatic potential, LeRoy did not reshoot the scene with lights.

As Harry had envisioned, Muni's searing portrayal shattered audiences' sensibilities and made the public aware of the brutality perpetrated by corrections officers. The outcry following the release of the film forced improvements in prison conditions. The movie was one of Warners' most successful of the year and it and Muni were both nominated for 1933 Oscars.

New stars were arriving and a few old ones were passing from the Warner studio scene. Rin Tin Tin died in 1932 at the age of sixteen. The films the German shepherd had made from 1923 to 1930 had earned over $5 million for his owner, Lee Duncan, and kept the studio alive during those first bleak years. Jack Warner, saddened that the studio's "mortgage lifter" was gone, dramatically (perhaps too dramatically) recalled:

> Rinty's gallant heart was tired and old, the strength had long since ebbed from the massive shoulders and legs. He was barely able to crawl to his master's side, and Duncan knew at once that no power on earth could help. He phoned across the street to his neighbor, the lovely, shimmering Jean Harlow, and she came running. And she cradled the great furry head in her lap, and there he died.

The eulogy read like a Hal Wallis release and no doubt made Jack muse, "What a way to go!"

The same year, Jack cast a young actor named John Wayne in an equine version of a Rin Tin Tin adventure, *Ride Him Cowboy,* in which Wayne saved a horse called Duke from being shot. Earlier in the year,

"Duke" Wayne had made his first picture for Warners, a horse opera, *Haunted Gold*, which costarred Ken Maynard.

Warner Bros. had also got into the cartoon business with the first Looney Tune animated short, "Sinkin' in the Bathtub," starring a Mickey Mouse knockoff named Bosko. (Disney created Mickey in 1928; Bosko made his debut May 6, 1930, preceding the Warner movie *Song of the Flame.*) Rudolph Ising, one of Bosko's animators, said the cartoon character "was supposed to be an ink spot kind of thing. We never thought of him as human, or as an animal, yet we had him behave like a little boy."

A year before Bosko was first shown to the public, Hugh Harman and Rudolph Ising had made a three-minute animated short called "Bosko the Talkink Kid" in hopes of riding the talkies' wave of popularity. "That's what we were selling—synchronized lip motion," Ising said. At the end of the cartoon, Bosko wiggled back into his inkwell, saying "So long, folks!" (a curtain line that was later given to Porky Pig, who changed it to "Th-th-th-th-th-that's all, folks!"). At first Ising and Harman were unable to sell their cartoon, so they teamed up with Leon Schlesinger, who had made a career of creating movie title-cards. The advent of sound almost put him out of business. Schlesinger sold the cartoon idea to Jack Warner under the title "Looney Tunes." They became so popular that Jack commissioned another series, "Merrie Melodies."

The cartoonists had more fun than the viewing public as they acted out the faces of their characters in front of mirrors. Originally the cartoonists drew in a studio next to where Busby Berkeley was shooting with 300 beautiful girls. That distracted them so much, they were moved to a little shack made of pressed paper that they dubbed Termite Terrace. There they created the 10,000 drawings needed to make each seven-minute cartoon. It took six months and $20,000 to create a cartoon. Mel Blanc did most of the voices: Porky Pig, Daffy Duck, Elmer Fudd—and Bugs Bunny's famous "What's up, Doc?" Mel hated carrots and tried crunching turnips and radishes, but nothing sounded like carrots. After each take he'd spit them out.

Jack Warner considered cartoons nothing more than a sop to distributors, whose customers wanted a full program of news, animated shorts, and feature pictures. One day Jack invited the cartoon staff for lunch and told them, "The only thing I know about you guys is you draw Mickey Mouse." After a stunned silence, the cartoonists assured their boss that they'd take good care of "Mickey."

Although each of the cartoon characters was loved by the public, Tweety Pie became a big hit when the flesh-colored canary uttered the line, "I tot I taw a puddy tat!" The ever vigilant censors thought the pudgy bird looked naked, so he became a yellow canary. What the censors never realized was that Porky Pig waddled around without pants.

A fan of Warners' cartoons, Steven Spielberg, said, "They were irreverent, antisocial, and never took any guff from anyone."

While Porky was being drawn without any pants, Harry, on a visit to Hollywood, found Jack ranting and raving about a new Production Code document from Will Hays's "voice of morality." Responding to pressure from the National Legion of Decency (a group composed of American Roman Catholic priests), Hays had issued a new Production Code governing the movies. Mae West's sexually suggestive films were one of the main reasons for the new code, as were several Warners films (such as the 1933 production *Baby Face,* starring Barbara Stanwyck as a speakeasy girl), which were said to contain too many "sex and violence" scenes. Studios now had to submit scripts prior to filming to the Hays office for review. If the script was deemed satisfactory it was given the Code's stamp of approval.

"It's all due to that bunch of tight-assed Roman Catholic bishops!" Jack screamed. "I mean, listen to this." Jack read from the document in his hand: "'Authority to exhibit films in America will be denied if there is any long tongue-involved kissing . . .'"

"Now, Jack . . ." Harry started to say in a placating manner.

Jack blurted, "Hell, I know, I'll just fire every actor with a long tongue."

"You're overreacting," Harry said.

"Am I? Listen to this one: 'If a man kisses a woman while in bed, he must have one foot firmly on the floor.' Why in the hell don't we just have them rub noses!"

Harry didn't say anything.

"No nudity!" Jack continued ranting.

"We don't have any nudity . . ."

Jack rolled up the Production Code paper into a scroll. "I said *no* nudity. *Not even on babies.* Good God, are they supposed to be born in Little Lord Fauntleroy suits?"

Harry just shook his head. Jack *was* overreacting. Both of the brothers were concerned about morality in the movies.

"You listening to me, Harry?" Jack slapped the papers on his desk. "What's happening to our rights? Is this still America?"

"These codes will not hurt our productions," Harry intoned. "Audiences will go to see films with artistry . . ."

"Yeah, and the next thing you know, they'll be saying that even *going* to the movies is a sin! If that happens, we're out of the movie business."

"That we'll never have to worry about."

The next day, Harry picked up the newspaper and was shocked to read the headline ROMAN CATHOLIC BISHOPS—GOING TO THE MOVIES IS A SIN!

At a luncheon honoring the Postmaster General, Harry made an impromptu speech:

> I had not planned on saying anything about this film cleanup move. . . . Many faults could be found with any industry, if one wants to look for faults, and it is an unfortunate mistake to judge an entire industry by the faults of a few. It's going a bit too far to make going to theaters a sin.

Harry added, more or less humorously, a suggestion that the Postmaster General "issue postage stamps bearing some of the film stars' pictures and charge an extra cent for them to create a fund to aid those that might be thrown out of work if the theater ban idea spreads."

The Catholic church, along with other religious groups, decided to give the industry another chance at self-censorship, and called off its threatened boycott.

Production Code or not, Warners had a full schedule of movies to make. And they continued to make them frugally. The *National Exhibitor* reported:

> Warner pictures are something like the Ford car used to be advertised, nothing going to waste. The Warner shows have been timely, topical, allowing theater men to cash in. The scripts are racy, attractive, move speedily. Two years ago Warners weren't setting the world on fire. Now, because the tempo of the organization has been quickened, the company is delivering. What Warners have accomplished is an inspiration to the industry.

Gangster movies, social dramas, biopics, cowboy shoot-'em-ups, and Looney Tunes cartoons were winning awards and selling tickets, but Jack wanted to make movies that could be viewed for their sheer entertainment value. That's when he decided to revive the movie musical.

> Come and meet those dancing feet . . .
> On the avenue I'm taking you to—
> Forty-second Street!

Until the 1933 triumph of *42nd Street,* the movie musical was scarcely more than a collection of production numbers with little, if any, story. Audiences were not going to put their hard-earned Depression quarters on the plate to watch a bunch of dancers parade around in a revue. (Rin Tin Tin had even got into the act, introducing—with an elaborate series of barks—the 1929 musical revue *Show of Shows.*)

In the early sound years, Warners had followed *The Jazz Singer*'s success by starring Al Jolson in musicals: *The Singing Fool* (1928), *Say It with Songs* (1929), *Mammy* and *Big Boy* (both 1930). John Boles appeared in *The Desert Song* (1929), best classified as a sound-film operetta, with sequences in muted Technicolor. The studio issued a moratorium on musicals until someone could come up with a story line that was more than a series of song cues. A story was finally found (borrowed from the plodding 1929 musical *On With the Show*)—and *42nd Street* was born. It created a sensation.

Budgeted at a mammoth $400,000 and directed by Mervyn LeRoy, *42nd Street* boasted the talents of dance director Busby Berkeley; staid, solid Warner Baxter in the lead; and Bebe Daniels, George Brent, Dick Powell, and Ginger Rogers.

It also introduced a new actress by the name of Ruby Keeler. Shortly before her death in February 1993, she reminisced about her arrival in Hollywood.

Other Voices

Ruby Keeler

"Ruby, how'd you like to go to Hollywood and work the picture houses for five days, maybe a week?" my agent asked. I had just signed with Mr. Ziegfeld to dance in *Whoopee* with Eddie Cantor. Rehearsals didn't start until September. Hollywood. It sounded like fun.

"I'll have to ask Mama," I said. I was only seventeen and Mama was with me day and night backstage, watching over me. Mama said she couldn't handle the train trip from New York, so I set it up with a girlfriend named Mary, who was five years older, to chaperone. On the trip I called her "Aunt Mary." At that time, the trains took three days to cross the country and it was such a fancy way to travel. At different stops the dining car chef would buy fresh trout and cook it for dinner. Fanny Brice, who was going to Hollywood to do her first picture, *My Man,* for Warner Brothers, was on the train, so there was a big welcoming committee at the Los Angeles station. My agent from the William Morris office was there to meet me. Standing next to him was a gentleman I had watched in the movies but had never seen perform on stage—Al Jolson.

Jolson—who I had heard was known as the "fourth Warner brother"—went up to Fanny Brice and gave her a big hug, to the delight of the publicity agents and photographers. Then Fanny introduced me to Jolson by saying, "This is Ruby Keeler."

Wow! Was I in awe.

"I know," Jolson said, holding my hand. "I saw you at Texas Guinan's place. You were that cute tap dancer?"

Double wow. He knows *me.* I had started working at Texas Guinan's speakeasy in New York as a chorus girl when I was thirteen. I celebrated my fourteenth, fifteenth, and sixteenth birthdays behind the stage. The dancers weren't allowed to sit with the guys between shows or drink the bootleg booze. If anyone touched us, Texas Guinan would usher him out the door. It was there that an agent spotted me and told Mr. Ziegfeld, "There's a kid over there who's a great dancer." What I didn't know was that Al Jolson had also seen me perform—and remembered.

At the train station he didn't waste any time. "How'd ya like dinner tonight at the Beverly Wilshire?" he asked.

I mumbled something about staying with "Aunt" Mary's relatives in Long Beach.

"I'll send my driver, Jimmy, to pick you up."

"*And* my 'Aunt' Mary," I stammered.

He smiled that big wonderful Jolson smile. I could feel his eyes watching me as I walked away.

Al Jolson and Ruby Keeler were married September 21, 1928. He was forty-two. Ruby was nineteen. Ruby went back to New York to dance in Ziegfeld's *Whoopee.* She was billed as Ruby Keeler Jolson. Her husband was furious, not because Ziegfeld had used his name, but because Ruby was billed below Eddie Cantor. "How can you have a Jolson below a Cantor?" he told his wife. Jolson took her out of the show the next day. Ruby was learning to live with a legend. Undaunted, Ziegfeld cast Ruby in a new Gershwin stage musical, *Show Girl,* which also featured Jimmy Durante. She was a smash.

Later in Hollywood, Jolson got a call from Jack Warner.

Jack said, "I'm planning a new musical, *42nd Street* . . ."

"Jake, I don't think you can afford me," Jolson cracked.

"Jolie, the picture's not for you. I want Ruby to do it. It's a great tap-dancing part for your leggy lady. She goes on, saves the show, you know, same old schmaltz, but a big production. We're budgeted for four hundred grand."

Jolson hesitated. "Okay, but only if I'm her manager."

"Sure."

Then Jolson jolted Jack by saying, "I want ten thousand dollars for her first picture."

"Ten thousand!" Jack screamed. "This is no gold mine."

"Yeah, but Ruby is." Hearing no response on the other end of the line, Jolson pressed on, "And one other thing—don't expect me to see her work. I don't want to watch her kissin' no other guys."

Ruby Keeler got her $10,000 and a 1930s partnership with Warner Bros., Busby Berkeley, and a new performer uprooted from Pittsburgh, Dick Powell.

"I was like a scared rabbit in *42nd Street,*" Ruby later admitted. "Dick Powell was too. I knew I wasn't an actress but figured all I had to do was say lines like, 'What?' 'Who?' 'When?'"

She also had some real corny lines, such as the one she purred to George Brent: "Remind me to tell you I think you're swell." (Not all the dialogue was bad: Ginger Rogers as "Any-Time Annie" got the classic chorus girl line, "She only said 'no' once and then she didn't hear the question.") The scene where Warner Baxter as the Broadway producer tries repeatedly to get Ruby to say the line, "Jim! It was grand of you to come!" and when she finally gets it "right" the line sounds no different from her first reading, was true to life. No, Ruby Keeler wasn't a good actress, but she was a wonderful dancer. Jack cast her in nine more musicals, mostly with Dick Powell.

Other Voices

Ruby Keeler

I remember the chorus girls saying, "Well, I would have got that part if I was married to Al Jolson." Al never did get me parts. Sure, there were hundreds of other girls who could have done what I did. The only reason I was able to do it was because of Busby Berkeley, who was in charge of the dance production numbers. Buzz himself couldn't dance a step, but he had such a fertile imagination. He'd say, "Try it, you'll do it great." Then if it wasn't right, he'd tell me. I knew that he was trying to improve me, so I would try.

Several months after *42nd Street* was finished, I did *Gold Diggers of 1933,* then *Footlight Parade,* which starred Jimmy Cagney. With Berkeley directing, I never knew whether I'd be sprouting out of a flower or dancing on a piano. One morning I walked onto the set of *Footlight Parade* and was shocked to see a huge pool of water with girls diving into it. I said, "Buzz, I can swim a little, but I don't like to be underwater. And I can't dive."

He said, "You'll be able to do it. Get in the water with the kids; get used to it."

"Yeah, but after that, what do I do?" I asked.

"For the first shot you'll go down to the other end of the pool and do a porpoise dive . . ."

"A porpoise dive?" I knew I wasn't a fish.

"Well, you dip your hands and you dive, then you swim underwater to this end of the pool with your eyes open because there's a window here with a camera. You have to time it so when you pop out of the water you're smiling."

Wow. Nothing to it. I said, "All I can do is try."

Buzz had a lot of champion Olympic swimmers diving off rocks. It looked dangerous to me, with everyone swooping off different levels on top of one another. I was supposed to dive off this high rock, maybe twenty feet, and he had someone do it for me. I'd get under the water where the diver was supposed to land, then pop up, smiling. I didn't like that. There were a lot of takes because I couldn't come out of the water on time. I had this long black wig they made me wear and it was always getting into my face. Buzz kept saying, "I can't *see* you with all the hair in your face." He finally got one take that looked pretty good.

To set up the diving scene, Dick Powell crooned "By a Waterfall" to me. Most moviegoers thought Dick and I were married, but the closest I got to him was looking in his face for seventeen choruses of "I Only Have Eyes for You."

In *Footlight Parade* I also had a big production number with Jimmy Cagney, "Shanghai Lil," which critics called one of the greatest musical numbers in film history. Cagney searches for his Lil in this sinister Oriental setting and I finally pop up out of a barrel in Chinese makeup and say, "I miss you velly much a long time." Then Jimmy lifts me up on a bar and I dance. It was great fun to dance with Jimmy.

One day Jack Warner, cigar in hand, dropped by the set while Ruby was filming the "Shanghai Lil" number, mostly to joke around with Busby Berkeley. But he had something else in the back of his mind. He stuck around long enough to have his picture taken with Cagney, Berkeley, Dick Powell, and Ruby. Looking Ruby over carefully, Jack decided he would eventually star her and his buddy Jolie in a musical together: *Go Into Your Dance* (1935).

"There were a lot of good musical numbers in it," Ruby remembered, "'About a Quarter to Nine,' and 'A Latin from Manhattan.' It wasn't any different working with my husband than it was with anyone else."

Jolson didn't feel the same way. The success of the movie called for a sequel, but Jolson feared comparison to his young wife and didn't want to end up as a husband-and-wife team. His ego wouldn't allow it. As Ruby said, "Al was called the greatest entertainer in the world. I know that was true because he told me so—many times."

One of her favorite musicals and the last one she did for Warner Bros. was *Ready Willing and Able* (1937) in which she danced on huge typewriter keys. "It was difficult jumping from key to key doing wing-and-taps, and I'd get charley horses in my legs from dancing on those 'footstools.' There was a solid black background. I had no idea that girls, lying on their backs and using their legs to tap out a letter, would be added to the final print."

Although Dick Powell and Busby Berkeley continued to make musicals for Warners, the old fizz was gone. Audiences clamored for the swashbuckling films of Errol Flynn and the melodramatic "weepers" of Bette Davis. Flynn's first movie, *Captain Blood* (1935), in which he appeared with another new Jack Warner find, Olivia de Havilland, made him an instant star. Jack Warner later said, "I knew we had grabbed the brass ring in our thousand-to-one-shot spin with Flynn. When you see a meteor stab the sky, or a bomb explode, or a fire sweep across a dry hillside, the picture is vivid and remains alive in your mind. So it was with Errol Flynn."

Once again, Jack must have had his publicity people working overtime to come up with such accolades. Actually, Jack did admire Flynn's hard-drinking, womanizing, devil-may-care lifestyle. The same couldn't be said for Harry. While he was on a visit to the Burbank studio, with Jack showing him around (and hugely enjoying his older brother's awkwardness and shyness around the stars), Errol Flynn bounded out of a dressing room and put an effusive arm around the horrified Harry.

"So you're Sporting Blood's better brother!" Flynn said, giving Harry an extra big hug. The actor turned to Jack: "Hey, ol' boy, I've got it all lined up for tonight—plenty of hot women and a hot game of poker."

Harry cringed.

Although Harry knew that Jack was having continuing affairs with a series of women, he had, for the last two years, reluctantly kept his mouth shut. His tirades against Jack's womanizing had done nothing but raise his own blood pressure. But Harry fretted silently that the family—especially Irma—would discover the truth about Jack's nocturnal absences from home.

What Harry didn't know was that Jack had taken up with a full-time mistress.

And she was pregnant.

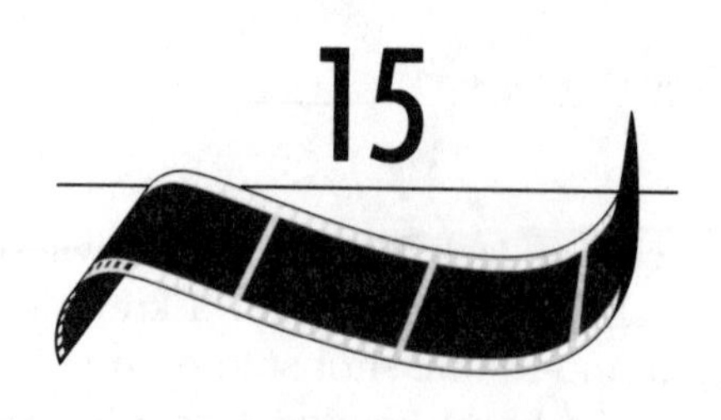

15

Family Affairs

Jack Warner was at a boring Hollywood party in the autumn of 1932 when his roving eye landed on a ravishingly beautiful brunette. He stood transfixed by the woman's raven hair, her dark exotic features, her slender figure. The curve of her body gave the silk material of her dress movement and life.

Her name was Ann Paige.

Several years earlier, Jack wouldn't have given her a passing glance. At that time she was a little brown mouse of a person who wore dull gray outfits and rarely spoke a word. She had come to Los Angeles with her Jewish parents, Sol and Sarah Boyer, in 1920 in the hope that the climate would help her mother, who was suffering from tuberculosis. Two years later her mother died. Ann's father soon remarried, but Ann and her stepmother didn't get along.

My great-aunt Lina Basquette told me that when Ann Boyer was a teenager, she got involved with a socialite, Eddie Sears, who took her around the world and taught her how to dress, how to use makeup, and how to style her thick dark hair. She came back a year later a glamour girl. When she showed up in Hollywood, everyone said, "Who is that girl?" She started to pal around with Mary Astor and Billie Dove, two of the most entrancing women in town. In 1924 she met and married Don Alvarado, a Valentino clone who had starred briefly for Warners as a Latin lover, when she was only sixteen.

Jack had given Alvarado his stage name. When the actor had auditioned as Joseph Paige, Jack said, "You look Spanish, you need a Latin name." The only Spanish Jack knew was Alvarado, the name of a Los Angeles street. Don Alvarado did a Rin Tin Tin movie for Warners and a few other parts.

A year later a daughter, Joy, was born to Don and Ann. Alvarado's voice didn't match his Latin looks and his career quickly faded when talkies took over in the late '20s.

By the time Jack spied Ann at the Hollywood party, her marriage to Alvarado was breaking up.

Jack wrote a note and slipped it to one of his "beards" to give to Ann. A moment later, Ann smiled at Jack over her cocktail glass.

Jack got Ann a Malibu beach house, a love nest that she and Jack shared whenever he could break free from the studio. She was intelligent and exciting, and the most exotic woman he had met. As his interest in her intensified, he helped her obtain a Mexican divorce from her husband.

It was easy for Jack to make excuses to Irma. He was in a business that kept crazy hours. If the studio was late on a movie release date, he'd have to stay up all night to get it on schedule, or he'd have to go to the laboratory and correct a print, or rerun the dailies, or reshoot a scene. Many nights he had legitimate reasons for not coming home, but with Ann waiting in the Malibu house his excuses became more frequent.

Irma was a good Jewish wife who did her own cooking. She also played a little golf, went horseback riding one day a week, and chummed around with Virginia Zanuck. She felt her home was a happy one. She tried hard to believe her husband when he made excuses for being away. She wasn't sure he was having an affair. Not yet.

In New York, Harry got an anguished predawn phone call from Irma. She told him she was sure something had happened to Jack. They had come home from a party, and he said he had to go to the studio to edit a film that had to be shipped out the next morning.

"Harry," she cried, "I just called the studio to see if I could reach him, but the guard at the gate said he never arrived. He'd been drinking pretty heavily at the party . . . I'm sure he's been in an accident."

Harry's voice on the other end of the line was steady, calming. "Don't worry, Irma. I'll get someone on it, and call you back as soon as I know something."

Harry immediately telephoned the studio's private investigator.

"Where's Jack!" It wasn't a question.

"Uh, I'm sorry, Mr. Warner, but I don't know."

"You know your job?"

"Yes sir."

"Then I'll give you ten seconds to tell me where Jack is or you're out the gate."

Harry got the number. And the name—Ann Paige.

Jack had to hold the receiver away from his ear as Harry yelled, "You no-good son-of-a-bitch! Irma's worried to death, thinking you're dead in a ditch somewhere. For all I care, that's where you should be. Not with some whore."

Jack glanced at Ann, lying on her side at the head of the bed. "Harry, she's not . . ."

"Whore!"

"Don't be so damned self-righteous." Then for Ann's benefit, he added, "Look, you should try it."

Harry was apoplectic. "You . . . you run around with your fly unzipped . . ."

Jack sighed, "Harry, I do what I want."

"Get out of there and get home—now!"

"Harry, you have no damn business . . ."

Harry shouted back, "I have all the goddamn business in the world." The ferocity in his voice took on a knife-blade edge. "I'm talking about family. Think of what you're doing to Mama and Papa, and what other people think." When Jack didn't respond, Harry added, "You hear me, get your prick back in your pants and get out of there!"

Jack heard the loud click of the receiver being banged down, then as he got dressed he thought up an excuse about being called to the film laboratory—an important negative had been ruined . . . He'd tell Irma he never made it to the studio.

Other Voices

Jack Warner Jr.

Sometimes my father stayed all night in his office, sleeping on the couch, or so he told my mother, who believed him, fully understanding why he had to devote so much time to the growing business.

Ann was at the tail end of my father's extramarital affairs. My mother had known about his liaison with Marilyn Miller in 1930, as I remember hearing them shouting about it through the heating vents in our home. My father's infatuation with Marilyn Miller waned as her movie career declined. Although her vivaciousness and petite beauty had enraptured audiences on the Broadway stage, that same exuberance could not be recaptured on the screen. Five years after their affair ended, Marilyn Miller succumbed to blood poisoning. She was thirty seven.

My father wasn't too careful. He'd come home with lipstick on his handkerchief—not the right shade to match my mother's. I once found on his dressing table this strange cylindrical packet, shaped like a cigarette. Unrolling it, I discovered a transparent balloon. I was only ten at the time and thought he'd been to a party. I asked him, "What's this?"

He yanked it out of my hand and snarled, "Shut up, kid!"

As Jack's marriage headed toward destruction, Harry's attention was diverted by his oldest daughter, Doris, who had announced her engagement to director Mervyn LeRoy. Hollywood reporters intimated that Harry was searching for a surrogate son to replace Lewis.

Harry and Rea had done their best to show off their daughter by taking her on long cruises to Europe, where she had had the chance to meet exciting eligible men. On a trip to England she had even met and danced with the Prince of Wales. But Doris's mild flirtation with royalty was the stuff of which fairy tales are made, not weddings. Not that she wasn't attractive and interested. At twenty, Doris had blossomed from a gawky teenager with toothpick legs and arms that were all elbows into a bright, beautiful woman.

Although Doris was definitely smart and able enough to have replaced Lewis herself, a woman executive was not acceptable at the time. Understanding this, she agreed to take a lesser job in product acquisition, reading and choosing scripts for possible movies. This is something she had done very successfully in New York. She had personally chosen and produced the stage play *Men in White,* which drew critical raves.

Jack was itchy around Doris. She was young, intelligent, and had the Warner chutzpah. He was relieved when her engagement was announced. LeRoy, who had once acted in vaudeville, was one of Jack's cronies, and they would exchange snappy dialogue for hours. No, Mervyn LeRoy was not a threat.

When Doris accepted an engagement ring from LeRoy, her younger sister Betty asked why she agreed to marry him. Doris replied, "I liked the ring."

"Do you love him?" Betty asked, hoping for a romantic story.

"No—but he's rich."

At that point Betty knew she had a different person for a sister than she'd thought. "Doris had developed into the very thing that my mother and father wanted," Betty said. "They had brought her up to marry well. They had clothed her, groomed her, and traveled with her around the world. She was brought up to be a princess. My parents created her."

Doris was married in New York on January 3, 1934. Betty, now thirteen, and Lita, seven, were flower girls. Jack and Abe attended, along with Paul Muni, Will Hays, and Adolph Zukor. LeRoy was pleased with the homage paid by these industry leaders, remembering that only a few years back he had been just an extra in such films as Cecil B. De Mille's *Ten Commandments.* His ascent as a director at Warner Bros. had been meteoric.

Ed Sullivan devoted his entire "Broadway" column to the wedding, reporting in part:

> The Princess of the Warner Empire, a dynasty of celluloid and actors, is the bride of the ace director of Warners. Doris Warner is now Mrs. Mervyn LeRoy, and when you read this the newlyweds will be preparing to embark on a 'round the world cruise. Boarding the same boat when it reaches the west coast will be another heiress, Barbara Hutton, sailing with Prince Alexis Mdivani. You will forgive me if I think the Warner princess, marrying a fine director, fared better than the Woolworth princess.
>
> The significance of this ceremony cannot be overestimated. Adoption of Mervyn LeRoy into the Warner family is freighted with

> importance. He now becomes the Crown Prince of the Warner Empire, a valuable addition to the manpower of the celluloid dynasty. Mervyn LeRoy becomes therefore the successor to the boy who passed away. There was a world of meaning in Harry's voice the other night when he addressed LeRoy as "Son."
>
> I predict that within five years these two kids, Doris and Mervyn, will be tremendously compelling figures in the talking picture industry. Not all of the other picture families have been so fortunate. For it is not every daughter who elects to place her pert nose at the grindstone which her father and uncles have worn so smooth. The Warner Empire is to be congratulated, doubly.

Harry read the column with an embarrassed sigh. LeRoy certainly was not the "successor to the *boy* who passed away."

Jack scoffed at the words "Crown Prince of the Warner Empire." He was confident Doris's consort would never ascend the Warner throne.

Other Voices

Betty Warner

My father had come to realize that Doris did not have the desire to be president of the company, yet he continued to groom her as a studio executive. She appeared to be sharp and shrewd—requirements for a successful career. I did not envy her. She didn't like the pressure of having to live up to Dad's expectations for her in the business world. Although Doris was the heir apparent, the mantle was too heavy to wear. She did accept the role of princess, and used her title for fun and games on the social scene.

Doris became the "queen of charity," and threw outlandish parties for movie stars. Six months after she was married, she sent me a letter in which she said:

> The men played tennis until eleven last night. Such lunatics. Mervyn played with Clark Gable as his partner, but they got took as Clark didn't play so well!
>
> Tonight Dolores Barrymore is coming for dinner and Dick Powell and we have the picture *Caravan* to show a lot of people dropping in after dinner for a spot of movie. We showed *Belle of the Nineties* the other night and liked it very much.

And so dear, I'm off to do nothing and will be thinking of you.

Lovingly,
Dot

In hindsight, it was lucky for me all the pressure and expectation was put on Doris and not me. I was permitted to be a normal child—going to public schools, playing with neighborhood friends, and doing athletic things like swimming and tennis, having dates with my school boyfriends, going to the local movie theater and going out for hamburgers and sodas after.

My family didn't have a lot of time for me. Mother and Dad traveled on business. This is the only reason Dad left his home. He *hated* to travel. Mother loved it and until she died she would go anywhere in the world shopping, to the theater, and to see friends. They were gone for three to six months of the year.

Lita and I were left in the care of Dad's secretary, who reluctantly moved in with us. I led her a merry chase, as I was used to doing what I wanted with my friends. Innocent though we were, she was in constant panic.

I might have felt neglected, sometimes, but I never felt unloved, and being left to my own devices had many pluses, although it was hard to deal with authority when my parents came back from their travels.

They were generous to me with gifts, but time together was at a premium. On May 3, 1932, Dad wrote from his New York office to our home in Mount Vernon:

My darling Betty,

You certainly are growing, and here I am congratulating you on your twelfth birthday. It seems like but yesterday that I thought you were a big girl at two, and ten years have passed and you are as sweet as ever.

I hope that I may be here to congratulate you ten years from today.

Love and kisses to one of the sweetest girls living.

Your
Dad

Meanwhile, Jack Warner's relationship with Ann had reached a critical point. What Irma had hoped was only another flirtation had now become a serious affair.

Bill Schaefer, who went to work for Jack in 1933 as his personal secretary at $25 a week, remembered that on one of his first nights on the job he got a call from Irma asking to speak to her husband. "Just a minute," Schaefer said, and went into Jack's office, where he really got chewed out. Schaefer remembered Jack telling him: "*I don't ever want to talk to her.* Don't you dare tell her I'm in the office whenever she calls!"

Other Voices

Jack Warner Jr.

Later when my father began to devote more time to his new mistress, Ann, his secret leaked out. Gradually and painfully my mother learned most of the truth about my father's late evenings. The arguments began in earnest and the noise level at our house went off the dial.

My father always resolved these scenes by rushing out of the house and slamming the heavy front door on the way to the garage. I would hear the roar of his yellow Jaguar as he took off down the hill, through the gates, and off into the night—not to return for a day or two. I don't remember his ever sitting down to talk with me about something he knew I could hear. There was only the slam of the door and the vanishing noise of his car as it left the scene of the battle. The rest of the long night would be marked by my mother's sobs. Finally, sometime in late 1933, the car roared down the hill and did not return. In a few days my father's masseur, Abdul Maljan, came by to pick up all of his clothing and personal effects.

I could see that Abdul was sad about the whole thing, but he didn't say a word. My father stayed away for weeks and I didn't know what was going on until my mother, between crying spells, filled me in, or at least gave me her version of events. Yes, my father had a female "friend," but he would soon return home. The word "divorce" was not mentioned.

Then, in January 1934, Ann told Jack she was pregnant.

Mike Levy, an agent and old buddy of my father's, took me aside and said, hesitantly, "Your father asked me to tell you this. He, well, he wants a divorce."

I was shocked, but the first question I had was: "Why doesn't *he* tell me this?"

Levy's shoulders sagged. "Please try and understand." He went on to voice a veiled threat that if my mother refused, then my father would "somehow get the divorce himself."

I asked, "I'm supposed to tell my mother that my father wants to end their marriage? Why doesn't he talk to her?" There was no answer to the question.

For a man who made his living packaging other people's lives and throwing their shadows on the screen, my father was a coward when it came to getting his own story across. That was a revealing side to him, his stark fear of direct confrontation. He had to use surrogates to face and resolve unpleasant problems. Whether it was telling his son about the breakup and divorce, or the firing of an associate, Jack L. Warner would invariably arrange to be out of town (usually nobody knew where) and the dirty job of delivering bad news would fall to somebody else, usually one of his yes-men.

My father and mother went back to Youngstown for a hometown celebration with Ben and Pearl and it was there that he told her.

There was a loudspeaker hookup from Youngstown to Los Angeles so all of us in the family who could not make the trip could be in on the celebration. My mother, obviously not realizing she was "on the air," started talking to one of the family, sobbing, "He wants a divorce. He's leaving me . . . what shall I do!" In Los Angeles I sat there listening to what was a real-life soap opera.

Shortly afterward, I graduated from high school with a mixture of joy and gloom. I remember the event mainly because I played the tympani in the orchestra, then had to dash out the side door immediately after my fourteen-bar solo roll in "Finlandia" and throw up on the lawn. I'm not sure if it was something I ate—or something I thought. My father was not present for my graduation.

At last I received a message to be at my father's office at the studio. I went past the assistant secretary, then past the executive secretary, then through the door, and walked the long approach to the big black desk behind which sat my father.

His hands played nervously with some pencils and he snapped one of them between his fingers with a crack that made me start. At the time, I thought he was terribly angry—later I knew he was terribly desperate. My mother was refusing to give him a divorce. He was going to use me to force her to do so.

It was not a father-son conversation about a difficult situation, but rather the ultimatum of the head of a studio to an employee. I was told that my mother had better give him his divorce—or else! The "or else!" came with a strong implication that if she didn't he would sue her himself for divorce and that he had plenty on her and she'd better know what was good for her.

I wanted a closeness, a heart-to-heart talk during this terrible time in all our lives. What I got was the opposite: a desperate man using his power as a club to crush opposition.

At one point, my father promised my mother that if she would give him a divorce, he would "work things out with the other woman." Meaning marry Ann, give their baby a name, then come back to Irma and remarry. This now sounds like a fairy tale, and if my mother believed it, then she was very naive. But she desperately *wanted* to believe. She loved Jack Warner. She wanted him back no matter what the cost. No matter how bad the embarrassment.

Irma, now totally humiliated and embittered, realized she could no longer hold on to her husband and sued for divorce, charging desertion. She testified that her husband had left home and refused to live with her. (That was no big secret; everyone at the studio was aware that Jack was now living with Ann.) Irma was granted an interim decree that gave her custody of Jack Jr.; they moved out of the Beverly Hills mansion, with its ugly memories, and temporarily took an apartment. The decree also stated that neither Jack nor Irma could remarry for a year. This was a last-ditch effort on Irma's part to preserve her marriage. Perhaps in a year

Jack would come to his senses and return to her. Jack and Irma's divorce would become legal in January 1935.

Ben and Pearl Warner knew little of Jack's estrangement from his wife. Pearl had been suddenly stricken with a cerebral hemorrhage on August 26, 1934, and was taken to Cedars of Lebanon Hospital in Los Angeles. There she and Ben quietly observed their fifty-eighth wedding anniversary. Death parted them before the next dawn. She was seventy-six. At the funeral, held in the Wilshire Boulevard Temple in Los Angeles, Rabbi Edgar Magnin intoned: "She was a character greater than any the films could have ever shown on the screen."

Pearl Warner, who had given birth to twelve children—and a film dynasty—was in reality a quiet, dutiful Jewish mother, a stout-hearted woman who loved to cook for her family and took pride in the accomplishments of her "boys." She was simply a peasant from Russia who wore fur coats in the warmth of Hollywood, and shopped for bargains in a limousine.

The shock of his wife's passing was so great that Ben had to be placed under a physician's care. He returned to Youngstown to visit his daughter, Annie, who still lived there with her husband David Robbins. While there, Ben Warner suffered a stroke. He died on November 5, 1935, at age seventy-eight. Harry was informed of his father's death by telegram and flew to Youngstown to make the funeral arrangements. Ben was laid to rest beside his wife in the Warner family mausoleum at the Home of Peace in Los Angeles.

Ben Warner, the Russian cobbler who had come to America to find streets of slush, not gold, the man who had hocked his watch and his horse to buy his boys a Kinetoscope projector, thereby launching them into the movie business, was gone. With Ben no longer on the scene, the last buffer between Harry and Jack had been removed. Harry, now more than ever, was the patriarch, a notion Jack found impossible to accept.

Shortly after his parents' deaths, Jack shocked the family by moving Ann into the mansion in Beverly Hills.

Other Voices

Jack Warner Jr.

I recall a rare talk with my father during this time of turmoil. We sat together in my car and I asked him point-blank if he was going to marry

Ann. He paused, looked me in the eye, and in a low voice answered, "No, no . . . I am not."

Not very long after this, in January 1936, he and Ann went to Armonk, New York, and were married. Their wedding luncheon was at the "21" Club and made all the newspapers, surprising me and upsetting my mother, who still had hopes he would return to her. Now I had a stepmother and a stepsister, Joy, and, as I was to shortly find out, a half-sister.

On September 30, 1934, Ann had given birth to a girl whom she and Jack named Barbara.

None of Jack's relatives attended the wedding. In Harry's mind, Jack and Ann's marriage was a personal insult. Perhaps Harry should have been more understanding. After all, Jack was a human being; he had human emotions and desires. He was no longer in love with Irma—he had fallen in love with someone else. Harry was blind to these emotions. It was the "family" that counted. He wrote Jack, saying, "The only thing good that has come on this day is that Mama and Poppa did not live to see it."

Betty Warner said: "Because of the way Ann felt she had been treated by Jack's relatives, particularly Harry, both before and after the wedding, she wanted nothing to do with her husband's older brother. This hatred was fed by the way Jack and Harry treated each other, and I'm sure she sympathized with Jack's anger at Harry. Quite simply, Harry was not letting Jack live his life."

Other Voices

Jack Warner Jr.

Jack never brought Ann to family gatherings for fear of her being insulted. Perhaps the sisters would have walked out of the room in Ann's presence. Even when Jack went to family dinners alone, it was like having a banquet on Mount Vesuvius, waiting for the lava to flow. Harry would hold back from voicing his hatred, and Jack, angry as hell at his brothers and sisters, and wanting to blast away at them, kept his hostility in check.

Many years after my father's marriage to Ann, I heard from my uncles and aunts that some of them had gone to my father and told him he had done a despicable thing. Yet getting a divorce and remarrying wasn't the worst thing in the world, even in those years of conservative morality. They could have understood and forgiven—he was a grown man and those things happen—but the woman for whom he left his wife and child had, in the eyes of his brothers and sisters, brought disgrace on him and the whole family.

They went so far as to tell their brother Jack just what kind of woman he had married, words that seared the ears and were the first cracks of a chasm that would widen and become unbridgeable, utterly dividing the brothers and sisters who for so long had stood together. Although they remained in business together, beneath the surface there broiled hostility and jealousy.

It would never again be the same.

16

Celluloid Dreams

The years 1934 and 1935 had been difficult ones for Harry Warner. Nineteen thirty-four had ended terribly when a massive fire swept through the Burbank studio, quickly igniting the wooden and paper facades of movie sets. Studio fire chief Albert Rounder collapsed and died of a heart attack after fighting the blaze, and when the embers had cooled the physical damage was assessed at close to half a million dollars. The studio's net loss for the year was over $2.5 million. Money Harry could borrow, sets he could rebuild, but the prints of early Vitagraph, Warner Bros., and First National movies that were consumed in the blaze were irreplaceable. The acrid smoke from twenty years of celluloid dreams lingered in the night air momentarily, then vanished.

In those two years, Jack had divorced Irma, Ben and Pearl died, and Harry was indicted by a federal grand jury in St. Louis on charges of conspiracy to violate the Sherman Anti-Trust Act, an offense punishable by a prison sentence of two years and a fine of $10,000. Ten other major film companies, including Paramount and RKO, and six individuals were also named. Harry M. Warner was the big name on the list. These companies and industry leaders were alleged to have "canceled franchises under which they agreed to furnish films and sought to coerce owners of buildings," and in general "conspired in the restraint of trade."

Harry had been through all this before when several years earlier he had narrowly escaped being convicted on similar charges brought by a financier named Harry Arthur who felt he had tried to "freeze him out by sealing up first-run movies." With his acquittal, Harry thought the issue had been put to rest, but now, in 1935, the government was bringing suit against him. Harry had originally appealed to President Roosevelt to call off the Justice Department and not prosecute. Roosevelt refused. Infuriated by the betrayal (he felt the Warners had helped Roosevelt win the election in 1932), Harry said, "The New Deal pays off its friends with the Sherman Act and causes them to lose theaters." Harry changed his political allegiance: He was now a staunch Republican.

In the face of this new legal action, Harry quickly issued a statement: "The charges are groundless and I welcome the opportunity to test the matter in court. We have done nothing but conduct business in a fair and honest manner."

The judge, a recent appointee hearing his first case, charged the jury in a vein that seemed to expect a conviction. Harry spent months on the witness stand. In November 1935, after an extended session in St. Louis, he wrote a letter to his two younger daughters.

> Darlings
>
> Betty & Lita, I am hopeing [sic] that this case will be over so I can be home with you and your mother next Saturday. It is favorable that I may. It is so lonesome here without you all. Otherwise Daddie is feeling fine. Love and kisses to both of you and Mother kiss her for me.
>
> Your loveing [sic]
> Dad

Three weeks later, Harry was still in St. Louis. He sent a postcard to Betty and Lita in which he noted:

> It seems like years since I saw you two and I hope it will soon be over but it looks like another ten days. It is just one of those things we have to go through in business.
>
> Dad

As the trial progressed, Harry began to realize that his defense was in jeopardy. How could the jury, themselves suffering through the dismal years of the Depression, condone what they were led to believe was the money-grabbing manipulations of a movie magnate?

Harry pointed out to his lawyer that one of the jurors was a "Negro," a man who had, no doubt, suffered through a lifetime of prejudice. "Play up the point that I'm Jewish," Harry told his lawyer. "Stress the persecution that Jews have suffered in America."

When Harry's lawyer rested the defense's case, and the jury left to deliberate, Harry noticed that the black juror gave him a slight smile.

Two days passed. The jury wasn't able to reach a verdict. Eleven had voted "guilty" and one, the "Negro," had voted for acquittal. The judge declared a hung jury and the case was set for a retrial. In the interim, Harry sold Warners' interests in the movie theaters in question and the case never went to trial again.

But Harry's troubles weren't over. He received a telephone call from a man who did not identify himself, saying only, "Watch out! Within forty-eight hours we'll have both your daughters."

Harry hired bodyguards for Betty and Lita. Betty, who was fifteen at the time, remembered: "Someone was with me all the time, to guard me, and I'm sure they even followed me to school. It lasted maybe six months."

Harry felt that the kidnapping threat came from the newly formed projectionists' union which was engaged in a struggle for survival with Warner Bros. Harry and Jack and Abe believed the unions were controlled by underground elements and refused to agree to their working and wage demands. Eventually Warners negotiated and accepted many of the union's proposals. Once that problem was solved, the kidnapping threat disappeared.

The only thing that eased Harry's mind in 1935 was the profit the studio showed: $674,158. The economic crisis brought on by the Depression was abating. That same year, two of their films were nominated for Best Picture. One of them, *A Midsummer Night's Dream,* was Warners' burnt offering to culture. In it, a young lad by the name of Mickey Rooney made a nimble appearance as Puck, until he broke a leg and had to be wheeled around behind the bushes on a tricycle by unseen stagehands. For this effort in filming one of Shakespeare's classics, Harry, as Warners' president, received an honorary doctorate of humanities from Rollins College.

In 1935, with the studio no longer marking its account books in red ink, Harry decided to move his family full time to California. He bought a home in Beverly Hills, and a beach house in Santa Monica that was formerly owned by Jesse Lasky. Most of the movie moguls had homes on Ocean Front Avenue. The houses shared a tennis court, but each had its own small swimming pool. An open gate led to the beach, which Betty called "the sandy front yard." It was an idyllic setting for Betty and Lita. They loved the barbecue picnics and romping at the water's edge with their pets.

Around the same time, Harry bought a 700-acre ranch in the valley northwest of Hollywood in Calabasas for the studio to use for shooting Westerns. He fell in love with the land and bought adjoining property for himself, about 3,000 acres altogether by the time he was done.

Harry enlarged the ranch house that rested on the east side of a hill. West of the house was a hill that cut out the sun two or three hours before he would've liked it to. He bought a thirteen-ton carryall, a Caterpillar tractor, and a man to move a million tons of dirt into the meadow. It took many months, but he finally got those extra hours of sunlight that he wanted. Harry had a mystical feeling about the land, which he associated with "hard work and the salvation of the soul."

Several years later he sold the Calabasas property and bought 1,100 acres in the San Fernando Valley, which was closer to Hollywood. He wanted to build a racehorse establishment and grow his own alfalfa and fruit trees, and be as self-sufficient as possible. He built four homes, one for himself, two for his horse trainers, and one for family and guests.

Harry also raised chickens—laying hens. He'd deliver eggs to everybody in the family, driving up in his car and dropping off a dozen. He would joke, "Enjoy them. Each egg cost me five dollars."

His move to California (which left Abe as the only Warner brother left in New York to handle the business) was prompted by his feeling that he could be of better use where the action was: at the Burbank studio. It was a mistake. He was more powerful as the mysterious figure in the East that no one could argue with. The move also put him in close proximity with the one person he could not deal with rationally: Jack.

In September 1936, Jack casually announced at a studio luncheon that he and Ann had adopted a two-year-old girl. Producer Sam Bischoff leaned over to his luncheon companion, an attractive studio secretary, and said, "I'm not going to believe that. I'm sure that's Ann and Jack's child."

While Harry was settling in on the ranch, Jack had set himself up in palatial style. When he moved Ann into the Spanish-style mansion on

Angelo Drive, she said she found it impossible to step over the threshold of a home built by another woman—unless it could be redecorated. She decided to design a home so luxurious it would be the envy of every star, producer, director, and studio head in Hollywood.

Ann studied architecture, then hired architect Roland E. Coate, who enlarged and rebuilt the mansion in the Georgian style, with an impressive Greek-revival portico. Interior designer William Haines, known as the decorator-to-the-stars (his clients included Douglas Fairbanks Jr., Carole Lombard, and Norma Shearer), outfitted the rooms in Georgian style to complement the new exterior. He helped Ann select the finest and most expensive European antiques, adding imported dark wood paneling to a study. (The wood, unsuited to the Southern California climate, popped and cracked painfully for years to come.)

The front door opened into a two-story hall with a cantilevered staircase that directed one's eye away from the extraordinary parquetry floors with their intricate wood mosaics to the upstairs bedrooms. Although each room was filled with antiques, such as George III mahogany armchairs and cut-glass chandeliers, Jack favored the library, for it was there he edited the films he brought home from work. The bookcases held a bound collection of every script Warners had produced. Written at the base of each binding was "Jack and Ann Warner." The movie screen was raised with a water pump and the front sofa would flip-flop for viewing.

Florence Yoch, a set designer and landscape architect, created magnificent grounds that were the envy of nearby stars. Jack, who had become an ardent tennis player, had had a court built when he lived there with Irma. Now he started inviting the top players to drop by for a round with him and his Hollywood guests. The Wimbledon crowd soon learned that it was best to hit the ball softly enough for Jack to return it. In fact, it was prudent to let him win the game. On Jack's court the game was called "Good shot, Jack!" His personal secretary, Bill Schaefer (and later Solly Baiano, tennis pro and talent scout), would work out the list of players, careful to put his boss on the winning team.

Terraces and garden-lined stairways led to the tennis court. Jack would have refreshments served courtside by Roget, the butler, who had to haul the food and drink on the long trek from the kitchen, then water to wash up, a task he disliked. Ann asked Jack if she could put in a dishwasher and a sink. He agreed. The next thing he knew, there was a pavilion with marble columns. Jack once said of Ann: "She is the woman I owe everything *for*!"

After passing through the estate gates and along the sycamore-lined drive, guests arrived at a circular brick-paved driveway in the center of which was a large fountain with a statue of Cupid riding a sea horse. The finished mansion was dubbed "San Simeonette" after the Hearst castle. It was into this transformed home that Jack invited his son, Jack Jr., to meet Ann for the first time.

Other Voices

Jack Warner Jr.

The day came when my father called me into his office and told me he wanted me to come to the house to meet Ann and their daughter, Barbara. At the time, I was living with my mother in an apartment on North Rossmore and had embarked on the life of a college freshman at the University of Southern California, so my contacts with my father had been few.

My palms were sweating as he began to give orders as to exactly how I should behave when I met Ann. I must be warm . . . above all I must be warm. I must throw my arms around her. I must make her feel wanted . . . accepted, loved . . . I must be nice, nice, nice!

My mother was terribly upset about my agreeing to visit with Ann, but I felt it was important that I get to know her and that my relationship with my father not be destroyed. How would I act at the meeting? I did not know.

I drove up to 1801 Angelo Drive in Beverly Hills. Even the address had been changed. When my mother and I lived there it had been 1871. After I announced myself into the little box set in the wall, the gates swung open and I went up the sweeping curve of the road past the tennis court, around the lovely waterfall where I used to climb rocks, then past the swimming pool where I had loved to swim deep under water to see the tile octopus, and then finally on up to the house with the circular court and the tremendous white columns lined up to greet me like frozen sentries.

My father stood waiting at the door. Behind him, deep within the big reception hall with the lovely inlaid parquetry floor he had brought from Europe, stood a dark figure harshly outlined by an uncurtained window.

Details seem vivid when I recall that meeting: my father watching me like a hawk for any possible future criticisms; Ann hovering over him and then suddenly asking my birth date and exact time of delivery. I told her, and, horror of horrors, I had been born under a sign in direct and violent conflict with hers! She judged that the two of us could never get along.

After a few meaningless words back and forth came a question from Ann asking why I did not seem to show some kind of warmth to her. I know now where my father had gotten that word he used so often—warmth. I stammered out something that told them both I knew about their long relationship, even before the marriage, and while my father and mother were still husband and wife. I gulped deep and told them how it had affected me and my mother and how I knew about everything else . . . about their child . . .

I remember seeing the sudden flush rising in their faces, then my eyes went down to the pattern of my argyle socks. Years later, it is that pattern, garish red and blue and white, I see most clearly when I think of that day which for me was filled with fear, dread, and anxiety.

We parted with the tight little words "So what . . . that's the way it is . . . it happens every day . . ."

I never felt comfortable in Ann's presence after that. Nor did she in mine. Later I returned several times to the house on Angelo Drive. What I recall is the almost complete withdrawal of Ann. As I drove up, parked my car in the circle, and walked toward the house, I could occasionally see her peering down through the curtained window of her bedroom on the second floor . . . yet she was usually "out" when I asked for her.

I did meet my half-sister, Barbara, a lovely child who looked very much like me at the same age, only with Ann's dark hair, while mine had been the light blond of my mother's. It was strange and not a little disturbing to see her playing in rooms that had once been mine and were now redecorated almost beyond recognition.

Of course, Ann and my father knew my feelings of discomfort and I could sense that there was a poorly concealed element of guilt on their part whenever we were together. The fact that he had kept this woman as his mistress, perhaps for four years while still married to my mother, and met with such terrible disapproval from all his family made it inevitable that, as I was later told by my father's lawyer, "Whenever he sees you, you remind him of Irma."

This feeling of guilt coupled with Ann's paranoid belief that my horoscope was in direct conflict with hers would eventually drive a wedge between my father and me.

Jack Jr. continued his university studies not knowing what his future with his father, or the studio that carried his family name, might be. Only Jack L. Warner could decide that—and he was too busy being a movie mogul.

Bill Schaefer had become his closest and most dedicated aide, following "J. L." around with a notepad, watching and recording whatever the Chief said as he went through his daily routine at the studio.

Other Voices

Bill Schaefer

I always planned to be at the studio no later than 9 A.M. so I could go through the mail and sift out anything that was important to tell J. L. about when he called in around 9:30. I would read him letters that might require action as well as the headlines of the *Hollywood Reporter* and *Daily Variety.* Sometimes he would respond to a letter over the phone and I'd have a typed answer for his signature when he got to the studio.

J. L. would arrive between 11:30 and noon and would first go through the mail, then talk to the production head, Hal Wallis, or Mel Obringer, who was head of the legal department, or Ed DePatie, the business manager, who, with Harry Warner around, had become a very important cog in the wheel. Then we waited around to go to lunch, biding our time until we got a report from DePatie that Harry wasn't in the executive dining room.

The Warner commissary was divided into three sections: the biggest part with hundreds of hardwood tables and a long serving counter was for movie crews and extras; a smaller area, called the Green Room, was set aside for the stars and directors; and a final secluded section, the executive dining room, was for the Warner brothers and the studio's executives. I was allowed to sit near J. L. with my notebook and record any decisions that were made at the table. Occasionally, featured players like Bette Davis, Edward G. Robinson, Paul Muni, James Cagney, and Errol Flynn were invited to sit at J. L.'s table, but there was only one star who needed no invitation—Al Jolson. He was J. L's buddy.

After lunch J. L., accompanied by Wallis, would go to the projection theater and look at the dailies from the previous day's filming. I'd sit next to J. L. in the theater and make notes of his comments. He had what was called a fader, a control that he turned to raise or lower the sound. When I first started to go with him to the dailies, I would sit behind him, because he wanted the film editor at his elbow. The editor had a clipboard with a light built into it, but the flash of light bothered J. L. and he made him turn it off. I started taking notes in the dark. Finally, one day, J. L. turned around, saw me scribbling away in the flickering light, and asked, "What the hell are you doing?"

"Well, I'm taking down the comments you said to the film editor."

He said, "You mean you can do that without a light?"

"Yes," I said.

"Can you read them back?"

By that time I had become adept at writing in the semidarkness so I said yes. From then on, I sat next to him while the editor sat to my right.

J. L. had the reputation of being one of the best film editors in the business. If he saw a scene that showed an actor walking along a street or crossing to another building, he'd say, "Why can't you just cut from the previous scene to the actor walking *into* the house?" He had a natural sense for keeping the story line moving. No one argued with his decisions about what to delete.

After the dailies were shown there would follow a discussion in the theater manager's office with the editor, producer, and director, and I would put these comments in the same chronological context as J. L.'s notes. (The next morning I would type seven copies, an original and six carbons, and pass them out to the various interested departments: sound; music; the film editor; the producer who worked with the director; Hal Wallis, whose office was across the hall from the Chief's; a copy for our files; and the original to J. L.)

Each day after watching the dailies J. L. would end up back in his office. If he didn't have an appointment, he would read the trade papers, then the newspapers, from which he occasionally got story ideas. The story department had a dozen readers and they would send one-page synopses taken from books, magazine articles, and the newspapers. It always amazed me that the readers could type a synopsis so that they'd end up with a page so full they couldn't get another line on the paper.

After J. L. read through this stuff, he would go into his private barber shop for a shave and a hot towel. The manicurist would fuss around with his nails and polish them. The barber was a funny guy and one day I heard him say to the manicurist, "Ida, stop playing with the Chief's fingers, you're giving him a hard-on." J. L., who didn't like anyone else to be the joke-maker, laughed at that one. After the barber had made a few snips at the boss's balding head, we'd go back to his office and I'd mix J. L. a Jack Daniels and water.

If he planned on running a picture in the evening, I'd order a steak and baked potato and some kind of vegetable from the Brown Derby and have one of our drivers pick it up. I'd make him a pot of coffee somewhere between 7:30 and 8 P.M., and he'd say, "Well, Bill, you can go home." Sometimes Jack Warner wouldn't get home until midnight.

Jack's biggest challenge as studio head wasn't selecting scripts or editing films, it was handling the stars he had under contract. His "children" were constantly whining, eventually to the point of rebelling against the studio's autocratic rule. Warners was considered one of the tougher studios to work for. Jack had a habit of patrolling the lot and seeing that no unnecessary lights were on and that everybody was at work. An actress who forgot to return so much as a muslin handkerchief to wardrobe would discover that its carefully catalogued value would come out of her next paycheck. Everyone learned to be careful with costumes, as each one would eventually drape some other player or extra in another film. But what raised the most serious growls from actors was typecasting.

James Cagney got so sick of being typed as a gangster, he expressed his dissatisfaction by growing a mustache and swearing at Jack in obscene Yiddish. Jack gave in and started casting him in detective and even Western parts—as the hero. The one actor who was satisfied with his contract was Paul Muni, but only because it gave him a veto on his roles, an option that he exercised more than once.

Jack's most difficult problem child was Bette Davis, who had won an Oscar for Best Actress in the 1935 movie *Dangerous,* a prize that she felt gave her the clout to demand better parts. She had appeared in a couple of dreadful movies in the interim between filming *Dangerous* and winning the Oscar, and, after being given another terrible script, decided it was time for a confrontation with the Chief.

"It's a piece of tripe!" Davis said, slamming the script on Jack's desk.

Jack, startled, picked it up, glanced at the title, *God's Country and the Woman,* and said, "Hey, what's wrong? You play a lumberjack in the north woods . . ."

"I asked for a *good* property. Something better than that last crap you forced me into."

Davis had just finished filming *Satan Met a Lady,* an early version of Dashiell Hammett's novel *The Maltese Falcon* in which the bejeweled falcon had been rewritten into a gem-filled ram's head. The inept script, which was played for laughs by producer Henry Blanke, bombed at the box office. Davis cringed when she watched the finished movie and considered it the low point in her career.

Seeing the bitterness in Davis's eyes, Jack added quickly: "We'll do it in Technicolor, costar you with George Brent, film it in Washington—you know, where they have all those pine trees . . ."

Davis stood up, glaring at Jack. "Get yourself another lumberjack!"

Putting on his best crooked-riverboat-gambler smile, Jack pulled his hole-card out of his sleeve. "Bette, baby, if you'll do this picture—just for me—I'll get you a great role when you return from the woods."

Davis looked at him warily. This she had heard before. "What role?"

Jack stepped around the desk as if he was going to give her a fatherly hug. "Uh, I'm optioning a great novel, one that hasn't hit the bookstores yet, it's called *Gone With the Wind.* The heroine's name is Scarlett . . . you were born to play her."

"Yeah, sure," Davis said. "I'll bet it's a pip."

Jack was not planning on producing *Gone With the Wind.* Doris Warner LeRoy had read the novel and told Harry she thought this was too good an opportunity to miss. Doris met with Jack, novel in hand, telling him to buy the book. His response was, "It's some kind of Civil War crap, cast of thousands, costumes . . . We can't put up that kind of cash."

When Doris countered with, "It's the chance to make one of the best movies of the decade," Jack replied, "We're not in the business of making big movies, we're in the business of making big money." He then told her: "Get the hell out of my office!"

Doris went ahead and offered Margaret Mitchell $50,000 for the screen rights. (Milton Sperling said at the time that it was like "buying the Arabian oil fields for only a million dollars.") Unfortunately, Doris's husband, Mervyn, told a group of directors at a party that night that his wife was buying the movie rights, and David O. Selznick overheard. He offered Mitchell $55,000 and, of course, the movie ran away with the 1939 Oscars, leaving Jack holding one statuette for a twenty-minute Technicolor short subject called *Sons of Liberty,* starring Claude Rains and directed by Michael Curtiz. It was a story of Haym Salomon, the Jewish immigrant from Poland whose financial assistance to the feeble American government during the Revolution enabled it to survive.

After the confrontation with Jack, Bette Davis fled to England and filed a suit in English court to break her contract with Warner Bros. Davis said, "I could no longer bear the scripts I was being given under my Warner contract. I was forced into some very definite action for the future of my career."

Jack was deeply concerned. If a studio star could dictate what picture she wanted to make, it would start an avalanche of similar suits from other stars.

When, during the trial, Davis called her contract a form of slavery, Sir Patrick Hastings, Warners' barrister, replied, "This slavery has a silver lining, because the slave was, to say the least, well remunerated." He noted that her seven-year contract (which would not expire until 1942) would reach the "sum of $2,400 a week." The barrister added, "If anybody wants to put me into perpetual servitude on the basis of that remuneration, I shall prepare to consider it."

Jack, when called to the stand, stated, "The studio has gone to great expense to build Davis into the star status that she now enjoys."

The next day, in the summation, Davis's barrister, Sir William Jowitt, said, "Miss Davis is a chattel in the hands of the producer. I suggest that the real essence of slavery is no less slavery because the bars are gilded." It was to no avail. The verdict decreed that the studio won a "three-year injunction, or the duration of the contract—whichever was shorter."

Davis returned to the studio and Jack welcomed his errant star back into the Warner fold with open arms: "Let's forget about all this, Bette, and get back to making movies." Yet the trial had made an impression on Jack, as he realized Bette Davis was serious about her work. He began to give her parts that were commensurate with her growing stature and cast her in such movies as *Jezebel,* for which she won the Best Actress Oscar for 1938.

Harry once gave the following advice to studio producers who were having difficulty with stars: "Take a death scene immediately. If you have the troublesome star killed in the picture, the writers can find a new ending and the star gives you no more trouble."

New actors were constantly being funneled into the studio system of star-making. One of those bright new faces was a radio announcer who broadcast games for the Chicago Cubs. The Cubs, owned by the Wrigley chewing gum family, did their spring training on Catalina Island just off the coast of Los Angeles. In 1935, the announcer, Dutch Reagan, got an assignment to cover the training session. In Hollywood he met an agent who called Max Arnow, a Warners casting director, and said, "I have another Robert Taylor sitting in my office."

Arnow answered, "God only made one Robert Taylor." Yet he decided to take a look at the young announcer.

Other Voices

Ronald Reagan

After appraising me like a slab of beef, Arnow—who did like my voice—gave me a screen test on June 1, 1937, reading a few lines from *The Philadelphia Story* with a Warner starlet. He told me to stick around Hollywood until Jack Warner had time to look at the screen test. Out of sheer ignorance, I did the right thing. I said, "The Cubs are leaving for Chicago tomorrow and I'm going with them to broadcast their games." They hate to hear anyone say no, so after I had been back in Chicago two days, I got a wire saying I had a seven-year contract with Warners—at a salary of $250 a week, three times as much as I made as an announcer!

The first thing they did was send me to the public relations department to give me a screen name. They sat around staring at me, thinking, What name does he *look* like? My mother had wanted to call me Donald, but her sister had a son first and used that name, so I was tagged with Ronald. I didn't care much for it. When I was a child I had a hairstyle that curled around my face with bangs, so my father started calling me the Dutchman. I preferred that name and was called Dutch Reagan in high school.

None of the publicity guys asked me what I thought I should be called, so I said, "At the radio station my name was well known as a sports announcer."

"Dutch Reagan? On a theater marquee? No way."

I sat there for a minute and said, "Well . . . Ronald?"

"Ronald . . . Ronald Reagan . . . hey, that's all right!"

I was one of the few actors who got to use their real names.

Warner Bros., like all the major studios, churned out two types of pictures: "A" movies, which featured their major stars, and low-budget "B" movies, which featured newcomers. I was assigned to the Warner Bros. B unit. My first film was *Love Is on the Air* (budgeted for $119,000), in which I played a radio announcer. I had had no training as a screen actor and I was terribly nervous when I showed up, but once the director said "Action!" I forgot all about the camera and crew and concentrated on saying the lines. To my surprise, when the director said "Cut" he was satisfied with the first take. I got to thinking: Maybe I can make it here.

I remember my first kiss in a picture was with lovely little June Travis, who had helped me with my screen test. When it came to the scene where finally boy gets girl, I *got* her, and *kissed* her. The director yelled, "Cut!" and came over to me and said, "You kissed."

I said, "Well, the script said to kiss her, so I kissed her."

"Yes, but you don't."

I said, "What are you talking about?"

What I didn't know was that the movies had designed a system of kissing. And that's why those love scenes in old movies are remembered, because you embraced but you barely touched lips, so that you didn't push the actress's face out of shape. That made the most beautiful love scenes. Today, a kissing scene in a movie looks like they are trying to eat their way through each other.

Anyway, after I did the kiss correctly, and the director yelled, "Cut! Print!" I turned to the crew and said, "I got news for ya, it was a helluva lot more fun at the high school picnic."

Ronald Reagan made thirteen "B" pictures in his first year and a half at Warners. He was finally invited up to Jack Warner's office, where he met both Harry and Jack. Reagan remembered that Harry never called him by the right name: "If he passed me on the lot, Harry would say, 'Good morning, Phil.'" (Phil Reagan had made a few musicals for Warners, the last being *Broadway Hostess,* completed two years before Ronald Reagan was signed by the studio.)

The "B" movies Reagan acted in were being made in black and white, mostly for financial reasons, and secondarily because color had not been perfected. In 1935 RKO had filmed *Becky Sharp,* the first feature in three-color Technicolor. One critic wrote that the actors looked like "boiled salmon dipped in mayonnaise." Even Harry had his reservations about the quality of Technicolor. He was quoted as saying, "Colors can't improve a picture. The best-painted woman is not necessarily the prettiest."

Harry had also decided to stop making gangster movies, and not because the studio was being criticized by journalists who blamed the films for crime. Harry had lashed back at that accusation by saying: "I would guarantee there are at least as many criminals who have never or seldom been to the movies as there are criminals who are moviegoers. I don't think there is any connection between moving pictures and crime. . . . The causes of crime are what they always have been, poverty, neglect, and bad environment."

In 1935 Warners made *G-Men,* starring James Cagney as its heroic lawman. Warners' ad copy stated: "Public Enemy Becomes Soldier of the Law." Headlines in trade papers proclaimed: Hollywood's most famous badman joins the G-Men and halts the march of crime. Will Hays praised the picture for placing "healthy and helpful emphasis on law enforcement."

Instead of gangsters, Harry continued his plan of bringing to the public the lives, struggles, achievements of great men like Louis Pasteur, Dr. Paul Ehrlich (who discovered the cure for syphilis), and Voltaire. Of the Pasteur film, Harry said, "When we made this picture, we didn't consider money. We know that it is difficult to get people to see and take an interest in that which educates them, but in spite of this we took up this film. We must show the people the good and noble things in life."

Jack was unconvinced that historical subjects were best for Warner Bros. When Mervyn LeRoy, noticing that dozens of fellow passengers on a cruise ship were reading Hervey Allen's blockbuster novel about the Napoleonic era, *Anthony Adverse,* cabled Jack PLEASE READ ANTHONY ADVERSE. WOULD MAKE A GREAT PICTURE, Jack wired back: READ IT. I CAN'T EVEN LIFT IT.

Jack figured educational features about historical characters to be box office suicide. The only thing Jack liked about *The Story of Louis Pasteur* was the critical acclaim and the Best Actor prize to Paul Muni. What Jack really wanted was an Academy Award for Best Picture, a goal that the studio had come close to but never achieved. He finally got it with *The Life of Emile Zola* (1937), once again starring Paul Muni. The movie was a classic that emerged as that *rara avis,* a film that was prestigious as well as popular. The *New York Times* called it "the finest historical film ever made and the greatest screen biography. It illustrates how injustice is combated by an idealist, how truth finally becomes victorious."

Harry, watching the film on opening night, stared intently at the screen as Zola says to his wife at the end of the story:

"I see it clearly now. The Cause and Effect—the roots and the tree . . . What matters the individual if the idea survives? It's not the swaggering militarist—they're puppets that dance as the strings are pulled. It's those others—those who would worthlessly plunge us into the bloody abyss of war to protect their power."

Harry had begun to see, as Hitler's military buildup got under way in Europe, that the world was about to hurl itself into destruction. Harry mused through Muni's voice that "thousands of children, sleeping under the roofs of Paris, Berlin, London, were doomed to die on some titanic battlefield, unless it could be prevented. The world must be conquered, not by force of arms, but by ideas that liberate. Then the world can build anew—build it for the humble and wretched."

Although Harry cherished his ideals for a better world, movie audiences were at first reluctant to spend money to see a film about a bearded Frenchman who spouted idealistic monologues. S. Charles Einfeld, the studio's publicity manager, had the star billed as *Mr.* Paul Muni. The columnists jumped on it right away, noting that even Sarah Bernhardt had not been billed as Mme. Sarah Bernhardt. It was utterly preposterous. But the journalists went on writing about the movie, devoting more space to it than they might have had "Mr. Muni" not been billed.

Exhibitors were also reluctant to endorse the movie at first, but soon rushed out to buy the film. Warners discovered there was a gold mine in the word "Mister."

Whereas Jack thought the movie would lose a bundle, it made $3 million, as well as capturing the Oscar. *Zola* also won for Jack the Legion of Honor award from the French Republic for Warners' service to the "glory of France, of science, of men of good will throughout the world and to the enduring art of the cinema." Jack began to build his trophy room.

The trophy room was next to the executive dining room and sported tables loaded with all the Academy Awards, plaques, certificates, prizes, statues, and whatever else had come to the studio. From *The Jazz Singer* to Looney Tunes, all the awards were there to astound, amaze, and impress visitors and VIPs who had just eaten in the dining room. The walls were covered with glass sheets protecting the closely packed photographs of practically everything and everybody with whom Jack or the studio had been involved.

Jack began to think that Harry was right about this "educate and enlighten" method of making movies. He even issued a statement that said, "Every worthwhile contribution to the advancement of motion pictures has been made over a howl of protest from the stand-patters, whose favorite refrain has been, 'You can't do that.' And when we hear that chorus now, we know we must be on the right track."

Certainly the annual profit/loss statement agreed. The studio listed a net profit for 1937 of almost $6 million. Jack felt that *he* was the reason for the profit margin. He said, "In 1937 the average Warner picture cost $400,000; the highest film cost was $1 million. This was done by never buying unnecessary stories, rarely making retakes, and always knocking temperament in the head." The Warners were known for getting more production money onto the screen than any other studio.

Some directors learned how to get more out of less. William Dieterle, who directed *Zola*, wanted 400 people milling in the street outside the courtroom during the trial scene. He was told he could have only 200. So he changed the lighting to make it a rainy day and got the prop department to break out 400 umbrellas, only 200 of which had people under them.

Jack was quoted as saying, "I'm running a factory. Making movies is like any other kind of factory production requiring discipline and order rather than temperament or talent."

Hal Wallis, who in the late '30s was at the peak of his reign as executive producer and dealt with the day-to-day details of production, used the motto "Volume is efficiency."

What kind of executives were Jack Warner and Hal Wallis? Agent Dick Dorso, who worked in Hollywood during this period, formed vivid impressions of them.

Other Voices

Dick Dorso

They knew nothing about being executives. They knew nothing about handling people. They ruled by fear. This was characteristic of everyone in their position. They used intimidation. The standard procedure was: "Jack, this is a very hot property, and you have till five this afternoon to make a decision because Metro is very interested in it." Jack would respond in kind, so there was warfare all the time.

That's why they'd yell. They had to intimidate you before you intimidated them. Nobody said, "Sit down, and tell me what the problem is and let's deal with it." They wanted to establish the edge right away, because they knew they couldn't handle it.

All the big studio execs were all the same—they yelled at you. So much of it was visceral. If you had a good stomach, good seat of the pants, and reacted properly to actors or properties, you'd be successful.

Jack was scared stiff all the time. He did a very peculiar thing. He went for jokes. When he was frightened, he told jokes. I watched him very carefully. The most accurate way to describe him was being a fool. He acted foolish all the time, because he was in a constant state of wild anxiety. I think he never had a true sense of self-appreciation—a real hard core of self. And the more successful he became, the more anxious he became. Because he really wasn't that person—he wasn't entitled to that success.

In person Jack Warner was overbearing, he was arrogant, he was ignorant, he was coarse, he was brash.

Yes, but he did make good pictures. And he made them economically.

Bryan Foy, who was known as the Keeper of the "B"s, bragged that he made one picture eleven times, using the same plot, just slightly changing the next version. He also cast the same actors over and over again in these movies, a technique that unsettled Ronald Reagan, who, although happy being a "B" movie star, longed for a juicy role in a major production. Three years after first signing with Warner Bros., he got his chance to play football hero George Gipp.

Other Voices

Ronald Reagan

When I was a kid the "Gipper," who had played for Notre Dame, had been a hero of mine. I worshiped football and had played it for eight years in high school and college. With my sports background, I began to think that a movie on the life of Knute Rockne, the famous coach of Notre Dame, would make a great picture. Jimmy Cagney, Pat O'Brien, and myself would sit together in the dining room at the Warners commissary and I kept telling them that I had a great movie idea.

One day I picked up *Daily Variety* and read where Warner Bros. had bought the life story of Rockne. I told Cagney and O'Brien and they said, "You talk too much. You've been all over the lot blabbing your story."

"I don't want to sell a story," I said. "All I want is to have the picture made and play the Gipper."

Cagney and O'Brien looked at each other, then me. "Well, you'd better start talking to the producer, because several actors have been tested for the part already."

I stormed into the producer's office and said, "Listen, this is the greatest football hero that ever lived and I want to play the part."

He looked me up and down and then said I was too small for the part.

"What you mean is I should weigh two hundred and twenty pounds," I said. "Would it interest you to know that I weigh five pounds *more* than the Gipper did when he played?" That wasn't getting me anyplace. Then I remembered what a cameraman had told me: "The only thing the fellows in the front office know is what they see on film." I rushed home, grabbed my college football picture—there I was in full uniform—and hurried back to the studio. I shoved the picture in the face of the producer. "Here, this is me playing football."

He looked at it and said, "May I keep this for a while?"

I wasn't home fifteen minutes when the phone rang telling me to be at the studio at eight in the morning for a screen test. Pat O'Brien volunteered to do the test with me and he ended up playing Rockne and I got the part of the Gipper.

I wasn't in the picture a lot, but it had a very emotional scene. Gipp is in a hospital bed, dying, and he says to Rockne, "Someday when things are tough and the breaks are going against the boys, ask them to go in there and win one for the Gipper." I watched the preview and heard the audience sniffling in their handkerchiefs. Was this the acting breakthrough I had been wishing for?

When I arrived home, the phone was ringing, telling me to report to the studio early the next morning. "You've been cast as George Armstrong Custer in *Santa Fe Trail.*" Errol Flynn was the star, but I was second lead in another "A" picture!

When I was fitted with the uniforms for the movie, I noticed another rack of uniforms with the name "Dennis Morgan—Lt. Custer" tagged to each of them. A wardrobe man came in and took them off the rack and tossed them in the corner and replaced them with the gold-braided uniforms made for me. I thought, "That can happen to me someday."

Errol Flynn—it was the funniest thing—he had a complex about his acting and was constantly worried that someone would steal a scene from him. Character actors didn't bother him, but the leading-man type—like myself—did.

There was a scene in *Santa Fe Trail* where a newly commissioned group of officers from West Point were gathered around a campfire while an Indian woman made designs in the sand telling us what our fortunes in war would be. I saw Flynn go up to the Hungarian director, Mike Curtiz, and heard him whisper, "Same uniform." At the first rehearsal Flynn and I had stood side by side, but now Curtiz called out in his heavy accent, "I must line you up all again."

He took me and placed me behind two of the tallest fellows. I was standing on a downslope so the only part of me that showed over their shoulders was my forehead. As I was standing there waiting for the shot to begin, I realized I was in loose gravel so I started raking it into a pile. By the time Curtiz said, "Action," I had a gopher mountain built. I stepped on it—my head was now way above the other actors' shoulders—and said my one line. Yes, you always had to watch out for Errol Flynn.

Michael Curtiz was not one of Flynn's favorite directors (the Hungarian liked to annoy the actor by calling him "Earl Flint"), but he was one of the top directors the studio hired to meet the production demands of the '30s.

Curtiz's command of English never did improve. He didn't just try to speak it, he attacked it, and in the process created a series of "Curtizisms" that kept everyone on the lot, including Jack Warner, laughing.

Curtiz once described Bette Davis as "the flea in the ointment."

Working with a child actor who was having difficulty doing his lines, Curtiz prompted, "Ad-lib, ad-lib. Say, 'Jesus Christ, you ruined my castle.'"

To a group of extras he gave this confusing command: "Separate together in a bunch."

Perhaps his most famous saying occurred when he cued the entrance of a herd of riderless horses for a scene in *The Charge of the Light Brigade* (starring Errol Flynn) by saying, "Okay, bring on the empty horses!"

Curtiz also directed the James Cagney classic *Yankee Doodle Dandy,* which he called the "pinochle of my career." Showing Cagney how to do a scene, Curtiz said, "Don't do it the way I showed you, do it the way I mean."

Other Voices

Milton Sperling

Michael Curtiz was a wonderfully flamboyant character. He could do any type of picture, a musical, comedy, drama, costume . . . He was like an actor, took on that personality. Whatever script he was given, he did. Of course, directors were given scripts only a few days before shooting, delivered to their doors like morning papers, with a note that said, "Start shooting Monday." Nobody sat down with a director and said, "How do you see it?" It was always, "Here's your cast, get moving." Not a line was changed without permission from upstairs. That was the studio system. And it worked.

In 1936, when Betty Warner turned sixteen, Doris decided to play matchmaker for her younger sister. She lined up actor Robert Taylor, and Milton Sperling, who had left Warner Bros. and had wound up working at Twentieth Century Fox as Darryl Zanuck's assistant.

"Betty was very shy, very shy and withdrawn," Sperling later remembered. "She spoke in a very soft voice and I could scarcely hear her. But she was pretty and intelligent."

Other Voices

Betty Warner

Milton was there when we arrived at the Pasadena/Glendale train station. He was the first "older man" that I dated. (He was twenty-four.) I thought he was worldly, funny, and talented. Milton and I went together off and on—he had another girl—for three years. When I turned seventeen we decided to get engaged. Milton went to the ranch to talk to my father and Harry took him out for a walk.

Before they left the house, Dad picked up a rifle and tucked it under his arm. They had a nice talk and both agreed that the marriage would work. Then suddenly, as they were passing a pond, Dad lifted the rifle and took a potshot at a frog. The strange thing was that I had never seen my father carry, let alone shoot, a gun. I guess he used it for emphasis.

I was going off to school at Mills College in Oakland and my father took the overnight train with me. We shared a drawing room and stayed up all night talking about my getting married at such a young age. He didn't disapprove of Milton, but thought I was being hasty. He never said, "Don't marry Milton," but he did keep repeating, "You're only seventeen." It was sweet and it was a loving experience I had with him on the train trip. His gentleness with me surprised me; I believed he and Mother liked Milton. That made it much easier for me, as Doris had come up to me before I left and said that I was too young to marry and that Milton was not right for me. She suggested I date the glamorous movie stars she knew and enter into her social scene. I had my own social scene and hers was not appealing.

In June 1939, Louella Parsons reported the following news:

> One of Hollywood's sweetest romances yesterday reached the stage at which rumors of the engagement were confirmed.
>
> Betty Warner, 19-year-old daughter of Mr. and Mrs. Harry Warner, is to wed Milton Sperling, 20th Century Fox writer. Sperling is one of the young writers rapidly making a name for himself.

Betty and Milton were married Thursday, July 13, 1939, in one of Hollywood's most beautiful weddings. The social columns reported that the bride wore "a white mousseline de soie wedding gown and a veil of white old lace. She carried a corsage of white orchids and lilies of the valley. Harry Warner gave his daughter away and Mrs. Mervyn LeRoy acted as the Matron of Honor. Lita Warner, the bride's younger step-sister, was the Maid of Honor."

I suppose I can say that this marriage meant a great deal to me, as I would eventually become Betty and Milton Sperling's third daughter.

Other Voices

Betty Warner

We got on a train for the honeymoon along with some of the family, including Jack's wife, Ann. Milton told me later that Ann snuggled up to him at the bar and said, "You know, you made a terrible mistake."

He was shocked and started to pull away from her.

"You shouldn't have married Betty, you should have married *me.*"

I couldn't believe it when Milton told me: She was actually flirting with my husband on our honeymoon. I was soon to find out that she flirted with every man that showed up on the scene.

In 1941 a liaison was reported between Ann and a young Warner contract player by the name of Eddie Albert. Jack discovered the affair and was outraged. Determined to destroy Albert's career, Jack kept him under contract but forbade any of the studio's producers or directors to use him. Soon after, Albert entered military service.

Early in 1939, several Jewish inmates of Nazi concentration camps escaped and brought back stories of Nazi atrocities. Few Americans believed them to be true. Few but Harry. During his European travels, he had seen the evidence of Hitler's brutal anti-Semitism.

THE SCENE:

Berlin, 1940.

Late at night—a glass-windowed office door with the Warner Bros. logo on it. Smeared over the glass in brown ugly paint is the single word JUDEN.

Suddenly a mallet shatters the glass, revealing a man, Joe Kaufman, inside the office, working at a typewriter. He looks up in horror.

Four German thugs grab Kaufman and ram his face into the typewriter. They destroy every object in the room, using Kaufman as a battering ram, then drag him through the broken glass and dump him in the garbage can in the alley, leaving him to die.

The scene fades out.

No, this wasn't a sequence filmed at the studio, it was an actual event, described to Harry in a telegram. After reading it, he immediately gave orders to close every Warner office in Germany. Then he called Jack. "We're going ahead with *Confessions of a Nazi Spy.*"

Confessions of a Nazi Spy was in many ways similar to the first major film Warners made before World War I, *My Four Years in Germany.* At first, Jack and Harry were wary of making another war exposé film. The murder of Joe Kaufman changed their minds.

The story line was based on the revelations of a former FBI agent, Leon G. Turrou, whose assignment had been to ferret out German subversive activities in America. Edward G. Robinson was cast as the head of this team of G-men who uncover a network of Nazi agents. The result was a well-documented indictment of the Nazis and the condemnation of the officially "friendly" nation of Germany.

At the 1939 premiere of *Confessions of a Nazi Spy* there were almost as many policemen and special agents in the audience as there were movie fans. Before the film was released, Harry and Jack had received threats that they would be "executed" if the movie made it to the screen. They were also told that the theater where the film was to be shown would be blown up.

Harry fought back. At a St. Patrick's Day dinner in 1939 he gave an impassioned speech decrying anyone "glorifying a dictatorship." In a clear, firm voice he said: "I wonder what is happening when I hear of American radicals wearing a foreign emblem on their sleeves and drilling and marching in a foreign style—with the goose-step!"

He paused before adding:

> And so our producing company is making right now a picture revealing the astonishing length to which Nazi spies have gone in America. We are making this—and we will make more like it, no doubt, when the occasion arises. We have disregarded, and we will continue to disregard, threats and pleas intended to dissuade us from our purpose. We have defied, and we will continue to defy, any elements that may try to turn us from our loyal and sincere purpose of serving America.

The personal threats against the Warner brothers were never carried out, but Harry's troubles were not over. *Confessions of a Nazi Spy,* as the first major motion picture to deal with Nazism and to warn the world of its threat to democracy, brought down the wrath of Congress. Ironically, the studio was accused by a Senate investigating committee of "creating hysteria among the American public and inciting them to war."

"We nearly hit the ceiling when Harry was summoned by the committee," one of his associates said. "Not that we were afraid of the investigation. We knew Harry would come through with flying colors. What really worried us was what he would say when he discovered the shortage of accommodations in Washington. And sure enough, at the end of the first day in that national capital he began toying with the idea of building a penthouse on the top of a Warner theater there."

To Harry, the summons was not a joking matter, and as a witness he defended the film as factual, testifying that it was based on reliable sources and that the events depicted had been documented by the FBI. Harry stated that he knew America was going to have to go to war against Germany and voiced his bitter anti-Nazi feelings.

Although several members of the committee came to agree with Harry's views, one senator, a rabid isolationist, could not forgive the Warners for making the anti-Nazi film. He kept repeating, "I don't mean to question what you say, Mr. Warner, but I must remind you that your testimony conflicts with the president of Metro-Goldwyn-Mayer. *They* don't seem to be operating the way you do. Why is that? Aren't you in the same business?"

"We are, Senator," Harry said, calmly wiping his glasses, "but there is one difference between us. The gentleman you refer to *inherited* his business. I built mine up from nothing."

Harry then summed up his life in one sentence: "Gentlemen, I am of Hollywood, and yet I have managed to remain married to the same woman for the last thirty years."

The gathered senators chuckled at this, and the "subversive propaganda" allegations were dropped.

But Harry was still concerned about Hitler's threat to Jews, and spent huge sums of money during the latter part of 1939 getting Warner relatives and employees out of Germany. The war hysteria continued as Hitler invaded Poland, then attacked France's Maginot Line. Paris fell. Dunkirk . . .

Dismayed about events in Europe and concerned about how little material assistance the United States was giving the British (Harry didn't feel President Roosevelt's new "cash and carry" system of providing help to the Allies was enough), Harry and Jack sent a blunt and passionate telegram on May 20, 1940, to the White House in which they stated, in part:

> The least we can do, it seems, is to supply all the material help we can command, short of actual troops to help destroy the barbarian gang that is overrunning the world. . . . We cannot stand by and watch while others die for the civilization which is ours as much as theirs. . . . We cannot contentedly sit still out here and do nothing while the whole world echoes with the march of savages to destroy everything we hold dear. We would rather die in an effort to be helpful than live to see barbarianism triumph. Will you please tell us, Mr. President, what you think we should do?

President Roosevelt did not respond.

In private, Harry mused, "If these many dogs were being killed, then we, the United States, would have come to their defense."

As the war shadows in Europe continued to darken, Betty Sperling gave birth to a daughter, Susan. It was Harry and Rea's third grandchild. (Doris had Warner Lewis LeRoy in 1936 and Linda LeRoy in 1939.) On the day the new baby was born—December 4, 1941—Harry passed out cigars to Warner employees. Before their smoke had cleared from the studio's sound stages, the United States had been thrust into World War II.

Part IV

1940s

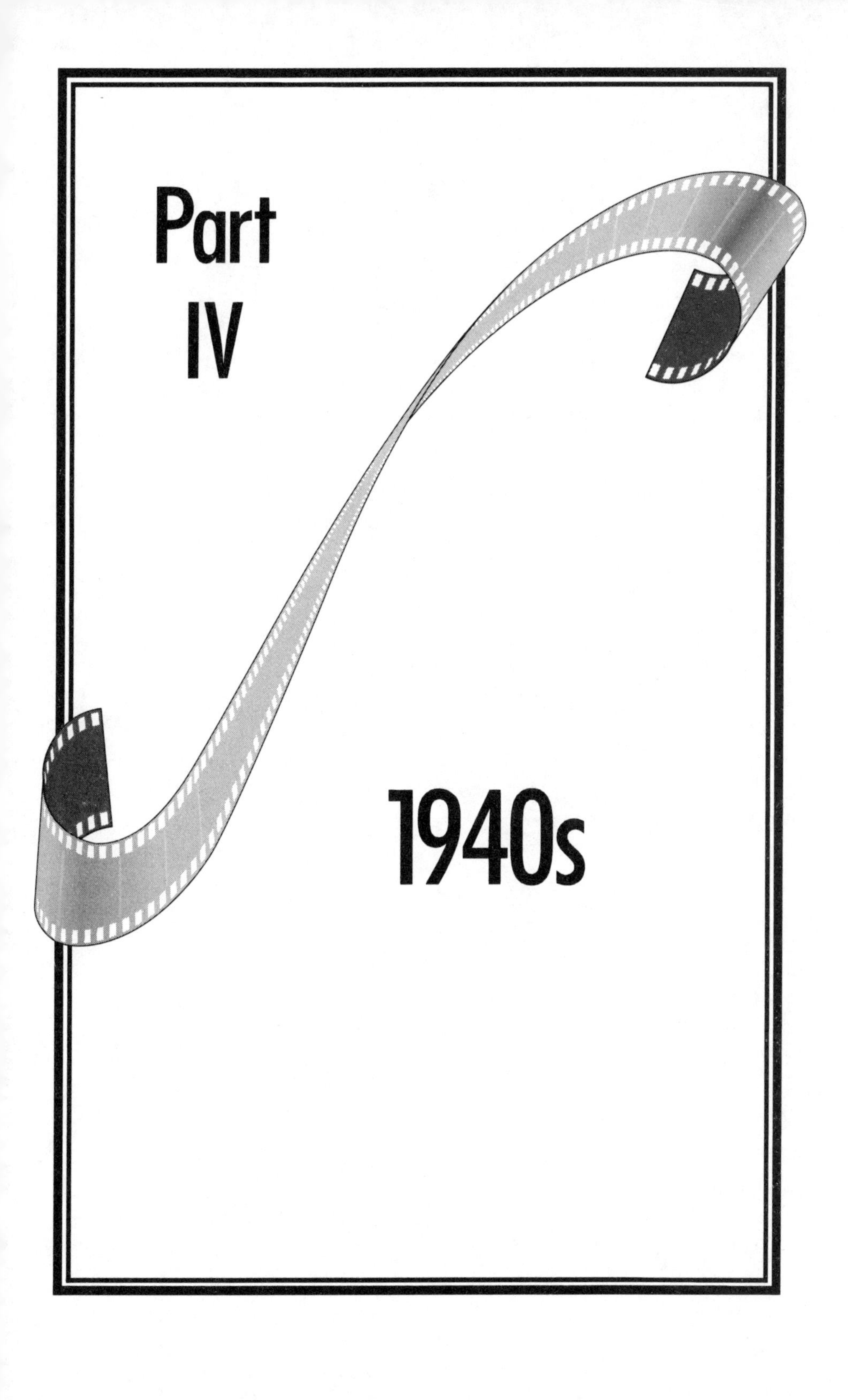

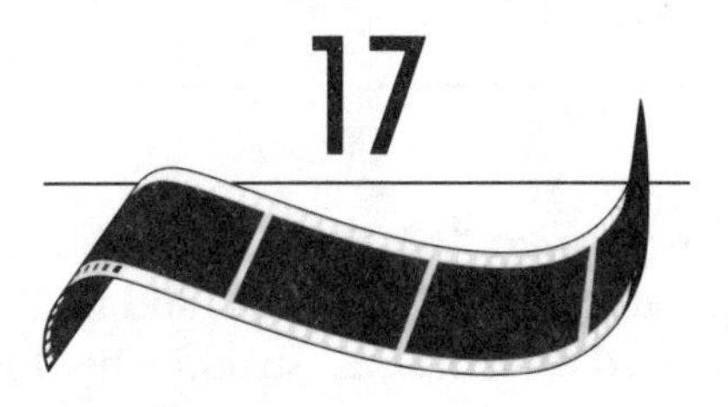

Warners Goes to War

"Shut up!" Louis B. Mayer said as filmdom's funniest mute, Harpo Marx, was about to tee off. Al Jolson and Jack Warner leaned on their drivers and laughed. Both wore plus-fours and V-neck sweaters in the mild California winter, and their golf caps were pulled tightly to their eyebrows to ward off the brilliant sun.

Harpo waggled his driver and wiggled his rear. Whack! He hooked his shot into the trees. Harpo turned to L.B. Mayer and mouthed the words, "Fuck you."

Still laughing, Jack lined up his tee shot. As he started his back swing, two men burst out of the pro shop shouting:

"It's coming in over the radio!"

"They've just bombed Pearl Harbor!"

"Oh, my God," Harpo said aloud, his face stricken. With Mayer and Jolson, he started walking toward the men who were now running toward the eighteenth fairway, calling out the news.

Jack continued his swing, whacking the ball in a straight line toward the three-par green. It plopped onto the surface . . . and rolled into the cup. "I sank it!" Jack screamed. "A hole-in-one!" He turned around to stare at empty air. "My greatest shot of the year, and no one saw it!"

He walked toward the others, shaking his head. "No one'll believe me."

Still mad, he said, "Pearl Harbor. Where the hell is that?"

With the outbreak of war, most studio bosses panicked and declared a period of austerity. But Harry jumped into the war effort with both feet by offering the major commodity he had available: films. He put the great resources of the studio at the disposal of the country.

Warners had gone to war.

In a speech to studio department heads, Harry stated:

> Our company is about to start the largest program of pictures for the government that has ever been undertaken to be made by any company in the industry. We have agreed to make from four to five hundred reels of training pictures in the coming year. We are also going to make Irving Berlin's *This Is the Army.* In making these pictures, we want them made at absolute cost. When I say absolute cost, I mean exactly that. I don't want to make a single dollar of profit out of these pictures.

At least one-fourth of the male employees of the studio joined the military service. Jack Warner Jr., twenty-five when the U.S declared war, was the only eligible male from the Warner family, and he had joined the Marine Corps Reserve five years earlier.

Other Voices

Jack Warner Jr.

During my first two semesters at the University of Southern California in 1935, I worked as an assistant director at the studios, as I was deeply interested in the business, wanting more than anything to be part of the company my father and uncles had built. One picture I worked on affected me deeply. It was a musical called *The Singing Marine* and starred Dick Powell. When recruiters from the U.S. Marine Corps Reserve came to the campus, I passed the physical, raised my right hand, and suddenly was PFC Warner, USMCR, and on my way that summer to train in San Diego. My father seemed sort of proud and pleased as he told people about his crazy son.

I spent two summers, more like vacations than training camp, and finally came the day I graduated from college with my degree and a second lieutenant's commission. I invited my father and Ann to the ceremonies as well as my mother and her husband-to-be, Al Rogell, a well-known movie director. To avoid complications, I got seats on one side of the L.A. Coliseum for my father and another pair on the opposite side for my mother. I was quite pleased to have both of my parents present, my mother with Al, my father with Abdul Maljan, his masseur.

It was 1938 and I took off with a college friend, Alan Gordon. We loaded our car aboard the *Normandie* and sailed to Europe. We got as far as Copenhagen, where we got stuck while the world teetered on the brink. Hitler was just about to send his armies across the Sudeten and the American consul advised us to head home fast! As we were homeward bound in mid-Atlantic, the meeting in Munich preserved "peace in our time."

Back to the studio, where for a while I worked in the production office. Then my father suggested I go to the home office in New York and I agreed, as it was time to learn all I could about distribution and exhibition of film. During the next couple of years I visited nearly every film exchange and Warner circuit theater in the United States and Canada. Later on I figured my father had done me a good turn sort of by accident, as his real objective was not my education but getting me physically far away. The closer I was, the more I reminded him of things he preferred to forget.

My mother came east to tell me she had set the date for her wedding to Al Rogell. I had gotten to know Al and felt happy for them both. We all met in Las Vegas, where they were married.

On December 7, 1941, the radio told us all of the attack on Pearl Harbor. I had resigned my Marine Corps commission several years before to accept a commission in the Army Signal Corps to assist with a program to recruit and train experts in film production in the event of mobilization, so I got right on the phone and called Major Nathan Levinson, the Western Electric sound technician who was now head of the Army Reserve unit, to ask where and when to report. His answer was the classic Hollywood "Don't call us, we'll call you."

My next call on December 7 was to my father. I drove up to Angelo Drive and the two of us spent two hours walking and talking on the estate grounds. Now and then he shot a worried look up at the house and the curtained window on the second floor. When finally we parted, the radio was blaring the latest details of the horror in Hawaii and the national emergency we all would live with for many years.

Jack Jr. did get his call from Major Levinson and a few months later he was ordered to set up and train signal photographic companies. Jack Warner Jr. (now an Army second lieutenant) had the kind of experience the Army Signal Corps was looking for in an effort to form a motion picture unit that would make training, propaganda, intelligence, and any other kind of film requiring skilled production personnel. The night before he left for Camp Robinson, Arkansas, the studio employees' club held its annual dinner dance at the Biltmore Bowl in downtown Los Angeles.

Other Voices

Jack Warner Jr.

I was there in uniform saying good-bye to old friends. Hal Wallis and I were standing with a group of people when my father and Ann came down the stairs and toward us. She clung tightly to my father, shot a black look at me, then shouted, "Stay away from my house! Stay away from my husband! Stay away from me!"

I was deeply shocked, as were all the people around us. Everybody stopped talking. My father went crimson and looked stricken. Ann yanked her arm loose and walked away. I felt deeply sorry for my father, who just stood there all alone in a crowded room, but I did not know what I could possibly say. Finally he shrugged his shoulders, tried a feeble joke, and slowly walked off in the direction Ann had taken. I did not see them again that evening, and when the train left the next morning for Arkansas he was not there to say good-bye.

My father, Milton Sperling, who still worked for Zanuck at Twentieth Century Fox and had seen his boss join the Army Signal Corps, enlisted in the U.S. Marines. Betty recalled that Harry had said to Milton as early as 1940: "You'd better join the Army, so you won't get drafted into a dangerous job." Betty, who was looking forward to having children, said no to that idea.

"When the war broke out, Milton decided he wanted to go to Europe and fight the Germans," my mother remembered. "He was sent to intelligence school and was commissioned a second lieutenant in the Marine Corps. To his dismay, he was shipped to the Far East, filming documentaries about Marine landings on Japanese-held islands."

General Hap Arnold, commander of the Army Air Force, lost no time in enlisting the aid of Jack Warner, who was summoned to Arnold's office in Washington and told, "Jack, you're a great filmmaker. How would you like to make some special training films for my new Air Force Signal Corps unit?"

Jack looked at the spray of silver stars on the general's shoulder and asked, "Uh, you know, Darryl Zanuck has been made a colonel in the Army, and, well, I wouldn't mind being a general."

Arnold grinned sardonically and agreed to commission Jack as a lieutenant colonel, one rank below Zanuck. A bit disgruntled, Jack returned to Hollywood, where he was fitted for a uniform by the studio wardrobe department. Admiring himself in the mirror, he had his secretary, Bill Schaefer, read officer etiquette from an Army manual.

Leaving the office, Jack walked the studio lot when two newly commissioned second lieutenants approached him. Seeing the silver clusters on Jack's shoulders, the officers raised their arms in a pair of snappy salutes. Jack, not knowing what to do, waved happily at them, saying, "Hi-ya, boys."

Other Voices

Milton Sperling

Before I left for Marine Corps training, I got a funny telephone call from Jack one day shortly after he was appointed light colonel.

"They want to send me to camp!" Jack cried. "Some fuckin' camp in Missouri."

So I said, "Well, you're a colonel; you should go."

"Hell, I've only been a colonel for five minutes."

"Maybe you'll enjoy camp."

"Shit, then they'll want me to go overseas. Who's going to mind the goddamn studio?"

I had just received orders to the Marine Corps intelligence school, so I didn't say anything.

"Look, Sperling, I gotta see you right away. You gotta help me."

I was still new to the family, but I went over to his mansion and Ann's there and she's hysterical. "They're going to send Jack overseas—he'll get killed!"

Jack sulked, "Why do I have to go to some camp?"

"Listen, Jack, when you put that uniform on, you made an obligation . . ."

Jack shook his head. "Hap Arnold never mentioned a word about this."

So I said, "Resign the commission. You're overage. You don't have to go to war."

He flicked an invisible speck off his uniform. "Well, jeez, what would it look like if I resigned the commission? I've only been a colonel for ten minutes."

Two hours earlier he had been a colonel for five minutes. At this rate he'd never be promoted to general. I said, "I'll tell you what to do. Get on the phone with Hap Arnold. He probably doesn't know this has happened. It's the bureaucracy: They saw your name, punched a button someplace, and out came 'Camp Missouri.'"

He got on the phone and told the general the story. I could hear the roar of laughter on the other end of the line as Arnold said, "Hell, I'd love to see you in camp when you report for duty, Jack. But you'd better not go. It would demoralize the whole damn Army. Go to Washington from time to time and advise me where the broads are."

Arnold called Jack again, to tell him the Army Air Corps needed pilots, college graduates. Jack quickly put together a training film, *Winning Your Wings.* Arnold credited the film with helping to recruit 100,000 pilots. One of them was Jimmy Stewart, who narrated the film. Stewart, a longtime private pilot, went on to win honors as a B-17 plane commander bombing German targets.

Warners began making a wide variety of short one- and two-reel films, some for public consumption, others for training newly enlisted men. Harry liked to say they were films that "made Americans proud to be Americans." They had titles like *March On, Marines; The Tanks Are Coming; Meet the Fleet; Wings of Steel,* and *Here Comes the Cavalry.* Harry said, "I don't want us to be known as the studio that made the best musical comedies during the war."

In fact, a *New York Times* review commended the studio for its "combining good citizenship with good picture making," a slogan that was subsequently emblazoned on a billboard in the traffic signal island across from the studio lot.

As the fighting with the Axis powers intensified, Harry became a frequent spokesman for the movie industry, expressing his feelings about the "violently changing world" by saying:

> The glamour era has vanished. Glamour belonged to a decade which brought us this war. We ought to leave it far behind and forget it as quickly as possible.
>
> The days of swimming pools and fast cars and bathing-suited cuties chasing rich young men into honorable marriage have been wrenched away from us and I, for one, am grateful for it. The cuties, if any, are in overalls these days, standing over airplane parts, or welding and riveting. Or they are living quietly in a small room existing on a soldier's pay and thinking night and day in terms of how to win and win quickly.
>
> Another decade of glamour and frolic may come back to us. If it does, I hope this monstrous war will have taught man to leaven his fun with some thought.

Jack, who still loved to throw lavish parties at his Beverly Hills mansion, could only hold his head in his hands and shudder with an embarrassed groan.

Harry reiterated his wartime philosophy to the press:

> In the hands of motion picture makers lies a gigantic obligation, honorable but frightening. We must have the courage and the wisdom to

> make pictures that are forthright, revealing and entertaining, pertinent to the hour and the unpredictable future. . . .
>
> Several years ago, against all well-meant advice from high places, we attacked the Nazi Swastika in *Confessions of a Nazi Spy.* We made the now renowned series of Warner Bros. patriotic shorts, telling graphically the lives of great Americans. If arms were taken up, we wanted to show the youth of America that they had something worth fighting for. In the same spirit we put into production *Captain of the Clouds* with James Cagney, a story about the Royal Canadian Air Force, and did the same for the RAF in *Desperate Journey* with Errol Flynn and Ronald Reagan. *Across the Pacific* dealt with Japanese treachery before Pearl Harbor, and is one of the first films to depict our Oriental enemies as cold, calculating and ruthless, completely efficient and not at all the nearsighted, weak and stupid enemy who can be "knocked off the map in six weeks."
>
> We also made *Sergeant York,* a story of a simple mountaineer, opposed to killing on religious grounds, who eventually knows he must fight, kill and even die, for freedom.

When *Sergeant York* was released in 1941, Americans were still undecided about their involvement in the European war. The true story of Alvin York's single-handed capture of 132 German soldiers in the Argonne in 1918 was an inspiration to young American men and many of them began answering the call to arms.

What Harry didn't say was that Sergeant York's inner conflict, "to kill or not to kill," was the invention of a young Warner studio writer by the name of Howard Koch—who *was* opposed to killing.

Koch had first come to the attention of Warners on October 30, 1938, when his radio script for Orson Welles's "War of the Worlds" caused thousands of people to believe that an interplanetary battle between Earth and Mars had begun.

Koch was hired by Warners for the princely sum of $300 a week—and discovered he didn't have anything to do. His secretary told him, "Don't worry, they always do that to start. They want you to get familiar with the place." So Koch hung around the Writers' Roost, the building where the studio writers were housed, and waited.

Occasionally Koch would catch sight of William Faulkner, who was there on assignment. Faulkner was a quiet man, not social. He had an office in the Writers' Roost where he worked for a while, until he asked Jack Warner if he could work at home. Jack, wanting to please the great novelist, said, "Sure." Weeks later, after no one had heard from Faulkner, the studio discovered that he had indeed gone home—to Mississippi.

Koch finally got his chance when studio producer Henry Blanke told him to script a seafaring adventure for Errol Flynn titled *The Sea Hawk*. It would be the first film to show off the studio's enormous new sound stage, where two full-scale ships had been built. Koch, an untried writer, was awed when Blanke told him the production cost was set at $1.7 million and was to be a topnotch film directed by Michael Curtiz. The movie was a huge success.

The next film Koch worked on was *Sergeant York*. Alvin York liked Warners' treatment of his life story so much, he made an appearance at the premiere. *Sergeant York* was nominated for 1941 Best Picture, but the Oscar went to Twentieth Century Fox's *How Green Was My Valley*. Gary Cooper did win the Best Actor award for *York*.

Not all of Warners' movies were of the caliber of *Sergeant York*. *You're in the Army Now,* the studio's last release of 1941, was an embarrassing slapstick comedy starring knockabouts Jimmy Durante and Phil Silvers. It gained some notoriety as having the screen's longest kiss, one that lasted 185 seconds.

With *Sergeant York* behind him, Koch was assigned to collaborate with two other writers, the Epstein brothers, on a film that made cinema history: *Casablanca*.

Casablanca started out quietly enough when Jack Warner bought the rights to an unproduced play by Murray Burnett and Joan Alison titled *Everybody Comes to Rick's*. Hal Wallis, the film's producer, called it a potboiler, but could see good cinematic values in the story.

On January 5, 1942, Warners announced in the *Hollywood Reporter* that Ann Sheridan, Ronald Reagan, and Dennis Morgan were to star in *Casablanca*. Morgan was to play Rick, Sheridan the part of Ilsa. For the idealistic antifascist leader, Laszlo, Reagan was penciled in on the casting chart. (Reagan does not remember being told he was to play the part.) On April 2 Jack sent a note to Wallis: "What do you think of using [George] Raft in *Casablanca*? He knows we are going to make this and is starting to campaign for it." Wallis replied, "Bogart is ideal for it, and it is being written for him, and I think we should forget Raft on this property." Casting the film became a merry-go-round ride, but this time around the studio caught the brass ring—Humphrey Bogart, Ingrid Bergman, Paul Henreid, Claude Rains, and a superb supporting cast. It would be difficult today to imagine any other actors in those roles.

All of the casting changes had a devastating effect on the script being written by Julius and Philip Epstein. The Warner time clock was ticking away, and another studio writer was assigned to the script—Howard Koch.

"*Casablanca* almost put itself together," Koch remembered.

> All of us who worked on it were on a conveyor belt. The Epstein brothers had done very solid work on the script, but when they were called to Washington on a war service project, I was told to finish it. It just sort of grew. Perhaps that was part of its vitality—one scene demanded another. With the production deadline creeping up and with director Mike Curtiz imploring me for more pages than could be mimeographed, I felt under more pressure than on any other previous assignment. The cast knew there was no finished script, as they were getting it in pieces.
>
> One day Ingrid Bergman came up to me and asked, "How am I supposed to play this love scene? I don't know who I'm going to end up with."

Paul Henreid in his autobiography, *Ladies Man* (St. Martin's Press, 1984, p. 121), wrote:

> Shooting started before there was a finished screenplay, in fact before there were much more than opening scenes. Since we were shooting in chronological order, we were able to do it. The pages usually came out of the writers' typewriters the day before they were to be shot. As a result, there were often different versions of scenes. Nobody knew what would come next, including Mike Curtiz, the director, who did a beautiful job, especially under the wild circumstances. Mike would apologize to us, saying, "Excuse me, I don't know any more than you do about what follows or goes ahead. I'm directing each scene as best I know how!"

Jack Warner asked Curtiz how the film was going and Curtiz answered, "Well, Jock, the scenario isn't the exact truth, but we haff the facts to prove it."

Henreid reported that one day Curtiz asked a prop man for a "poodle."

The startled prop man asked, "What do you want with a dog?"

"I want a poodle in the street!" Curtiz shouted. "A poodle, a poodle of *water*!"

Curtiz bulldozed his way through the piece-by-piece script and created a superb screen entertainment. The film, released during the

Casablanca Conference of Anglo-American leaders at the beginning of 1943, captured the mood and intimacy of a wartime situation and moviegoers loved it, making it one of the year's highest-grossing films. *Casablanca* won the Oscar for Best Picture and Michael Curtiz was honored as Best Director, and the three writers took the Best Screenplay award. Because of wartime rationing, painted plastic statuettes were handed out at the ceremonies.

Trouble had been brewing between Hal Wallis, who was *Casablanca*'s producer, and Jack. It reached a crisis the night of the Oscars. Wallis recalled:

> After it was announced that *Casablanca* had won, I stood up to accept, when Jack ran to the stage ahead of me and took the award with a broad, flashing smile and a look of great self-satisfaction. I couldn't believe it was happening. *Casablanca* had been my creation; Jack had absolutely nothing to do with it. As the audience gasped, I tried to get out of the row of seats and into the aisle but the entire Warner family was blocking me. I had no alternative but to sit down again, humiliated and furious.

Later that year Wallis would leave Warners for Paramount.

As an added reward for winning his Oscar, Howard Koch got a one-time invitation from Jack Warner to eat at the studio executive table.

"It wasn't much of a reward," Koch recalled. "All I did was sit next to Jack and react to his lousy jokes."

Koch later said:

> I always looked upon *Casablanca* a little mystically. It had its own reason for being. It was a picture the world audience needed. What it said was "there are values that are worth making sacrifices for," and it said it in a very entertaining way. If *Casablanca* had ended, in my opinion, with Bogart going off with Bergman—the romantic ending—this wouldn't have happened. There wouldn't have been a *Casablanca*.

Joy Paige, Ann Warner's daughter from her first marriage, who was eighteen, auditioned and won the part of Annina Brandel, a young married girl who has decided to sleep with Captain Renault (Claude Rains) to get an exit visa for her and her husband, Jan. Rick settles the matter by letting Jan win enough at the roulette table to buy the visa.

Jack Warner was surprised when Metro signed Joy to a contract, and he teased her about trying to be an actress. No one in the Warner family (other than Lina) had appeared on the screen. When Metro didn't use

Joy in films, she asked to be released from her contract. Besides, she was in love with an actor named Bill Orr.

Orr had been put under a Warner Bros. contract after a talent scout saw him in the 1939 play *Meet the People.* Orr was introduced to Ann Warner at a party Elsa Maxwell put on. "Ann fluttered in in a beautiful dress," Orr remembered. "She was stately and beautiful." Ann invited Orr to the house to see a film, saying, "I have a rather pretty daughter." Orr thought Ann looked very young and guessed she meant a nine-year-old daughter. The actor was surprised when he met Joy, whom he found to be a "charming young girl, with a special quality of sweetness."

In front of Jack, he asked Joy to dinner. She demurred, "I don't think so." Jack blurted, "Why not? You don't have anything else to do." Joy was so mortified, she could have killed her stepfather.

The young couple dated for six months before Orr proposed. Ann thought it would be ostentatious to have a big wedding in wartime and had the ceremony performed in a small chapel. But Jack called in the Air Force and when the couple left the chapel, ten officers were lined up, swords crossed. Orr couldn't believe it: swords were for heroes. Jack picked up the tab for the honeymoon.

Several years after *Casablanca* was released, Loew's scheduled a Marx Brothers movie called *A Night in Casablanca.* Threats of legal action by Warners' lawyers prompted Groucho Marx to fire off a "brothers-to-brothers" letter:

> I had no idea the city of Casablanca belonged exclusively to the Warner Brothers. However, it was only a few days after our announcement appeared that we received your long, ominous document warning us not to use the name Casablanca.
>
> It seems that in 1471, Ferdinand Balboa Warner, your great-great-grandfather, while looking for a shortcut to the city of Burbank, had stumbled across the shores of Africa and named it Casablanca.
>
> I just don't understand your attitude. Even if you plan on re-releasing your picture, I'm sure the average movie fan could learn in time to distinguish between Ingrid Bergman and Harpo. I don't know whether I could, but I certainly would like to try.
>
> You claim you own Casablanca and no one else can use that name without your permission. What about "Warner Brothers"? Do you own that too? You probably have the right to use the name Warner, but how about Brothers? Professionally we were

brothers long before you were. We were touring the sticks as The Marx Brothers when Vitaphone was just a gleam in the inventor's eye, and even before, there had been other brothers—The Smith Brothers; the Brothers Karamazov; and "Brother, can you spare a dime?"

Now Jack, how about you? Do you maintain that yours is an original name? Well, it's not. It was used long before you were born. Offhand, I can think of two Jacks—there was Jack of "Jack and the Beanstalk," and Jack the Ripper, who cut quite a figure in his day.

As for you, Harry, you probably sign all your checks, sure in the belief that you are the first Harry of all time and that all the other Harrys are impostors. I can think of two Harrys that preceded you. There was Lighthorse Harry of Revolutionary fame and Harry Applebaum who lived on the corner of 93rd Street and Lexington Avenue.

I have a hunch that this attempt to prevent us from using the title is the brainchild of some ferret-faced shyster, serving a brief apprenticeship in your legal department. I know the type well—hot out of law school and hungry for success. This bar sinister probably needled your attorneys in attempting to enjoin us. Well, he won't get away with it! We'll fight him to the highest court! No pasty-faced legal adventurer is going to cause bad blood between the Warners and the Marxes. We are all brothers under the skin and will remain friends until the last reel of "A Night in Casablanca" goes tumbling over the spool.

Sincerely,
Groucho Marx

After Jack and Harry stopped laughing, they told their legal department to lay off the Marx Brothers.

Casablanca was followed by Irving Berlin's military musical, *This Is the Army.* Although Harry had said he didn't want Warner Bros. known as "the studio that made the best musicals during the war," Jack decided to take the risk and paid $250,000 for the Broadway revue, which he felt was suited to the mood of the country. In this flag-waving screen version Jack used the 350 fighting men who were in the Broadway production and added such Warner stars as Ronald Reagan, Joan Leslie, George Murphy, Dolores Costello, Kate Smith, and even heavyweight boxing champ Sergeant Joe Louis. Harry stated that Warners would not take any of the profit from the picture and eventually donated $7 million to the

Army Emergency Relief Fund. A second war-theme movie, *Hollywood Canteen,* was also immensely successful and Harry donated 40 percent of its profits to the Canteen and for the rehabilitation of servicemen after the war.

Then came the most controversial film in Warner Bros. history: *Mission to Moscow.* It would become one of the few films Harry and Jack wished they had never made. It would also be a film Howard Koch would regret writing.

On Christmas Eve day, 1942, Koch was getting ready to leave the Writers' Roost to visit his ailing father on the East Coast when he got a telephone call from Mr. Warner's secretary to come to his office. It was like getting—as Koch said—"a summons from Olympus." Except for the one time he had invited Koch to the executive luncheon, Jack had never talked to the writer.

Then the secretary added, "Mr. *Harry* Warner wants to see you." Koch was baffled. He knew Harry ran the financial affairs of the studio but rarely had contact with the artists, least of all writers. Up to now, Koch had seen Harry Warner only at a safe distance. Jack, resplendent in his colonel's uniform, was more in evidence. He could often be observed from Koch's office window on his way to the executive dining room, followed by an entourage of assistants, secretaries, and agents. Koch noted that Jack's "jaunty stride suggested a bantam cock asserting his authority over the roost."

Harry and Jack wanted to talk to Koch about his refusal to write the screenplay for *Mission to Moscow.* Koch was vaguely aware of the book by Joseph E. Davies, former ambassador to the Soviet Union. It was a sympathetic account of Davies's prewar experiences in Russia. Koch felt he didn't know enough about what was taking place inside the Soviet Union (nor did any writer) and had turned down the assignment.

In his autobiography, *As Time Goes By,* Koch recalled the meeting in Harry's office.

Other Voices

Howard Koch

I was ushered into an imposing office large enough to accommodate a princely court. Seated at his desk was Harry Warner and standing over his shoulder was Jack. Both Warners—my God, what had I done?

Their cordial greeting reassured me. Harry Warner began by thanking me effusively for my part of the work on *Casablanca*. The elder Warner was a soft-spoken, grandfatherly man in his sixties, his rather ponderous manner in contrast to his off-the-cuff, wisecracking brother. He delivered speeches rather than talked to you, making conversation somewhat difficult. His rambling discourse was larded with personal reminiscences that were not always relevant to the subject at hand. Jack, not a patient type, was in the habit of interrupting, although Harry never relinquished center stage without a struggle. The fraternal relationship suffered an uneasy balance of power, with one brother occupying the throne and the other being an ambitious pretender.

After a lengthy preamble, Jack finally got sole possession of the floor. He asked me why I had turned down the assignment just offered me, his tone implying that evidently I didn't realize its importance.

"Look, Koch, you can't turn this down," Jack said. I was surprised at his vehemence. He took a turn at irony. "You've heard there's a war on."

I smiled. "Yes, I heard."

"Well, Russia's our ally," he said, and then as an afterthought, "whether we like it or not."

Harry Warner was getting impatient. "Koch, what we're trying to tell you is that this isn't just a studio matter. We're doing this picture because the government wants it done."

"Our government?"

Mr. Warner allowed a pause to emphasize the importance of his revelation. "The President of the United States . . . go ahead, Jack, tell him."

Jack was more than willing. "I was invited to dinner at the White House. Ambassador Davies was there—the fellow who wrote the book. The president had it on the table beside him. He handed it to me. 'Jack, I see you're in the Army.' I had on my uniform. 'As one officer to another, I suggest you do a film based on this book. Our people know almost nothing about the Soviet Union or the Russian people.

What they know is largely prejudiced and inaccurate. If we're going to fight the war together, we need a more sympathetic understanding.' Or words to that effect. Then Mr. Davies offered to work with the screenwriter in an advisory capacity. He added, 'At no salary.'"

Koch could hardly refuse to do the script now and spent several months with Davies at his home learning all he could about the Soviet Union, which Davies referred to not as a Communist country but a "socialist state." Koch wrote the script, and as the filming got under way, Harry and Jack began to have capitalistic qualms about portraying the Soviet Union in a favorable light. The brothers asked whether the film couldn't be toned down a bit or "inject more sins into the characterizations of the Russian leaders."

But as Koch said:

> The production wheels were turning; it was too late to shift gears. In a sense the Warner brothers had trapped themselves into making a film with which they were not entirely in sympathy. When it was finished, director Curtiz said, "I make no more social pictures." Ambassador Davies was generous with his praise for the completed feature, and even showed it to Stalin in the Kremlin's projection room. Stalin showed surprise that a film on Russia had been made in America.

The movie, which many critics saw as a blatant attempt to bully Americans into thinking that the Russians were trustworthy allies, was competently made and absorbing and did well at the box office. Harry, in an attempt to mollify its pro-Russian stance, said:

> *Mission to Moscow* is realism; it is historically important. We believe it will do more to clear up the Russian picture for the American mind than anything thus far offered. . . . We owe an emotional gratitude for Russia's heroic stand, a cheering, warming appreciation of what it has meant to the cause of the Allies. It is this company's hope that *Mission to Moscow* accomplishes at least some of that.

Mission to Moscow, patriotic in the '40s, would be perceived as subversive in the '50s—and cause the Warners to bury it deep in their vaults.

By the middle of 1943, people in the film trade decided the public was becoming bored with war pictures. That piqued Harry, who insisted that the industry was under obligation to make them. He urged theater owners not to be "intimidated or coerced into not showing pictures dealing with the war effort." He added, "If the screen does not attempt to inform the public and explain the current struggle, there would be little justification for our existence."

The *Motion Picture Herald,* a trade journal, reported that studios other than Warners had 126 comedies or musicals either completed or in preparation and quoted a poll of 101 exhibitors as saying the public wanted "non-war, escapist pictures." Harry stormed back, saying, "Any arbitrary exclusion of war films, either to satisfy a small appeaser element or for personal reasons without regard to the general public interest, is equivalent to sabotage!"

Warners would keep making war pictures, although Harry continued to hedge his "no musicals" pronouncement by making entertaining, flag-waving films such as *Yankee Doodle Dandy.* James Cagney played the role of pint-size entertainer and songwriter George M. Cohan with patriotic zeal. Three premieres were held for the film, in New York, Los Angeles, and London, admission to each of which was gained only through the purchase of U.S. War Bonds valued from $25 to $25,000, bringing in $15.6 million to the governments of the United States and England.

One of Harry's happiest moments during the war was christening a Liberty Ship after his father, Benjamin Warner. During the ceremony he read a prepared speech in which he said:

> Benjamin Warner knew the meaning of liberty better than we can know it because he had the bitter experience of oppression. He also knew the value of unity. His constant advice to me was this. "Harry," he would say, "you are the oldest of my sons, and it is your responsibility to keep your brothers together. As long as you stand together, you will be strong." And that, I feel, would also be Benjamin Warner's message to our country today.

As the fury of the war progressed, Harry began to feel the advancing years and tried to spend more time at his San Fernando ranch. In a *Fortune* magazine interview he related this story:

> I haven't been feeling very well. I went to my doctor and explained how I felt. He asked me if I drank. I told him, honestly, I drank only an occasional highball or cocktail.
>
> "Stop it," he ordered. Then he asked if I smoked. I said I smoked a few cigars. "How many?" he asked. I told him.
>
> "Cut them out," he said. "What do you eat?"

> I explained that I ate what almost any normal human being eats.
>
> "I'll give you a new diet," he said. And he did.
>
> So now, I can't drink, I can't smoke, and I can't even eat what I like. But I'm a success.

Abe also had a farm, in North Carolina, which during the war became a working farm. He would send food to all the relatives. His step-grandson, John Steel, recalled that despite rationing, Abe would send "ham, eggs, bacon, a tremendous amount of food every week. Abe also liked to putter around the farm and fix things, like the concrete walk, not letting anyone help him."

Jack and Ann bought 22,000 acres of rangeland in Arizona not far from Tucson. Ann had had a dream (a kind of mystical experience, she said) that told her they should have a place where everyone could go if a crisis arose. Her first husband, who was still under contract to Warners, was sent to manage the property, a task he enjoyed, as he spoke Spanish and had grown up on a ranch.

As the war entered its final months in 1945, Harry began thinking about the future. What would be Warners' place in peacetime? He expressed his views by saying:

> The motion picture industry would be shamefully remiss if it were not looking ahead to its task in the postwar world. The essence of the task can be stated in a single phrase, "To interpret the American Way."
>
> We at Warners Bros. have sought for more than a decade to combine entertainment with a point of view. We have tried to say on the screen the things about America and democracy and Fascism that need to be said.
>
> The screen transcends language, time, distance, and custom. One of our chief aims now in the postwar world will be to show Americans how millions of Chinese, Icelanders, Indians, Eskimos, and Russians live. I can think of no clearer, surer way to achieve a community of nations—and certainly of no clearer, surer way to show the world what our democracy means. Once the world understands the blessings of democracy, the results are beyond question.

Harry, Jack, and Abe could look with satisfaction at the wartime accomplishments of the studio and the efforts of its personnel. More than $20 million in war bonds were purchased through the studio. Red Cross mobile units made twelve trips to the studio, collecting 5,200 pints of plasma. But most of all, they could take pride in knowing that 25 percent of the studio's employees had served in the armed forces, a total of 763 men and women.

Marine Captain Milton Sperling was returning home to his wife, Betty, and two daughters. The question was whether he would become part of Harry's plan to "interpret the American Way" in the postwar movie industry. He had been promised a job by his old boss Darryl Zanuck at Twentieth Century Fox.

Also returning was Colonel Jack Warner Jr. of the United States Army Signal Corps. He was now twenty-nine and felt ready to take his place as the heir apparent of Warner Bros. After all, he was the only surviving son of the four brothers. He was hopeful he would one day be president of the company.

All he had to do was convince his father, Jack L. Warner.

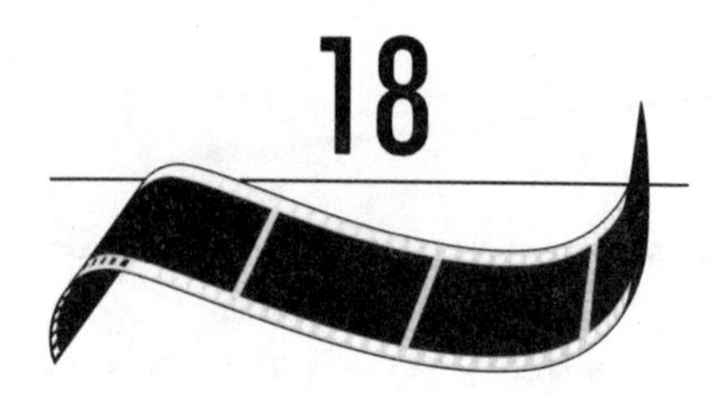

18

Family Entertainment

THE SCENE:

Warner studio lot, 1946.

The gray-haired man walks down a narrow street on the studio lot, his eyes scanning the ground in front of him. The street is jammed with Confederate soldiers, Nazi storm troopers, pirates, and a line of chorus girls in tap shoes and sequined costumes. Every half a dozen steps he stoops to pick something up. He examines the object carefully, then stuffs it into the pocket of his blue double-breasted suit.

"Who is that little guy?" a huge Gestapo man asks his companion, a stunning girl in a gold lamé gown and blond wig.

"Oh, him? He's nobody." She giggles. "Just the president of the studio, Harry Warner. Hunting for nails."

Harry Warner, the president of a motion picture company with a weekly payroll of over $600,000, was indeed on the lookout for nails. The double-headed nails used in making movie sets were expensive, and Harry, without realizing what he was doing, scooped them up whenever he was walking around the lot and sent them to the carpenter shop. At times Harry must have felt like that undersize kid who would keep a dozen cobbler's nails in his mouth so as not to waste time picking them up from the floor.

"I thought about making him a stick with a magnetic tip on the end so he could get them up easier," Milton Sperling recalled, "but I didn't want to embarrass Harry." Sperling had a good reason for not wanting to alienate Harry: He was now working for his father-in-law.

Other Voices

Milton Sperling

Darryl Zanuck, learning of my return from the service, sent a letter saying I could have my old job back at Twentieth Century Fox. Harry had asked me to come to Warners and work for them. "A member of the family can't work for a rival studio," he told me. "We'll give you your own production company and you can make whatever pictures you like." I said I was still under contract to Twentieth Century Fox, so Harry called Zanuck, told him how much he wanted me in the "family," and asked him to consider releasing me from my contract. "It's up to him," Zanuck said and abruptly hung up. He and the Warners weren't exactly kissing cousins.

I set up a meeting with Zanuck and he said, "Oh, hell, stay with me. You can't go to the Warners. They're all a bunch of fucking thieves."

I said, "Look, I'm part of the family. They're at my house all the time; I have to eat dinner with Harry and Rea. I have to talk to Harry. What else can I do?"

"You're making the biggest mistake of your life," Zanuck told me. "You'll be a whipping boy for Harry and Jack, you just wait and see."

"Darryl, I have to go."

He shrugged his shoulders. "Well, we were both in the war together, so . . . it's up to you." Shaking his finger at me, he added, "When you start taking too much shit over there, you can always come back to Fox."

As promised, Warners gave me my own independent production company, United States Pictures. I produced my first feature-length film in 1946, *Cloak and Dagger.* Gary Cooper played a nuclear physicist who is parachuted into Germany to rescue a scientist captured by the Nazis. Cooper was the best man I'd ever worked with. He didn't bother anybody, showed up on time, knew his lines, didn't get drunk. The only weakness he had was the ladies, and that sure didn't bother Jack. But Zanuck was right about Warners. Not Harry, as he turned out to be a straight shooter, but I should have known about Jack from that year I worked for him in 1932. Because I was married to Harry's daughter, he saw me as a threat to usurp his power.

My relationship with Jack made me feel like I was riding a pogo stick: one minute I was reaching for the heights, the next minute I was groveling in the mud. His temper was totally unpredictable and I never knew when he was going to chew my ass. Like the time he saw me driving a Rolls-Royce . . .

I was making more money than I ever had before, so I bought something I'd always dreamed of owning, a Rolls. The first day after I drove it into the studio lot, I noticed that Jack didn't speak to me. He saw me in the executive dining room and walked past me, totally ignoring my presence. The next day it happened again. This silent treatment went on for several weeks. He wouldn't return my phone calls, answer my memos. I had no idea what I'd done wrong.

One day, when I was driving out of the lot, I happened to see Jack ahead of me in his Cadillac. Suddenly he screeched to a stop and I almost banged into his rear end. Driving up alongside his car, I could see he was livid. He shouted, "Look, you miserable son-of-a-bitch, how can you afford to drive a Rolls when I'm driving a Cadillac?"

That's what he was pissed about. I tried to make a joke. "If you'd just told me you didn't like Rolls-Royces . . ."

He screwed his face into a smile that came out a smirk. "You just go on blowing your money, little man producer, and I'm gonna die a rich man, and you'll be buried like a fuckin' pauper." With that Jack drove away. I had been given a reminder that second-class executives didn't drive Rolls-Royces. I also knew that as long as Jack was the studio chief, I'd never make it to the top.

The next week Jack was talking to me again.

During the war, Jack Warner Jr. had had little communication with his father. The young man's letters were rarely answered. They did meet several times in New York when Jack Jr. was stationed there for a couple of months, and Jack Sr. did appear interested in what his son was doing. Jack Jr. went overseas in 1943 and worked on the staff of the Twelfth U.S. Army Group starting well before D-Day, under the command of General Omar Bradley, and after the war ended he was sent to Wiesbaden, Germany, as the staff Signal Corps photo officer. When Jack Jr. heard that his father was in a party of visiting film executives, he picked up a fleet of Jeeps and met the group at the Frankfurt am Main airbase.

The next week went fast, with inspections, briefings by General Eisenhower, a look at prisoner-of-war and concentration camps, a short trip up the Rhine to Cologne on Hitler's yacht, and a relaxing hot soak in the Wiesbaden spa.

Both father and son wound up in a little room in the Schawartzer Bock Hotel and for the first time, Jack Warner opened up to his son and talked about Ann and family relationships. The older man resolved that when they all returned to Hollywood there was going to be a "brand-new beginning."

When Jack Jr. drove his father to the airport he felt better than he had for many years. He was eager to get back home and have the life that had been denied him for too long.

Other Voices

Jack Warner Jr.

Several months after the war was over, I turned up in uniform at the studio. I felt something like a character out of one of the war films Warners had ground out, only with an upbeat ending. Or so I thought.

I had been released that morning from active duty and had been freshly promoted to full colonel. I went right to my father's office, where he grabbed my hand and seemed overjoyed that I was back safe and sound. Recalling that for a brief time he had been commissioned a lieutenant colonel in the Army Air Corps while setting up the first motion picture unit, I said it was a good thing he was out of uniform or he'd have to salute me. His face suddenly froze. I should have remembered something he often said in the executive dining room: "Around here I get all the laughs."

There was a picture of Ann in his office, and as he glanced at it, a chill settled over the room. He withdrew his hand from mine and sat down behind his huge black desk. From the enthused and buoyant man he had been the week before, he now seemed to have turned back to his old prewar version. He let me know that my presence created problems for him and that "peace" at home was being threatened. Then he told me I would have to go to Ann and "work things out" with her.

That was fine with me, but I wondered if she felt about things the same way my father and I had back in Wiesbaden, and if indeed he still felt that way.

He gave me no clues, but something about the way he told me how ill Ann was at the Scripps clinic in La Jolla, and how she might not be able to see me, made me feel as if nothing had changed. (Ann was a diabetic, and Bill Orr said that she had also been diagnosed as having a neurological disorder which affected the ends of the nerves, making her feel like a light breeze kept blowing over the skin. A doctor in Europe gave her a gold mist to take internally, which seemed to help.) I was determined to try and see her, so I changed from my uniform into clothing I hadn't worn for years and drove south to La Jolla. I found the inn where Ann was staying and reached her by phone. She said she was about to bathe and asked if I would please call her in an hour so we could meet. I walked the beach until an hour had passed, then called again, only to be told she had left the inn and nobody knew when she would be coming back. It would be several years before I would see Ann Warner.

Although I was under the assumption that I would go to work at the studio, my father felt my presence in town was a threat to his lifestyle. He told me there had been many important changes in distribution and exhibition and once again New York was the best place for me to go and study those changes firsthand. I wanted to work with him at the Burbank studio, but I knew better than to argue.

Another returning serviceman was Ronald Reagan, who hoped to be promoted to leading man in "A" pictures. Although he had been thrilled with playing George Gipp in the Knute Rockne story and had excelled in 1942 in *King's Row,* a movie he considered his finest acting effort, he was concerned about what his future might be at Warners.

Other Voices

Ronald Reagan

I always said this about Hollywood: If you were a sailor in a picture that made money, then buy yourself seasick pills, because you won't see land again for years. I had been pretty successful with the drawing-room comedies, the Cary Grant type of roles, but I wanted to do a Western. I started bellyaching all over the studio. Finally I went to see Jack Warner and said, "If you ever let me do a Western, you'll make me a lawyer from the east." The next thing I knew, I was cast in *Stallion Road.*

I was excited, because I was going to costar with Humphrey Bogart in a Technicolor film about horses. It wasn't exactly a Western, as I played a veterinary surgeon, but it did have horses! A week before shooting began, Bogart dropped out and was replaced by Zachary Scott. Jack decided to shoot the picture in black and white. I was back doing "B" movies.

I was next cast in *That Hagen Girl,* in which Shirley Temple starred in her first grown-up role. Sure enough, Jack cast me as a lawyer. I played Shirley's lover, but audiences weren't ready for the Lolita aspect of the story. My favorite role of that period was *The Voice of the Turtle,* based on John Van Druten's play. I portrayed a lonely soldier who woos Eleanor Parker. In that movie I was caught with my pants down.

Each year the studio gave a party and editors of the different films would save some of the funniest outtakes (clips of film that were ruined because of technical problems, or because the actor blew his dialogue). In this one scene with Eleanor Parker, I was supposed to change out of my uniform into civilian clothes, so I was directed to step behind a chair to make the change—but on each take the pants zipper stuck. I tugged at it behind the chair and finally walked before the camera, saying, "The damn zipper's stuck." That happened several times and the crew was in hysterics.

Mervyn LeRoy, who had remained at the studio during the war, felt like he had been in a war of his own. On August 12, 1945, Doris won a divorce from him, testifying that she had been subjected for several years "to a course of extremely cruel conduct." Doris added, "Life has been miserable, my peace of mind disturbed and health impaired."

My mother, Betty, thought that Doris's and Mervyn's personalities were like "oil and water," and added, "Mervyn was an incredible director, but in person he stuttered, was a hypochondriac, and was totally dominated by Doris."

When reporters asked Doris if she was going to marry Charles Vidor, the Hollywood director who had been linked with her romantically for several months, she replied, "This is not the day to speak of remarriage." Two months later, the trade papers reported that Doris, thirty-two, had married the Hungarian-born movie director, who was forty-five.

Harry hated seeing family problems aired in the newspapers and was embarrassed that his daughter had divorced. He was hardly pleased when he read the following headline in the July 1, 1947, edition of the *New York Daily Mirror*: LINA SEEKING CHUNK OF WARNER BROS. FORTUNE.

Lina Basquette, the former silent-film star, was breaking her silence twenty years after Sam's death with a Supreme Court suit to collect $15 million from Warner Bros. She charged that Harry, Jack, and Albert Warner

> willfully and fraudulently, for their advantage and in the breach of their trust to me, offered the will of Samuel L. Warner for probate in New York, although they knew that the residence and domicile of Samuel L.

Warner at the time of his death was the state of California, where state community property laws gave the widow a bigger share of a husband's property.

Lina wanted to know, she told the court, what had happened to the 62,500 shares of Warners stock, which currently were quoted on the stock exchange at slightly more than $15 million. She reminded the court that upon Sam's death she had been granted the interest income from a $100,000 trust fund, which amounted to only $3,000 a year. She said she was "just a child" when Sam Warner died and added that she knew nothing about legal matters at that time. She now believed she was entitled under California law to one-fourth of the estate.

When asked why she had waited so long to file the suit, she responded: "I have delayed bringing legal action because I wanted to spare my daughter, Lita, from the notoriety I have known."

Under the headline about the lawsuit, the newspaper ran a series of sensual photographs of Lina in a two-piece bathing suit cuddling a cocker spaniel puppy. She was—as the caption said—a "vivacious woman."

Harry could only shake his head. Five months earlier, he had been gratified to witness the marriage of Lita, now twenty-one, to Dr. Nathan Hiatt, a thirty-two-year-old Jewish doctor who was on the staff of Cedars of Lebanon hospital in Los Angeles. Lita, who had not seen Lina since she was four years old, asked to visit her mother in New York. Harry had agreed, but now realized that the meeting no doubt had prompted Lina to think back twenty years—and remember.

The Warner lawyers entered a denial of the charges, contending that it had been nineteen years since Sam's will was probated and there was no case because Lina had waited more than the six years prescribed by the statute of limitations before bringing action. The affidavits also ridiculed her because of her six subsequent marriages.

Lina responded that the brothers were trying to "prejudice the court against me because of my marital misfortunes," and added bitterly that their tactics were "neither relevant nor gentlemanly."

Lina Basquette had admittedly led a turbulent life. After her 1928 marriage to cameraman Pev Marley failed, she wed Ray Hallam, an actor, who she said was her "greatest love" and who died in her arms of leukemia three weeks after they were married. In 1931 she married Teddy Hayes, a one-time trainer for heavyweight boxing champion Jack Dempsey. Teddy and Lina divorced, then remarried and had one son, Eddie. In 1938 she married British stage director Henry Mollison. He joined the British

Merchant Marine, and was assigned to a ship that was blown out of the water by a German U-boat. He was rescued by the Germans and spent the rest of the war in a prison camp. Lina was granted a divorce in 1946, saying of her life with Mollison, "We fought like the devil but on stage and in bed we were terrific together." In 1947, she married Warner Gilmore, the general manager of the St. Moritz hotel in New York.

And now, here she was suing Warners for a percentage of Sam's estate. Harry wanted to make her disappear and quickly agreed to a settlement: Lina could have the $100,000 trust in cash if she would drop the suit. He also stipulated that she not write about the Warners for ten years. (At the time, Lina was working on a Warner exposé, a book with two different working titles: *The Inside Story of Life and Sex in the Movie Capital* and *Virtue Is a Dirty Word.*)

Lina told me she had a good laugh at that: "I guess they figured that with my lifestyle I wouldn't live past fifty. Either that, or someone would kill me. All I wanted was to be free of the whole thing and buy a piece of property to raise dogs. I never had much faith in the courts, so I took the hundred thousand, of which thirty-three percent went to my lawyer."

Lina Basquette, estranged from her daughter, Lita, was finally, and permanently, out of Harry's hair. With the bizarre mess over, Harry noted with pride that Jack Warner Jr. had become engaged to a New York girl.

Other Voices

Jack Warner Jr.

I met Barbara Richman while I was working with a film exchange in New Haven, and a year later I proposed. I immediately telephoned my mother and my father. When I told him Barbara's father was an attorney, he was less than pleased. (Next to agents and writers, my father disliked lawyers most, thanks to his bruising encounters with the profession.) He told me Ann was in New York, so Barbara and I went over to the Waldorf Towers where they had an apartment, and sent up our names. Ann sent word back asking us to come back in half an hour because she was getting dressed. We walked around the block, checked the time, then returned to the Waldorf, only to be told that Mrs. Warner had left the hotel and it was not known when she would return. Ten years were to pass before Ann met my wife.

When we mailed out wedding invitations, I made certain they were sent to Ann and my half-sister, Barbara, both of whom were now in France. My father flew to New York alone for our wedding and enjoyed himself immensely. I remember him dancing with my mother, Irma, and having a wonderful time.

A honeymoon in California followed a lovely dinner with my father at his house. There was a feeling of great warmth and he was very charming to my new wife. Ann remained in Paris. Somehow it seemed that the more distance between my father and Ann, the more relaxed and carefree he became.

I learned from my father that Warner Bros. was entering into a long-range coproduction deal with Associated British Cinema Ltd. of London in which Warners owned considerable stock. They needed a production liaison head to get the program started and I felt more than fully qualified. My father agreed, and soon my wife and I were on our way to London. The first coproduction was *The Hasty Heart* with Patricia Neal and Ronald Reagan.

We lived in London a year as I trained a replacement. I had done the job I had been sent to do and now wanted to return to California. I brought the idea up with my father by letter and cable, but he rejected it and ordered me to remain in England. I felt I had to sit down and discuss the entire matter in person, so I booked passage on the *Queen Elizabeth.* In Burbank my father refused to see me—and *barred me from the lot.* I read in the trade papers that Ann was in town.

For a while I worked with Al Rogell, my stepfather, on an independent film he was producing, then I set up my own company and made a feature starring Lee J. Cobb and Jane Wyatt. Whether I would ever return to Warners was a moot point.

What I did not know was that behind the scenes a battle was taking place, with Harry Warner demanding that my father take me back into the studio. He and Uncle Abe had learned the facts behind my leaving and were angry at my father for keeping me off the lot.

I heard that Harry told my father that if he didn't take me back into the company, then Jack L. Warner could leave it! Suddenly, the door opened and I was once again on the lot as assistant studio production manager.

I also learned that Harry had emphatically told his brother that Ann Warner was not running Warner Bros. so long as he, Harry Warner, was alive! In view of future events, this was a prophetic statement.

Harry, tired of the constant arguments with Jack, began to spend more time at his 1,100-acre ranch and his new passion: thoroughbred racehorses. A character in F. Scott Fitzgerald's unfinished novel *The Last Tycoon* says, "The Jews had taken over the worship of horses as a symbol—for years it had been the Cossacks mounted and the Jews on foot. Now the Jews had horses and it gave them a sense of extraordinary well-being and power." Harry's interest in racing was so intense that he was instrumental getting a new horse track, Hollywood Park, built.

Harry tried turning his ranch into a thoroughbred breeding farm, but no matter how much he spent on trainers, none of his horses entered the winner's circle. (He sold one horse for $100 and it won its next race and the $2,500 purse.)

His rival, Louis B. Mayer, had been raising thoroughbreds for nine years and had won purses totaling $1.7 million, a feat that racing experts described as "phenomenal." Hollywood columnists used other adjectives: "colossal" and "stupendous."

Harry was astonished when he learned that Mayer was going to auction off his stable, including Honeymoon and Stepfather, which had brought in $279,500 in purses. Mayer said he'd also put on the auction block Busher, a horse that had become the biggest money-winning filly of that time, with $334,000 in purses to her credit. Mayer liked to joke that his studio "received more requests for pictures of Busher than of Lana Turner, and Busher didn't wear sweaters."

The auction was held at the Santa Anita Park clubhouse, an event that drew 10,000 people. Bids for Honeymoon opened at $75,000 and ended four minutes later when Harry picked the thoroughbred up for $135,000. When Stepfather was led into the charmed circle, $40,000 was piped as an opening gesture. Harry gave the nod at $200,000 and the auctioneer's voice boomed, "Mr. Harry Warner—the new owner of Stepfather!"

Four months later, the best showing Stepfather made was fourth in the Santa Anita, a refund of $10,000 on his original investment. Harry had better luck with Honeymoon when the filly won in June at Hollywood Park. Harry was so happy, he kissed the trainer, Graceton Philpot.

Not long after this first win, Harry learned from his accountants that the studio executive dining room was operating at a $60,000 annual loss. He told his executives, "I don't know why it should cost so much for us to eat here. At my stables I know exactly how much the barley and oats fed to my racehorses cost."

Producer Jerry Wald whispered to his luncheon companion, "Just let one of those horses produce a picture, then you'll see a change in his eating habits."

While Harry and Rea were relaxing on the ranch, Jack and Ann were giving lavish parties at their Beverly Hills estate. It was a mecca for movie stars, and no actor on the Warner payroll refused an invitation. If they could get one. Anyone invited had to make at least $3,000 a week, unless it was a pretty starlet or handsome young actor that Jack and Ann used to decorate the room. Ann spent great sums of money on these parties and invitations were highly prized.

The artist Salvador Dali lived with Ann and Jack while he painted their portraits. Ann's portrait was an incredible likeness, capturing her faraway look. She was in a low-cut red evening gown, her arm resting on a stone cornice; the background was an arid scene with an oasis. Jack's painting showed him sitting on a cement wall with flowers in the background. Jack looked at it and said, "Wait a minute, I have six fingers!" He didn't like the painting and neither did anyone else, as it didn't look like a Dali work.

Dali lived at the house for three months while he did the two paintings, and each day he would paint, then go to the guest bathroom. He would sit there and doodle on the toilet paper, then pull it off the roll and stuff it into the wastebasket. Jack would go into the bathroom and retrieve it, then had it framed and hung on the wall—genuine artwork by Dali.

I chatted with Barbara Warner, Jack and Ann's daughter, and she remembered her life with her parents during these years.

Other Voices

Barbara Warner

When I was a teenager, my father was a bit strict. He was very suspicious of people, perhaps because he knew so much about the world. He was always distrustful about the people I was going out with. Sometimes he was right and other times he wasn't. He was very old-fashioned. He didn't like to see me in too much makeup or in a dress that was too low, or heels that were too high.

I loved him very much. He was fatherly, very attentive to what I was doing in school, who my friends were, and what time I came home. He was a good friend to many people, and was very generous.

My mother was very glamorous, five feet two inches tall and very fine-boned. She was more intellectual than my father. She wrote poetry. She was a lot of fun and had a great sense of humor. She *loved* astrology and was a very good astrologer. When I was in my teens, she'd be sure to get the birthdate of the guy I was going out with, where and when he was born. Then when I came home, we would sit up for hours doing my date's chart. My father would come in and see us and shake his head: He hated astrology and didn't believe in it at all. Anyway, she and I had a good time doing it.

My mother always felt rejected by the Warner family. She loved Jack, and couldn't understand why Harry and the Warner sisters wouldn't accept her. The sisters in particular weren't nice to her. Perhaps they were jealous, I don't know.

My father believed in family more than most people realized. He came home practically every night for dinner. We used to always go on wonderful family vacations during the holidays, to Florida or to Arizona, and once to Puerto Rico.

Jack bought a villa on Cap d'Antibes in the south of France in 1948. Nearby were the gambling casinos of the French Riviera. Jack stated that he planned on spending a lot more time there to get away from the hectic pace of the studio.

While Jack was turning his attention to the baccarat tables of southern France, Harry immersed himself in lofty ideas. Concerned about the problems in Palestine and the settlement of Jews there, he came up with a grandiose scheme that he believed would work. He even wrote a lengthy paper, titled "Land Without People—An Approach to the Problem of the Displaced Persons of Europe." On June 13, 1946, he sent President Harry Truman a telegram outlining his plan. It said in part:

> My Dear Mr. President:
>
> The statement that this country is pressing for the admittance of 100,000 Jews into Palestine because we don't want them here prompts me to lay before you a suggestion which might be a constructive step towards solution of this difficult and distressing problem.
>
> My suggestion is that the United States adopt emergency legislation to permit entry of displaced persons into Alaska, a territory with 72,000 population which could sustain many millions.

After offering this startling solution, Harry went on to add: "I understand that the principal objection to opening Alaska to immigrants is that such immigrants would not remain in Alaska but would quickly move to the United States. This difficulty could be overcome by providing appropriate terms and conditions through legislation." Harry stated that he was available to come to Washington and discuss the problem with the president, and a few weeks later received a telegram that said: "The President will be glad to see you Wednesday July tenth eleven-thirty A.M."

After the meeting with Truman, Harry wrote a memo about it.

> President Truman stated that I had arrived at an opportune time, as at eight o'clock that very morning, a mission had left for London to try and arrange for the admittance of 100,000 Jews into Palestine. . . .
>
> I told him that I had come to present him an idea which I, after thorough study, considered essential to the creation of better understanding in this world, and that I had studied the history of mankind and found that persecution and execution of man has existed since man has been on this earth and that the only thing that has saved the persecuted people was space. . . .

Harry related to his own family's experience by saying,

> Had we Warner brothers not stuck together, our company, Warner Bros. Pictures, Inc., would never have survived because when the Depression

> hit our company in 1932, we owed something like 170 million dollars. There is no reason why our company should have survived except we brothers were more interested in saving our name than in making money and, being interested in that cause, we united for the purpose of not having our company go into bankruptcy.
>
> If we started creating space through Alaska and some of the deserts of our own country, that would be followed by other countries. We must take the lead.

Truman ultimately rejected Harry's proposal, saying that an Alaskan program had been tried in 1935 with families selected from relief rolls from several states and hadn't succeeded. But that attempt, Harry replied, had been made with American people who were already living in a country of plenty, whereas those "living in barbed-wire fences would certainly be glad to be given the opportunity to have space where they could build homes for their families . . . No one can appreciate that more than I can, because I came to America fifty years ago, from Russia-Poland, which persecuted its minorities at that time and I know the opportunity I was given in America."

Shortly after this meeting, the United Nations voted to partition Palestine and carve out a Jewish state. Harry's dream of an Alaskan refuge for Jewish immigrants vanished in the smoke of Jewish-Arab conflict.

Back from France, and brimming with confidence at the success Warners was enjoying in the postwar era, Jack agreed to an interview with Louella Parsons. Dapper in a tailor-made suit, his pencil-thin mustache framing his toothy smile, Jack listened while Parsons stated, "You must be so very proud of what you have accomplished."

"Family pride," he responded. "My brothers and I are examples of what this country does for its citizens. There were no silver spoons in our mouths when we were born. If anything, they were shovels. But we were free to climb as high as our energy and brains would take us. If we sell that ideal to the rest of the world as a rule to live by, I think we can sell permanent peace."

For once, Jack sounded more like Harry than himself, but "peace" was far from settled at Warner Bros. Not only were Harry and Jack at each other's throats more and more—Jack Jr. once had to step between them and break them apart, only to have his father storm out of the room yelling for his son to "leave that madman!"—but the studio was experiencing labor unrest as well.

Although the war had ended, and a new prosperity was predicted, Harry and Jack had continued their stringent production budgets, cutting costs, rooting out waste. Many of the studio's union employees, most of whom had been totally loyal during the war, felt they were being underpaid. They held out for better pay and better working conditions.

In September 1946 a union strike and riots broke out at Warner Bros. as a result of a fierce struggle between two opposing unions. Thousands of pickets marched in front of the studio. Rocks, chains, and clubs were used as weapons, and the police had to intervene with tear gas. Studio writer Julius Epstein, noticing guards armed with rifles atop the studio buildings, commented, "That's Warner Brothers: combining good citizenship with good marksmanship."

Jack was in Harry's office when several studio workers tried to crash a pickup truck through the line of union strikers. The truck was stopped and the men jerked from the truck's bed and beaten with baseball bats and lead pipes. "Jesus, did you see that, Harry? They just beat the piss out of our guys. What the hell's going on? We're fighting each other."

Harry stared silently at the melee.

"Didn't you hear me, Harry? The Commies are doing this!"

Harry turned toward his brother. "That's what the red-baiters want you to believe. They're trying to turn Hollywood against itself!"

"All I know is that these guys are out for blood."

The strike officially ended in October, one of the first rounds in what would soon become feared as Communist-inspired riots.

In early 1947 ten filmmakers—writers, directors, and producers—were served subpoenas (the number would grow to three hundred) to appear before the House Un-American Activities Committee (HUAC) and testify about their activities and association with the Communist Party. It wasn't by accident that Hollywood was tagged as having subversive elements: This was where dreams were made, where ideas were generated and disseminated. It was time to suppress any wayward Tinseltown ideas and bring any dissidents into line.

What the committee wanted was names of suspected Communists in the movie industry. And the truth was, no one was forced to name names. Agent Dick Dorso, who lived through the ordeal, believes that had Elia Kazan, at the first meeting when he was called, said, "I do not have to do this, this is a violation of my constitutional rights, and I refuse to do it and you are wrong," it would have ended. But when Kazan talked, and Senator Joseph McCarthy realized the publicity bonus he would reap, the committee went wild. McCarthy got what he wanted: a list of ten writers and producers who were suspected of having Communist affiliations.

The hearings, the "Ten" (as they were now called) knew, would not be an impartial trial. This was an inquisition, designed to destroy lives. Headline-hungry congressmen such as McCarthy began to chill the very marrow of America with red-baiting stories about "pinkos" in Hollywood.

As the hearings progressed, Harry was shocked when he heard an impassioned speech on the radio given by a congressman from Mississippi: "They are trying to take over the motion picture industry, and howl to high heaven when our Committee on Un-American Activities proposes to investigate them. They want to spread their un-American propaganda, as well as their loathsome, lying, immoral, anti-Christian filth before the eyes of children in every community of America."

Harry was even more concerned when Jack Warner was called before the committee to testify. Warners had been under attack for producing the wartime movie *Mission to Moscow.* Now Jack had to defend its production to the committee.

On October 27, 1947, Jack went to Washington full of confidence. He knew he was glib, clever, and bright; he'd wow 'em. Instead, he made a fool of himself.

In the committee room the heat was oppressive, and the smell of sweat mixed with the acrid cigar smoke that lingered in layers over the heads of the people packed tightly into the airless space. As each celebrity was called to the long, low table to testify, photographers elbowed each other and raised their bulky press cameras to get an "exclusive" shot. The exploding flash bulbs were blinding. The congressional inquisitors leaned forward from their high perches, cajoling, intimidating, attacking each witness. It was into this mad arena that smiling Jack Warner stepped.

"*Mission to Moscow* was made when our country was fighting for its existence with Russia, one of our allies," he began, confident that he could sway the opinions of the committee. "It was made to fulfill the same wartime purpose as many other of our films did, such as *Air Force* and *This Is the Army* . . . We made *Mission to Moscow* out of patriotism. I was asked by the president, Franklin Roosevelt, to produce the film with the aid of Ambassador Davies."

Seeing the grim look on the faces of the committee members, Jack began to sweat under the collar of his shirt. His speech was not having any effect. Then, feeling the need to protect himself, he began to babble, talking off the top of his head, saying that he and his brothers would "subscribe generously to a pest-removal fund."

Grinning, he added, "We are willing to establish such a fund to ship to Russia the people who don't like our American system of government." Then he did a stupid thing: He offered to name a dozen screenwriters he had fired for being Communist. Surprised, the committee happily agreed to accept the list.

One of the names Jack gave was Howard Koch, who had written *Mission to Moscow* reluctantly at Harry and Jack's urgent request. Koch was already on the committee's second list of nine. Jack calmly stated that he had to get rid of Koch because he was "slipping Communist propaganda into films."

The committee also asked about the Epstein brothers, Philip and Julius, who had worked with Koch on *Casablanca*. Jack had never been fond of the Epsteins. The brothers had always resented having to punch the time clock every Monday through Saturday morning like factory workers. One Saturday Jack asked the brothers to take a script they were having problems with and look it over on the weekend. The brothers came in on Monday, went to the front office, and said, "Yes, we can lick this." Jack said, "Tell me how." They responded, "We can't. We thought of it on Sunday." When testifying, Jack said, "I don't trust 'em; they're brothers."

After exposing the writers, Jack retired from the witness table, wondering if he had done the right thing in trying to protect his name and the Warner studio.

Other Voices

Jack Warner Jr.

I went to Washington with my father when he appeared as a witness before HUAC. It was a frightening experience, and when he left the chamber he was soaking wet with sweat and angry as hell at himself for cooperating with the committee. He felt they had made him look and sound like a fool. But it was he who had made himself appear that way, not HUAC. He had sounded as if he were rambling on in the executive dining room at the studio, not in the halls of Congress. The whole thing was a very disappointing performance.

My father, Milton Sperling, told me this story about seeing Jack in the studio dining room after the hearings:

> *Jack had this gloomy look on his face when he sat down beside me and didn't respond when I said hello. I knew better than to say anything when he was in one of his moods so I kept quiet. Finally he said, "Sperling, you better not eat lunch next to me, or in the same room." I asked why and he growled, "One of us will not be eating here again." Uh-oh, I thought. He's mad at me for the petition I had signed to protest the committee. I got up and said, "Well, it's your studio, I'm leaving." Before I got two steps, he called me back. "We have to stick together." Then in a desperate voice: "They're after all of us."*
>
> *When Jack told director John Huston that he'd told the committee a "few names," Huston shook his head. Jack said, "I guess I shouldn't have, should I?" When Huston agreed that he had made a mistake, Jack added, "I guess I'm a squealer."*

At a second hearing, Jack told the committee he had been carried away naming names. "I was rather emotional. I have never seen a Communist and wouldn't know one if I saw one."

That hardly helped Howard Koch, who was exiled from Hollywood and forced to go to Europe and write under another name. Koch had hoped to rebut Jack's statement, but the committee canceled his subpoena with a last-minute telegram. Koch heard later that Jack had confided in a friend that he had "made a mistake" in naming Koch.

In late October 1947 a group of Hollywood stars traveled to Washington to attend a session of the HUAC hearings in support of others called to testify. Included in the group were such Warner stars as Humphrey Bogart, Lauren Bacall, and Paul Henreid. Jack was furious that some of "his people" went on the trip.

On November 24, several studio heads met secretly at the Waldorf-Astoria in New York and issued the Waldorf Statement. While the statement acknowledged that there was a "danger in hurting innocent people," and "creating an atmosphere of fear," the blacklist was put into effect and the first Ten were fired. The statement read in part: "We will not knowingly employ a Communist or a member of any party or group which advocates the overthrow of the government." The blacklist became one of the darkest moments in Hollywood history.

Years later, in March 1951, while the Korean War was in full fury, the HUAC hearings on Hollywood started up again. Mounting anti-

Communist feelings were principally directed at smoking out "Reds" and preventing possible subversive activity. Harry, to ensure that suspicions didn't center on Warners, called a special meeting to warn the 2,000 workers of "Communist conspirators." Everyone from manicurists to stars met on Stage 11 to hear Harry give a five-minute talk on the evils of Communism. The meeting lasted forty-five minutes.

"Don't allow these bullies to bully you," Harry said. "Get rid of them and any one of them who wants to go to Europe, where they can teach that kind of stuff. If it is Russia and the Russian way of life these people want, send them to us and we'll be glad to pay their way to that country.

"If my own brother were a Communist," Harry cried, "I'd put a rope around him and drag him to the FBI." Many there thought Harry would settle for simply putting a rope around Jack's *neck.*

Harry added that the studio would soon begin filming a new movie, *I Posed as a Communist for the FBI,* which would serve as a further blow at subversives. The meeting ended with the showing of a short subject, *Teddy, the Rough Rider,* which Warners had made ten years earlier and in which Theodore Roosevelt delivered a message on Americanism, and sounded a warning against subversive influences.

Harry's impassioned talk was just right for the times, and he was cheered by his employees and lauded by news journals.

One day Harry fired a writer for being a Communist, and the writer was horrified. Armed with papers and documents, he went to Harry's office and said, "Mr. Warner, there is a misunderstanding. I'm not a Communist. I've got all these documents to prove I'm not. As a matter of fact, I'm anti-Communist." With that, Harry interrupted and said, "I don't care what kind of Communist you are. Get out of here."

The House Un-American Activities Committee's hearings continued on into the early '50s with "Mad Dog" McCarthy screaming that Hollywood was a nest of "pinkos." As late as 1954, a judge dismissed a $51 million damage suit by twenty-three actors and writers who had lost their jobs as a result of the hearings. The judge noted, "The motion picture industry has a right to blacklist from employment workers who have refused to testify before the committee." Before it was over, 212 people involved in the film industry—both in front of and behind the camera—were blacklisted. The wounds inflicted by the committee hearings would fester for years.

The McCarthy era accelerated what was to be one of the major changes made in the movie industry. McCarthy had divided the movie "family." Now it was the government's turn to shatter it with the consent decree.

In late 1947 the Supreme Court began legal proceedings against the major studios, stating, "Motion picture studios who own theaters constitute a monopoly." Harry well remembered a decade earlier when Roosevelt's attorney general, Thurman Arnold, invoking the antitrust act, had tried to force the studios to give up their theaters, but the owners banded together and refused.

Jack grumbled, "If they say we can't own our own theaters, then tell the phone company they can't own wires to send calls. Next they'll be telling carpenters they can't own the tools to build a house. What the hell—is this the free enterprise system or not?"

In May 1948, the industry, realizing it was fighting a losing battle, agreed to a series of consent decrees stating that the major studios had to choose two of the three areas of operation: production, distribution, or exhibition. It was impossible to stop production and survive, and the movies produced had to be distributed, so with great reluctance the studios opted to sell off their theaters. The studios were ordered to phase them out over a five-year period. The line of theaters that Harry had so painstakingly collected were sold off at a great loss.

By depriving the studios of a guaranteed audience for their movies, the court decision was the first major step in bringing an end to the studio system.

What many would refer to years later as the Golden Age of moviemaking was coming to an end. The consent decree reduced the movie moguls from a position of absolute authority to one where they had to deal with actors, directors, and producers on a peer basis. The myth that movies could be made only by the major studios was now dispelled. Even the movie stars were beginning to realize this and acquired agents to represent them.

Jack tried to control the confusion by cracking down and barring agents from the lot. He felt they were continuously making his business dealings with his stable of talent exceedingly difficult.

Harry came in one morning and told him he was wasting his energy fighting these guys, that "he was bucking a stone wall." Jack's heated response was, "I'm not bucking anybody. They're bucking me, and by God they're not going to come in here and smash up everything we've built."

During the shooting of *Beyond the Forest,* Bette Davis again asked to be released from her contract and Jack, wearying of the struggle, agreed. Davis had this to say: "In the long run we respected each other. I'm truly indebted to Jack Warner, who during all these years gave me the opportunity to have a successful career. He was my professional father." Although Jack scoffed at his ex-star's response, he realized that the studio would no longer be able to hold on to their "children."

In the middle of all this, on March 8, 1948, my mother, Betty Sperling, gave birth to her third daughter, Cassie. I was named after the character Lana Turner played in a 1947 MGM movie, Cass Timberlane. *My second sister, Karen, had been born April 8, 1945.*

Despite the turmoil caused by the studio stars' new-found independence, Warners' 1947 net profit of just over $22 million was the biggest in its history. Yet it was the worst year for awards: the only Oscar winner was a Tweetie Pie cartoon, which made Jack shake his head in dismay.

In 1948 the company's profits fell almost 50 percent, in spite of such rewarding pictures as *The Treasure of the Sierra Madre* and *Johnny Belinda*, which won a Best Actress Oscar for Jane Wyman.

The following year, the net profits were once again cut to just a little over $10 million. Harry calculated that the total cinema admissions were decreasing, down to 60 million. Something was happening to change America's leisure habits. As the 1940s wound to a close, the studio began to realize that they would have to contend with a novelty on the entertainment scene.

Television.

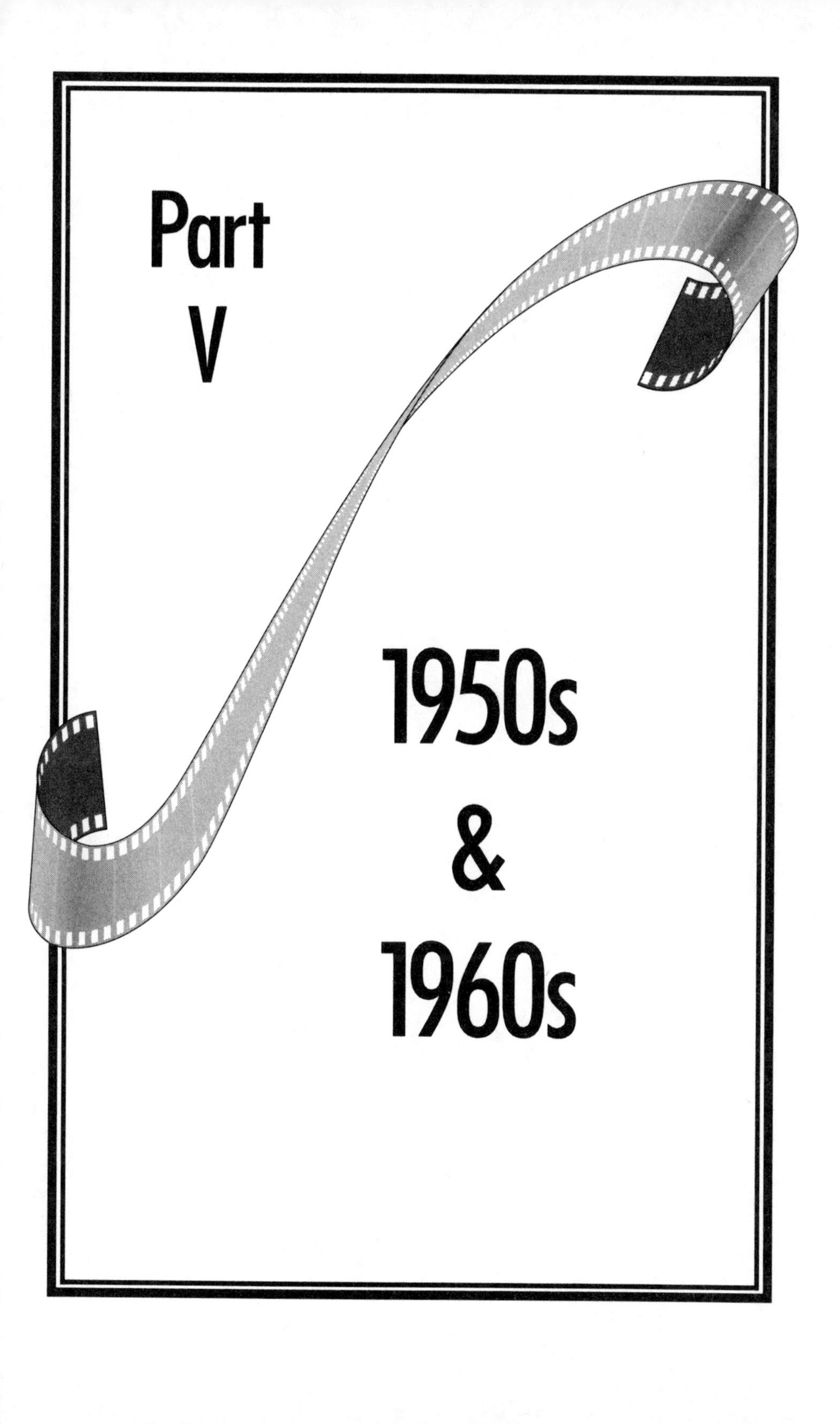

Part V

1950s & 1960s

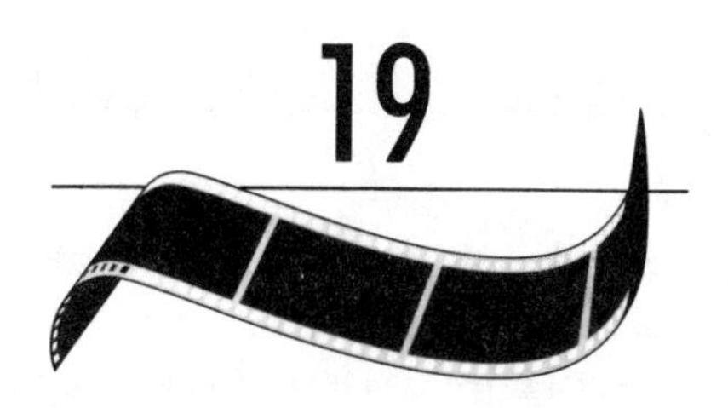

19

The War Between the Warners

"I'll get you for this, you son-of-a-bitch!" Harry Warner, raising a three-foot lead pipe threateningly over his head, chased his younger brother down the streets of the Warners studio lot.

Several showgirl extras costumed in silver sequined tights, their feathered headdresses swaying like knightly plumes, huddled against a wall, whispering as the two brothers charged past. Studio writers, who had heard the commotion, leaned out of their office windows, shaking their heads at the scene being improvised below. One of them nudged his buddy with an elbow. "Looks like old Harry is rehearsing a murder scene with Jack."

"That, my friend, is the real thing," the other writer said. "Truth is sometimes stranger than . . ."

"Yeah, I know . . . fiction." He pulled his head back in as the brothers churned around the corner of a sound stage and out of sight. "Just don't ask me to write it."

Harry, his breath coming in deep spurts, was losing ground. "Just . . . let . . . me . . . get . . . my . . . hands . . . on . . . you!"

Jack Warner slowed down, and called over his shoulder: "You old fart, you can't run as fast as I can!"

With one final burst of effort, Harry threw the pipe and watched in satisfaction as it spun like a boomerang toward his brother. Jack did a neat little side-stepping toe dance and the weapon clattered harmlessly a few feet from him.

Harry, his chest heaving, twisted his tie loose and glared at his brother. Jack glared back.

This terrible fight became part of studio lore. I asked my father whether there was any truth to the stories of these kinds of outbursts between the feuding brothers.

Other Voices

Milton Sperling

Yeah, they had terrible fights, the most memorable being that day Harry chased Jack around the whole lot with a lead pipe. So many people saw it. Can you imagine? The president of Warners, a dignified old gentleman who had won hundreds of personal awards, chasing the head of studio production through the lot saying, "I'll kill you, you son-of-a-bitch!" No, no, it wasn't done in fun. It was deadly serious.

Boy, did Harry and Jack fight. I spent most of my time on the Warner lot carrying truce flags back and forth between them, just to keep them from tearing the studio apart. They'd walk into the same room together and you could see them bristle, then jump at each other's throats. It was like watching a gamecock fight.

I'd say, "Listen, you're brothers, and you're trying to kill each other. But you're killing the studio instead." They'd both just stare at me. Harry was spending more and more time at his ranch, breeding racehorses, planting trees, and surrounding himself with his daughters and grandchildren, so I told him, "Harry, why don't you go into a kind of quasi-retirement and not come to the studio?" This time, *he* just stared at me.

My God, for ten years all I did—I tried to make a few pictures in between—was act as referee. I would come into the office and hold these two guys apart, then say, "All right, now, break clean and come out fighting." It was terrible.

Jack would call me into his office and say, "Listen, Harry's driving me crazy about the pictures I'm producing. You tell him that I'm letting you make these pictures—they are your pictures. I have nothing to do with them. Make 'em, do 'em, will you? For Christ's sake, you do them. *You do them!* Just tell Harry to get off my back. You fight him."

So I'd put this picture together and Harry would tell me, "Now that's what I call using your head. Why doesn't Jack do that?" Harry just couldn't see it, he couldn't understand what was happening with Jack, that his brother really was making good pictures. He didn't understand because he hated him so much.

As far as Harry was concerned, Jack was a first-class screw-up: He had divorced Irma, he ran around with anything in skirts, he had married what Harry referred to as a "whore."

Jack was foul-mouthed. I'll never forget what he said to a room full of people when he was trying to think of a synonym for a word. "I've been trying to think of a word all day," he said, then asked, "What's another word for cunthound?"

No one said anything, so I asked, "What do you need a synonym for?"

"Hell, I can't write that word in a letter!"

So I said, "How about Don Juan?"

"Naw, that ain't it."

"Seducer?"

"That ain't it. It's some other word."

"Why don't you look it up in the dictionary?"

"I did. 'Cunthound' ain't in there."

Harry had his own way of saying things. He had a prostate operation when he was in his sixties and it was very painful. Every time he took a pee, he'd say, "It had to happen to me instead of Jack!"

My mother's observations were similar to my father's. She said that when Harry raged about Jack, which was daily, a vein in his forehead stood out, throbbing. As Harry turned seventy and Jack sixty, the rows between them got more heated. And louder. Employees could hear Harry in the studio office berating his brother for wasting money, chastising him for leaving the studio months at a time to go to the south of France to gamble, and socialize with royalty.

In the past the obstacles the brothers had to overcome bound them together. Even united, however, they wouldn't be able to cope with the new threat: television.

In 1946 there were only 6,000 television sets in American homes. By 1950 there were 6 million. The viewing medium in a box was winning over the hearts and homes of America. There was no need to leave the comfort of one's living room, stand in lines, or hear ushers say, "Please get your feet off the seat." Television was far from perfect—it was in black and white, it was expensive ($500 to $800 for a set), and there was the hassle of wrestling with rooftop antennas, an aluminum tubing blight that had spread quickly across the nation's skyline. But after the original outlay of cash, television was *free.* And people who couldn't afford to buy a set could watch in a store window or a bar.

Harry had seen it coming. His daughter Betty remembers his fascination with television. "He said it would replace motion picture theaters," but also act as an outlet for the movies the studio produced. Jack said he didn't want any part of it. Harry argued that it was "the voice of the future." He even told Betty that he would have liked to quit the movies completely to go into television, but he felt he was too old.

In 1949 he had made preparations to ease the studio into the new medium by applying for a television transmission license. To his dismay, the Federal Communications Commission wavered in granting it. Harry decided he couldn't wait forever and said that his studio would shelve its plans to enter the field of television film production—which he said would have involved an expenditure of $50 million—unless the FCC made up its mind quickly.

Harry's real reason for not wanting to get into a prolonged fight with the FCC was that he was tired. He had fought too many battles—antitrust suits, the consent decree. He said, "I am sixty-eight now and I don't expect to live much beyond eighty, so I can't wait for the FCC. If we took as long in the business to make a decision as the FCC does, we'd fail."

He told Hedda Hopper: "After working forty-eight years I no longer know in which direction to go. Today one is punished by being too successful. I'm hoping we'll make the right decision, but if we do I doubt we'll know it."

Harry joined with the heads of the other major studios, such as MGM, Twentieth Century Fox, Paramount, Universal, Columbia, and RKO, in behaving as though television had to be a passing fad. The programming for the fledgling stations by independent producers across the country was shoddy at best. All the moguls had to do was hold on, they said, even though they saw more and more production companies providing the new medium with shows and more and more prominent stars joining the procession to television.

The major studio bosses had an unwritten law: They would not sell their high-grade productions to television, even if the movies turned to dust in the film cans while locked in studio vaults. They stuck to the theory that if a movie was good enough, the public would pay to see it at a theater.

Yet weekly ticket sales continued to drop dramatically. Strong measures had to be taken to win the audience back from the little box with its fuzzy images.

Producers tried new techniques such as Cinerama, which debuted in *This Is Cinerama* on September 30, 1952. Unfortunately, there were only two theaters in the United States equipped to show the three-camera projection on screens that curved a third of the way around the audience. United Artists thought the answer was in three-dimensional productions and released *Bwana Devil* on November 27, 1952. (The press dubbed 3-D "the deepies.")

Jack Warner countered with *House of Wax,* in which everything portable on the set was thrust through the screen. As the audience filed into the theater for this first Warner 3-D showing, they were given cardboard glasses with blue-red plastic lenses. The illusion of having rocks and spears hurtling out of the screen may have elicited a few screams, but the glasses induced more headaches than they did box office receipts.

Twentieth Century Fox tried wide-screen CinemaScope and released *The Robe*—"You see it without special glasses"—in September 1953. When the movie opened in New York, Abe called Harry. "There's lines around the Roxy, night and day. Maybe we ought to get into this thing." The Warners tinkered with the wide-screen technology in a system called WarnerScope, but the lens makers had miscued and almost half the image at the top and bottom was blurred. They went with CinemaScope and in 1953 made *The Command,* a loser starring Guy Madison, followed by *The High and the Mighty,* a winner, with John Wayne.

Paramount eventually came out with VistaVision, and Michael Todd produced *Around the World in Eighty Days* in 70mm TODD-AO, a new name for a process that had been around since 1930, when MGM tried it in a few specially equipped theaters to enhance a mediocre film, *Billy the Kid.* Audiences enjoyed these new productions, not because they were entranced by the illusion of 3-D, but simply because—as audiences always had—they loved spectacles, something that television had yet to offer.

Paramount's *Ten Commandments* and *War and Peace* did cause attendance to swell. Other movie companies thought they'd get smart too and upgraded their product, giving birth to megabuck movies. Although Jack fumed about the excessive costs involved in making these blockbusters, he okayed a series of epics: *King Richard and the Crusaders, Land of the Pharaohs,* and *Helen of Troy.*

Because these costume dramas were expensive to make, and seldom recouped their costs, producers had to search for a different way to counter TV, as the medium was now called. The answer was to film subject matter that was taboo on television because of its unacceptability to commercial sponsors. Drug addiction had never been addressed, so movies like *The Man with the Golden Arm* were made. Refreshingly adult movies were released, such as Columbia's *On the Waterfront* (the first film to use the word "damn" since *Gone With the Wind*), and Paramount's *Fear Strikes Out,* a movie about mental illness. In 1954 Columbia released *The Moon Is Blue,* which dared to use the word "virgin." Warners came out with the steamy Tennessee Williams play *A Streetcar Named Desire,* with Marlon Brando, then released Williams's *Baby Doll,* with its risqué theme of child sexuality. Yet nothing could turn back the tide of television. The movie industry during its peak year of 1946 recorded nearly 100 million admissions per week; by 1956, weekly attendance had dropped to 46 million. Feature film production fell from nearly 500 movies a year to 254 in 1955.

Harry thought there was too much loose talk in Hollywood about TV and jokingly threatened to put into effect the same ban he had instituted in the early 1930s when the Depression almost ruined the movie business. "At that time, all our employees talked about was how bad the economic conditions were," he said. "We decided that if the employees would concentrate on their work instead of worrying about the Depression, we might be able to produce some movies and pull out.

"So we banned all discussion of the Depression—except between ten and eleven A.M. on Thursdays. That was known as the 'worry hour' for the week, during which all of us could sigh over conditions. But promptly at eleven A.M.—no more talk about the Depression. Now we'll have no more talk about television."

There was one Warner Bros. employee who wanted to talk about television: Jack Warner Jr.

Other Voices

Jack Warner Jr.

When I worked in London after the war I visited the BBC studios many times and was impressed with their progress in this new medium, which was far in advance of ours. I firmly believed that Warners should continue its role as a pioneer and not wait for others to act. I compared it to the time the brothers grabbed the talking picture and ran down the field for a touchdown while the rest of the industry sat in the stands watching. Yet none of the Warner executives would take the ball and run. I decided to talk to my father about it.

Television was a dirty word to Jack Warner, and he refused to even allow a TV set to appear in a film scene, even after they had become common pieces of furniture. The distribution department had him scared stiff that theater owners would stop buying Warner pictures if the studio defected to what they saw as the "enemy." Also, because the major production companies had been involved in an antitrust case, the Department of Justice did not look kindly on an expansion into another medium.

When I started working at the studio again in 1950 I went to my father and suggested Warners get into television. His reaction was to jokingly order me to stop using the word "television" in his presence.

"If you say that word again, I'll have to wash your mouth out with soap," he said.

Columbia Pictures furnished me with the ammunition I needed in their annual report, which showed big profits from their TV subsidiary, Screen Gems. Harry Cohn at Columbia had been the first studio boss to break ranks when he signed with Ford Motor Company to make thirty-nine half-hour dramas for $25,000 apiece. I pointed out Columbia's bottom-line figures to my father. He only chewed on his cigar and grumbled.

A few days later he called me into his office and said, "Okay, you want us in television? All right. We're in it. You set it up and run it—but don't spend any money!"

It took time and lots of corner cutting, but Warners had a brand-new television department consisting of me, a secretary from the typing pool, a story editor loaned out by the writing department, and a young man by the name of Gary Stevens borrowed from the publicity department. The firm ultimatum prohibiting me from spending any money put tremendous limits on the operation. I could use only properties already owned by the studio and was forbidden to shoot any pilot films.

Unknown to me—I read about it later in the trade papers—Bennie Kalmenson, the head of distribution, worked out a deal with friends at ABC Television, securing an order for a weekly one-hour TV series to be produced by Warner. I was surprised to read that my part-time assistant, Gary Stevens, would be the executive producer of the series. I was angered, outraged is more like it, by the bypassing of my department and what I saw as a deliberate slap in the face administered by my father.

Warners, at my father's bidding, had agreed to do a one-hour weekly show for ABC, "Warner Bros. Presents," showing segments of some of their great movies like *Casablanca* and *King's Row.* Each hour included a ten-minute segment, "Behind the Camera," in which Warners previewed new studio productions.

Harry and Jack Sr. invited the ABC executives to the studio for a tour and Harry delivered one of his rambling speeches. Jack elbowed the executives on either side of him and cracked, "I apologize for my senile brother." Harry heard the laughter, but, fortunately, not his brother's comment. That is, fortunately for Jack.

Jack Warner Jr., who had felt like walking out of the studio when the ABC deal was made behind his back, stayed on, mostly because he felt his new job as head of television commercial production had great potential.

Other Voices

Jack Warner Jr.

I set up a division making television commercials and industrial films. That required a lot of negotiations with vice presidents of major corporations. I felt I would function better as the head of a corporate division—a vice president of the company.

I drove up to my father's house one morning and invited myself to breakfast. I explained the situation to him and he was cool and unresponsive.

"What the hell," he told me. "You don't need to be a vice president."

"Dad, I'm dealing with the vice presidents of American Telephone and Telegraph, of Ford Motors, General Motors, eight beer companies . . ."

"You want to be a vice president?" It was like he had never heard the request.

"Yeah . . . a vice president."

"Just so you can talk to people?"

I got burned and jumped up and hit the flat of my hand down hard on the table between us. He jumped almost out of his chair, looked at the spot on the table I had just hit, and kind of sheepishly said, "Okay, you want to be vice president . . . you're a vice president."

Then, thinking about charges of nepotism, he added, "I hope I don't hear nothing from the stockholders."

I said, "If you do, it will be compliments." I had finally spoken a language he understood. I should have hammered a few more desks.

Jack Warner Jr. had also decided to tour the studio on his own and evaluate its efficiency. He went so far as to write up his observations in a confidential seventeen-page letter to his father. In it he noted that there were many "internal problems" confronting the studio. He named names of producers and directors, and defined adverse personality traits. The lengthy report said in part:

> H. M. Warner, president of the company, is right in the production picture, yet his tie-in with production is rather unclear and undefined. My opinion is that he should be concerned with the broad picture rather than the tiny details of everything. A feud is going on between H. M. and J. L. Warner which has its roots in personal family matters. It is the old story about letting personal things slop over and mess up everything and it takes a damn mature person to keep things in their proper proportion. In too many ways these grown men, and the people around them, are *not* grown men.

Under the subtitle "Where Is the Boss?" Jack Jr. noted this about his father:

> In every business the workers like to see their Boss now and then. The soldiers in line like to know their general is right there with them—even though he may do most of his work in a house well behind the action. Here at W.B. the Boss rarely, if ever, gets on the back lot . . . I find that swinging around the sets twice a day has done a tremendous amount of good. I am still amazed that I am the first person from the "front office" ever to sit in on a technical department meeting. Of course, the Boss cannot take the time to swing around the lot, but now and then he should follow the advice of one of England's kings who said, "Show yourself to the troops."

Jack Warner didn't read his son's report for six weeks, but when he did, he called him into the office. Looking sternly at his son, he asked, "Has anybody else seen that report you made?"

"Only Steve Trilling," Jack Jr. answered, referring to the new executive producer Jack had appointed after Hal Wallis's departure.

"How many copies are there of it?"

"Only the original and the copy I gave Steve plus one for my files."

Jack jumped from his chair. "Don't show it to anybody! I want you to destroy the copies you have . . . get the one back from Trilling . . ." He slumped back into his chair, closed his eyes, and waved a limp hand. "Let's not talk about it further."

That was the end of the report—except the one copy Jack Jr. saved.

Even with the inefficiency Jack Jr. had tried to chronicle, Warner Bros. continued to crank out films in the uneasy '50s—246 of them, only slightly fewer than the 274 released in the 1940s (that number would have been higher if not for the war), but less than half the 572 produced in the '30s. The '50s would come to be Warners' least distinguished decade, despite such pictures as *A Star Is Born,* with Judy Garland, and the James Dean classics, *East of Eden, Rebel Without a Cause,* and *Giant. A Lion Is in the Streets,* a story about corrupt politics, starred James Cagney, and Anne Francis as a tempestuous blonde called Flamingo. Francis made five films for Warners during these difficult years, including *Girl of the Night,* in which she played a prostitute.

Other Voices

Anne Francis

After I made *Girl of the Night,* which was my favorite film, as I felt I did my best work in it, my agent sent me a script, *Claudelle Inglish.* I thought the story was trite and my part a caricature of a soap-opera character. I told my agent I didn't want to do it, and immediately got a call to appear before Jack Warner. I didn't want to go. Not because I was afraid of him (at least he'd never made a pass at me), but I just didn't want to get in a hassle. Of course, I went. Jack was in his huge office, sitting behind his huge desk, a casting director by his side.

"Anne, baby," he started, "this is a good juicy part: farm girl, abandoned by her boyfriend, becomes a man-hater."

I disagreed and argued that the script was dull, the language ugly.

"Anne, do this part for me."

"It's just not right for me."

"Aw, Anne . . ." Then he pulled out his billfold and slowly started slipping money out of it. He kept doing this all the time I was there. I guess it was his way of bribing me to do the part. I refused. The movie was made with Diane McBain, and bombed.

A couple years later, I went to a Hollywood event and found Jack sitting opposite me at the dinner table. He leaned over and said how much he loved the movies I had made for the studio. Then he looked me deep in the eyes and said, "You know, you're a wonderful actress, *Arlene.*"

Elsa Maxwell wrote a glowing article about Jack in her column around this time:

> Jack Warner is a man who improves with the years. I have received many letters asking what sort of a man he is, what he likes, eats, wears. Here goes!
>
> He is part clown, part comedian, whose greatest joy is to make people laugh. And he is probably the most public-spirited citizen of Los Angeles. There isn't a committee or a cause of any importance that Jack doesn't sustain or assist . . .
>
> He is a warm and wonderful friend, one of the kindest and most thoughtful men I have ever known. He is supposed to be the hardest man to do business within Hollywood, but since I do not have any business dealings with him, I don't know that side of him at all. But as a friend, I will put him against a thousand lesser ones . . .
>
> Ann prefers to remain cloaked in her anonymity, but her finger is on Jack Warner's pulse every moment.
>
> Jack likes gay-colored neckties, but his suits are sober in color, and well made; he loves fine clothes, good food, and has the strangest ideas of economy for a rich man. Every light in the house must be put out. If he comes into my sitting room and sees a picture a little on the bias, he is completely miserable until he has straightened it to his satisfaction. If there are loose pencils, bits of paper, or any element of disorder on your desk, he unconsciously straightens them; if there are ashes in your ash tray, he empties them into the wastebasket. He is the most meticulous man I have ever known . . .

I was six years old when my dad started taking me to the studio. I began to sense that he and my grandfather were treated with a special respect. They both had their own offices and bungalows and secretaries who called them "Mr. Warner" and "Mr. Sperling." They ate in a fancy dining room—the Green Room—where we were personally served and which was much less crowded than the commissary next door. There were two doors that led into my dad's office, so that when you opened one there was still another to open before you were in the office. The chairs smelled like a new saddle, and there was never any dust around.

Going to the studio was like going to a circus. I was immediately welcomed wherever I went. I loved it. The Golden Age may have been over, but there was a definite magic happening on the sound stages, and a buzz of activity swarming around the stars, all of whom knew they were getting their moment to be bigger than life.

Hollywood was in transition, and not simply because the studios were embracing television. Heavy U.S. taxes and escalating cost of film production in Hollywood drove producers to start making films in Europe where it was cheaper, because they could avoid the theatrical unions. By the mid-'50s, 30 percent of all Hollywood productions were being made outside the United States. By 1956 the major studio heads were looking for a way out of their empires, even to the point of selling their vaults of movies.

On February 13, 1956, the Warner Bros. board of directors agreed on the $21 million price offered by the Lansing Foundation for the world negative rights to 784 features and 1,800 shorts, all made before 1948. The contract was entered into with Elliot Hyman, president of Associated Artist Productions, a television film distributing organization.

On March 1, 1956, CBS made a cash offer of $25 million for Warners' backlog of features and shorts. Jack wanted to wiggle out of the Associated Artist deal, but Harry chose to stick with it because they had verbally agreed to it, thereby losing $4 million.

The announcement of the sale shocked the film industry, but the other major studios soon followed suit by selling their film libraries.

On February 16, the three Warner brothers turned down a bid for stock made by a Canadian investment trust.

Stories started circulating around Hollywood that the Burbank studios were up for sale. Errol Flynn sent Jack a telegram pleading: "Chief—Don't sell. Who will I fight with?"

At the same time, Paramount sold out to Gulf+Western. Hollywood—and Wall Street—waited to see who would be the next to go.

Abe Warner wanted to be the next. He was tired of working, and wanted to move permanently to his big home in Florida with his wife, Bess, who was ill. He hated to visit California because of his allergies. "Harry," he kept telling his brother, "I'm seventy-two, I want to relax, go to the horse races . . . Harry, I want out of the business."

Harry, whom top-echelon employees called the "Strategic Generalissimo," took great pleasure in his job and knowing everything that was going on at the studio, but any chance he got, he would head for the ranch to spend time with his beloved horses, as well as his grandchildren.

At the end of the windy road, and after what seemed like hours of sisterly squabbles and some car sickness, there was the serenity of Grandpa and Grandma's ranch waiting for us.

Ol' Prince, the St. Bernard, would greet us with his massive clumsy body and wet kisses. Grandpa used to let us ride him until we were about three.

By 1955, when I was seven, I had a favorite ritual which I always made sure I had time for. After stuffing myself with the usual goodies—poppyseed cake, lox and bagels, potato pancakes with applesauce, pickled herring in sour cream sauce, and fresh fruit salad—and getting bored with the adult conversation, I'd excuse myself and go outside and mount my favorite deer. So what if she was made of lead? That only meant she would always be there in the same place, resting on her haunches, legs tucked under her, staring out at the open fields.

Once I had taken in the panorama and opened all my senses to the familiar smells of sage, damp straw, and ranch dust, I'd fix my gaze just like my metal friend, and drift off into the comforts of my thoughts. I felt immortal, privileged, without borders, and forever safe. By the time I had indulged in this form of personal dessert, my food had digested so that I could get permission to swim. The pool bordering the knoll that the ranch house sat on overlooked the expanse of land below.

Going to the stables was the only thing that got me out of the pool. Grandpa's pride would come shining through as he walked us down the hill, sharing his love and respect for nature and the land, then he'd give us a tour of the sleek racehorses in their immaculate stalls. I remember the joy I felt when Grandpa told me he had named a colt after me, and that she was bred to be a prize racehorse.

Cassie was the most beautiful horse I had ever seen, and he told me he knew she was going to be a winner.

Then there was that room next to the hallway entrance that I was always drawn to. It had a presence to it, like a museum. It was dark and mysterious—a room full of treasures. This is where Grandpa would find me and explain the significance of some of the shiny objects I was so curious about.

What I didn't know was that his story, which had begun with Sam and Abe and Jack, and the first nickelodeon, was now coming to an end.

Jack Warner had begun to spend more and more time in the south of France, gambling. He loved to play the gentleman's games of chemin de fer and baccarat, in which he kept company with such luminaries as the deposed ruler of Egypt, King Farouk, and the Duke and Duchess of Windsor. Jack himself was treated as a monarch.

Ann preferred shopping trips to Europe and would precede him by six weeks. Jack felt she needed an escort, and selected Richard Gully, an affable and sophisticated young Englishman from a prominent family who had come to Hollywood to "be in the sunshine" and ended up working at the studio in 1940.

Other Voices

Richard Gully

When I first met Ann Warner she was just about thirty years of age—not only lovely but wonderfully vivacious and a terrific dancer. I used to see her continually at parties when I escorted Countess Dorothy di Frasso. I quickly discovered that Ann and I shared mutual interests. She was fascinated in astrology and had all sorts of mystics come to her house. That's one of the things she liked about me: I was a very good astrologer. I would do her horoscope, predict things. I thought she had an extraordinary mind.

The first time Ann ever invited me to be in her party, she and Jack took me to a restaurant with Lili Damita and Errol Flynn. Those were the days when Elsa Maxwell, Lady Mendl, Barbara Hutton, and Mrs. Evalyn Walsh McLean with her fabulous Hope Diamond were all in Beverly Hills, adding greatly to its elegance and excitement.

I remember in the early '40s Jack and Ann were so close, holding hands, flirting in public with each other. By 1948 Jack started having mistresses. I know, because I became his cover-up, making excuses to Ann. But she knew. Ann adored Jack, she just couldn't control him.

And then Ann started to flirt with others, and I had to cover up for her. She was glamorous and attracted men, all sorts of famous European men. She collected men easily. Ann had a crush on Ali Khan one year, and I had a hell of a time keeping him away from her before Jack would show up on the scene. Ann never had lovers. She had flirtations. Well, there was the affair in the early '40s when Jack found Eddie Albert in bed with Ann, and she *may* have hopped into bed with Ali Khan . . . and there was a very good-looking Italian, a certain Prince so-and-so. . . . They were never more than flings.

Ann Warner had a rare quality that I can say about very few women. She was exciting. Even the number one glamorous courtesan of our age, Pamela Churchill, who slept with every famous man in her day, wasn't exciting. Ann was glamorous *and* exciting. Norma Shearer and Mary Pickford gave good parties, but there was no electricity in the air. With Ann there was always electricity. For twenty years, from 1936 to 1956, Ann was the top hostess in Hollywood. She and Jack gave lavish parties, and received fascinating invitations in return. They went to Venice with Marion Davies and William Randolph Hearst, visited the White House and chatted with FDR, dined with kings and queens.

And that's where it began to happen.

They'd go to these elegant, important parties in Europe—and she'd be ignored. In Europe there's such a thing as protocol: If you are a duke, you are seated at the hostess's right; the duchess next to the host. That didn't work with Hollywood's nobility. In Europe they'd ignore the wives of the movie moguls. Everyone would fawn over

Jack, and ignore Ann. Jack was so happy basking in his own glory, hobnobbing with royalty, that he didn't notice what was happening with his wife.

The saddest case I remember was a party Elsa Maxwell gave for the king and queen of Italy. There were two tables, and Jack was seated at Elsa's with the king and queen, and Ann (and I) were relegated to the other. Ann was very sensitive about this treatment; after all, she was beautiful, and she was not a boring woman like Mrs. Zanuck or Mrs. Mayer. Because Jack overshadowed her, she stopped going to parties, saying, "To hell with it. I've had it!"

And she began to put on weight.

The first sign of danger I saw was when she refused to go to Grace Kelly's wedding. The hardest thing for a woman is to be a raving beauty and to suddenly lose her looks. In Ann's case it wasn't her face (she never had a facelift), or her hair, as she had beautiful hair; rather, it was her body. After 1956 she put on about eighty pounds. It bothered her terribly, but she couldn't do anything about it. She was a diabetic, and she was also a night person, raiding the refrigerator at three in the morning. She had no self-discipline.

She became a recluse. It's not that she wasn't interested in everything—she still talked on the phone to her close friends (I talked to her for hours at a time at least once a week)—it's just that *she didn't want people to see her overweight.*

It saddened me terribly, because Ann and I were so close. I was closer to her than I was to my own mother. (Ann and I must have known each other and lived many lives before, as we could almost read each other's minds.) It really bothered me that she wouldn't let me come to the house anymore. She said, "No, we've had so many glamorous times together, I don't want you to see me now." I couldn't get through the gates.

Harry saw his brother's more frequent departures to the south of France as a violation of their lifelong basic agreement as a team. "All for one and one for all" must still be the code they lived by. Harry desperately wanted to keep it a family-run business but was struggling with Abe's reluctance to continue. Jack seemed to be withdrawing his allegiance to the team by spending longer periods abroad, leaving his job to Steve Trilling and Bill Schaefer.

Harry was being urged to retire by everyone in the family who was concerned about his health. But life without "the business" was hard to imagine. If only there was someone to pass the crown to in the family. If only Lewis . . .

No, there was no one to take his place.

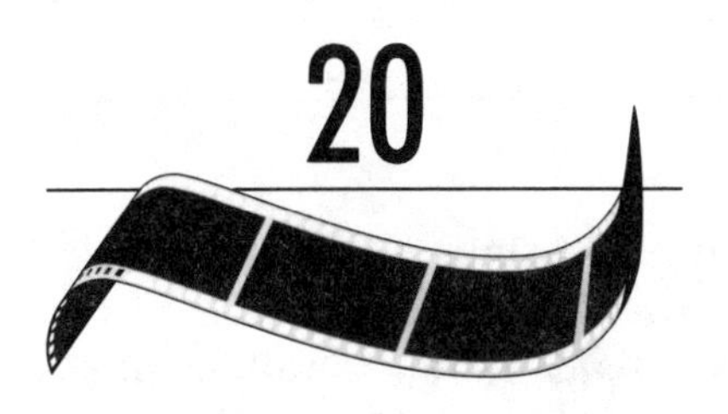

20

The Betrayal

Several big companies began to see that Warner Bros. was ready for a buyout. Secret talks began between Jack Warner and a couple of syndicates that were showing an interest in gaining control of the company.

On May 2, 1956, Toronto industrialist Lou Chester and his Canadian syndicate teamed with Charles Allen, a New York stockbroker, and announced that a deal "is on the verge of being wrapped up."

"I'm not ready to be put out to pasture," Harry yelled when he read of the offer in the trades.

Jack held his temper in check. He was playing the best hand of poker in his life: His fingertips could almost feel the sharp edge of his hole-card—an ace. In control of his voice, he said, "Harry, what the hell else can we do? It's time to sit on our ass and watch sunsets. I want out, Abe wants out . . ." Jack paused. His fingers were trembling; he hid his hands below the desk. "Uh, look, Harry, this 'united we stand' crap was okay for a bunch of kids from Youngstown, but not for three old bastards."

Harry felt the vein in his neck pulse as he slowly and deliberately said, *"I'll buy you out."*

Jack was startled. The smile on his lips faded. "Uh, Harry, you don't want to do that . . . Your family . . . you talked to Rea . . . ?"

Harry had talked to Rea. She was vehemently against his continuing with the company, as were his children. They had seen the deterioration of his health, caused in part by the terrible daily tensions between him and Jack. Rea especially wanted Harry, who was now seventy-five, to slow down, to enjoy his final years. Still, he didn't want to let go. The studio and its employees were his family, too. He loved the work, the challenge of making great movies, of staying one step ahead of the competition, of looking to the future through the medium of film. Being president of Warners kept him alive.

Just three months before, on February 24, 1956, he had been reelected president of Warner Bros. by the company's board of directors. President . . .

"You want it, don't you?" Harry asked, watching his brother carefully, trying to see the lie in his eyes. Or the truth.

"What the hell's that supposed to mean?"

"*You* want the title of president—you have always wanted it . . . the prestige that goes with it."

"Harry, you're talking in riddles. We're talking about selling, not arguing about who's president. Hell, you are. You always were."

"Were?"

Jack's fingers flicked the imaginary ace. "Were! Are! What the hell? I'm tired of fighting. Why don't we sell . . . enjoy the last years of our lives?" He leaned forward, elbows on his dark mahogany desk, wanting to wipe away the tiny beads of sweat that had formed on his forehead, knowing he couldn't. "This is a good offer from a prominent company. We'd go out in style."

Harry bowed his head and stared at the floor. He knew it was over. He *was* weary from all the years of quarreling with Jack. So weary . . . "All right," Harry sighed. "Let's sell."

The deal with the Toronto industrialist and the New York stockbroker was "on the verge of being wrapped up" for several weeks, when Abe called Harry from New York and said, "I got some bad news. You know they said they had a guy to take over Jack's job as head of the studio? Well, it's not true."

Harry pinched his nose between his eyes, feeling the pain. *Jack had prearranged this sale.* His immediate response to Abe was, "No deal. Forget it."

"Harry, it's not like Jack's taking over the company. He'll just be around in his old job until they find someone to take his place."

Abe wanted to sell. Not only was he tired and longed to totally retire, but he was very conservative and wanted to take his profit out of the company. He had convinced himself that if he didn't sell he would lose his money. The value of the stock would disappear if all of them, Harry, Jack, and himself, weren't running Warner Bros. "Look, Harry," Abe added, "let's say to hell with whatever Jack's trying to do. I feel we should go ahead with the sale."

"That's not what I agreed on."

"Harry, it's still a good deal."

"No! We all go or we all stay."

Jack, after a heated conversation with Harry, was forced to issue a statement to the press on May 3, 1956: "There is no truth to the rumor that a deal is now being made with Chester for the purchase of our stock."

May 11, the trade papers reported that a group of investors, headed by Boston banker Serge Semenenko, had offered the brothers $22 million for 800,000 shares (90 percent) of the Warner brothers' holdings. The group agreed to let Harry, Abe, and Jack remain on the board of directors.

Abe called Harry again. "I told Jack I'll go along with it if you do." When Abe didn't get any response from Harry, he added, "I really think we should both go along with it. It's what the family wants."

Harry was still cautious. "Who's going to be president?"

"They got a guy named Simon Fabian—they call him Si—he's a key figure in the new group and will be elected president. They say his participation has to be cleared by the Department of Justice. You know, the same old antitrust crap. It's just a formality." Abe said that Jerry Wald, who had produced such hits as *Mildred Pierce* (which won an Oscar for Joan Crawford) and *Johnny Belinda* (which won one for Jane Wyman), had been suggested to take over Jack's job. When Wald had been at Warners, many thought Jack had been grooming him for studio chief. Wald was now the studio boss at Columbia and it had been reported that he was negotiating a change in contract status.

"A couple of others have been mentioned," Abe added. "Bunch of bums . . ."

Harry paused for a long time, so long that Abe thought he was off the phone. "Harry, you still there?"

Finally he replied: "It's okay. I agree. But—*we all go together.*"

Abe had heard the "together" song too long. He just wanted out. "Okay, Harry, we all fade away into the sunset. Just like in the movies."

Harry didn't want to fade away into retirement. Retiring meant there was no longer a *game* to play. A high-stakes game called "Motion Pictures."

Harry sat in a large echoing boardroom behind a long cold table. Jack and Abe were across from him and in front of each was a folder of papers. The shadows at the day's end emphasized the sunken, tired features of the brothers—especially Harry's. A lawyer stood at the head of the table. When the four men spoke, their voices were low. It was like being in a tomb.

Jack, conciliatory, talked softly of the old days—of the fights they had fought together against all odds. Although tired, his mood was happy. "It's time to rest, get out of this cockeyed business while we still have our health."

The lawyer handed a pen to Jack, who quickly signed the document in front of him. Harry took the pen after Abe signed. The pen hovered over the paper for a moment, then touched the signature line. Harry sighed. "I never thought I'd see the day that Warner Brothers would be run by a corporation."

Jack watched Harry pause over the sale document.

Harry signed the document and mumbled to himself, "Sorry, Pops."

"What was that, Mr. Warner?" the lawyer asked.

"Nothing . . ."

Louella Parsons's column of May 14 reported: "All three brothers dined together at Perino's. After selling the studio for $20 million, they could afford it."

The next day she wrote:

> Jack Warner made Warners' studio employees very happy by calling a meeting of Warner department heads and announcing that the purchase of the studio for nearly $20 million doesn't mean that he is leaving as production head.
>
> "I'll be around as long as my feet hold out," Jack said, "and that should be many years. There's no truth in the rumor that I'm leaving in October and Jerry Wald is taking over as production head. Serge Semenenko has asked me to remain and I have every intention of continuing to head the studio."

"It's a dirty trick," Abe said to Harry, standing stiffly in front of his older brother's desk.

"It's just Jack." Harry was distressed by the news, but tried to stay calm. "I guess he'll be on some kind of salary to keep his old job."

"It's still rotten. Not one word, he didn't say one stinking word about staying."

"It doesn't matter, he sold his stock." Harry got up and looked at a huge aerial photograph of the studio complex, then made a sweeping gesture with his arm. "He doesn't own one square foot of the studio. Let him work his balls off."

Abe just shook his head. "He's never going to change."

"I don't want to talk about it," Harry said as he felt the familiar palpitations. He swallowed and lowered himself unsteadily into his chair. "Get the doctor in here." Harry gulped for air, trying to force the erratic heartbeats back down into his chest as Abe grabbed the phone.

A few minutes later the studio doctor hurried in. "Some patient you are," he said as he pressed two fingers on the vein on Harry's neck to slow his heart down.

A cold sweat had broken out on Harry's body, but he could feel his pulse weaken from the pressure the doctor was applying. "I'm all right."

"I told you, take it easy," the doctor said. "No more bullying your brother. The last time, you chased him with a pipe and almost killed yourself."

"I missed the son-of-a-bitch."

"You should go home and plant a tree on your ranch."

"I plant trees," Harry said, dabbing at the clammy sweat on his face with a handkerchief.

"Just don't worry about what your brother is doing," the doctor said. "Worry about yourself."

According to Neal Gabler in his book *An Empire of Their Own,* Jack reportedly told another executive after the coup, "I've got the old bastard by the balls at last. He can't do a goddamn thing."

Dennis Hopper recalled being introduced to Serge Semenenko while walking on the studio lot with his friend James Dean, shortly after the news of the sale had come out. "When Semenenko went to shake Jack's hand, Jimmy reached into the pocket of his pants and pulled out some change and threw it at Semenenko's feet and walked away. I ran to catch up with Jimmy to find out what that was all about. He told me that he liked Harry and that he didn't like what Jack had done to Harry. He knew that Semenenko was behind it."

Hollywood insiders were convinced that Jack would never take production orders from the Serge Semenenko group or any new member of it. Taking orders was not one of Jack's talents. Everyone thought he would hold on to his position only as long as it took to find a new studio storekeeper. Jerry Wald was still one of the leading candidates for the job, but the guessing game continued, digging back further into Warner history and turning up Darryl F. Zanuck, whose name and fame had been born at Jack's side. Another Warner graduate mentioned as qualified to fill the post was Mervyn LeRoy. After his divorce from Doris, he had come back to the studio and produced 1955 Best Picture nominee *Mister Roberts.* (The movie was James Cagney's forty-fourth and last picture for Warners.) Even Louis Mayer and David O. Selznick were mentioned as successors to Jack's forty-year reign.

Harry heard all these musings and was satisfied that Jack would not work for long at the studio.

Until a new studio chief was named, production continued under Jack's supervision. Alfred Hitchcock was filming *The Wrong Man* with Henry Fonda. Spencer Tracy was doing Hemingway's *Old Man and the Sea,* while newcomers Tab Hunter and Natalie Wood were featured in *The Girl He Left Behind.* Cutbacks were being made, however, and people were laid off in every department on the lot. Jack announced: "No more B's and Z's; no more run-of-the-mill A's. We are going with the absolute tops. If we can't get that, we'll lay off until we do, because nothing but the big ones are going to strike pay dirt."

On the last day of May, Harry read the startling news that Simon Fabian, who had been scheduled to take over as Warner Bros. president, had been denied the position by the Department of Justice because he hadn't disposed of his theater interests.

Who *was* the president of Warner Bros.?

Harry sat at the breakfast table in the ranch house, going through his usual ritual of reading the trade papers. He opened up *Variety* to see the headlines trumpeting the story of the sellout. Scanning the newsprint, he was startled to read the line: "Jack Warner retains substantial stock holdings in the company."

Then, in big bold print under that was the statement: **JACK WARNER THE NEW PRESIDENT OF THE COMPANY.**

Grasping the paper tightly, Harry continued to read: "Jack L. Warner, the new president, said, 'I am very happy that my brothers, the board of directors and the distinguished financial group have placed under my direction the perpetuation of the company which our family has pioneered.'"

Harry turned pale. He dropped the paper, grabbed the edge of the table, and fell to the floor.

Abe, when he read the news in New York, felt totally betrayed as well. His family reported that he wanted to kill Jack. Had he been in the same room with his younger brother, he might have done just that. From that moment, Abe never spoke to Jack again. (Abe later joined the company board, saying he did it only to watch Jack and "keep him from stealing the stockholders blind.")

The day after Harry collapsed at his breakfast table, *Variety* reported that "Harry M. Warner checked into Cedars of Lebanon Hospital for observation."

Jack Warner Jr. recalled:

> Barbara and I had planned on going to Hawaii for our tenth wedding anniversary when we heard about Uncle Harry's heart attack. We felt it would be wrong to leave at such a time and canceled our trip. I went to the studio the next day, where my father saw me and said, "What the hell are you doing here? I thought you and Barbara were off for Hawaii." I answered that with Uncle Harry so ill we just couldn't go.
>
> "Dammit! Why don't you leave!" he screamed. "He's not *your* father!"
>
> I had no words for a reply.

Harry had suffered a stroke, which impaired his walking ability. His doctor gave him a cane and told him to remain in the hospital for a week. Harry hated hospitals. He told Rea he wasn't staying. She practically had to hold him in bed, telling him that he would remain there as long as the doctors advised it.

His daughter Betty recalled visiting him in the hospital:

> I went to see Dad and it was there we made final peace. He took my hand and said, "I love you. You are a good person." I gave him a hug and whispered, "I love you." He then turned to the nurse in a panic and said, "I'm getting out of here! I'm going back to the ranch. Where are my pants?" He jumped out of bed, obsessed with wanting to leave. He believed one would die if one stayed in the hospital. He opened the closet looking for clothes. It took two nurses to put him back in bed. They said they would call the doctor. He angrily agreed to be in bed, if he could go home that night.

When Betty left, Harry crawled out of bed and limped to the closet and put on his clothes. A nurse caught him trying to sneak out the door and led him back to his room. Six days later he was released. He immediately sold forty-two of his thoroughbred racehorses. He was seen at the racetrack, still using the cane to steady his walk.

The details of the acquisition by Semenenko and his group quickly became clear to Harry and Abe. Jack had sold his stock, but had made an under-the-table deal with Serge Semenenko to buy the stock back—*after* Harry and Abe had signed away their shares. Jack had joined Semenenko in acquiring principal ownership with the purchase of some 600,000 shares of Warner Bros. stock. It was agreed that Jack would retain ownership of 200,000 shares—and the presidency. He was now the company's largest stockholder. Ben Kalmenson was appointed executive vice president, and Sam Schneider, Harry's New York vice president and assistant to the president, would retain his position and take over the added responsibilities as treasurer, Abe's old job.

Bill Schaefer, Jack's personal secretary, believed Ben Kalmenson (who was general sales manager of Warner Bros.; Harry had originally hired him because he felt he was a strong executive) was deeply involved in the deal. "I think there was a little chicanery involved," Schaefer remembered.

> I'm not sure Kalmenson was one hundred percent for the company. From the conversation I heard, he got a payoff from Elliot Hyman, when the deal was made to sell Warners. Elliot Hyman must have owed the Bank of Boston or Semenenko's company a lot of money. The only way the Boston bank figured they were going to get that money back was for Hyman to buy out Warner Brothers, which is why I think there was a little finagling going on. Kalmenson was a difficult person. He was kinda like a fireplug in his build and spouted obscenities. He appeared to have some kind of hypnotic influence over Jack, who agreed to his judgments.

Richard Gully had this viewpoint about the sale:

> It was a very deliberate double-cross. Jack hated *not* being president of Warners. When I first went to Europe with him, he would never sign hotel registers "Vice President of Warner Brothers." He'd sign "Head of Production." He wanted to be *head man.*
>
> Then he met Semenenko, who conned Jack into selling the studio. Semenenko was a big figure in Jack's life—a key figure financially—but

> he was a lecherous, vulgar man, who pinched ladies' fannies . . . One time I lost my temper and called him a cheap son-of-a-bitch for the way he was treating some ladies. Jack took it calmly, looked a bit bewildered, shrugged his shoulders and walked away.
>
> Jack wanted to be president and saw a chance to do it with Semenenko. Semenenko came to Jack and said, "Look, I can make a deal for you. We arrange it, set it up so you and your brothers sell—and for that my fee will be a million dollars." Semenenko pocketed the million bucks, but had to give it up when the government found out about it. Anyway, it was a crooked deal.

Whatever the inside reasons that led to the sale, the betrayal was now complete. Jack L. Warner was now President, Warner Bros. Pictures Company Inc.

Harry never did get over his brother's treachery. Rea told her daughter Betty: "Jack might as well have put a gun to Harry's head."

One day, after calling ahead and determining that Jack was not at the studio, Harry returned to his old office. He was told by a temporary secretary, Lois McGrew, that there was a young man waiting to meet him. The youth, dressed in a suit and tie, approached Harry nervously and stood in awe of the frail old man.

"Mr. Warner, I just want to say that I know the problems you're having, and . . . well, would it help if I . . . I said . . ."

"Yes?" Harry prompted.

The young man blurted the words: "I consider you a living legend."

Harry leaned on his cane, a wry grin on his face. "A barely living legend."

"Mr. Warner, I envy you for what you've done, what you've accomplished, the millions you've made."

"No," Harry began, standing as straight as he could, "it is I who envy you. For you have something money can't buy. *You have time.*"

After the young man left, Harry told the secretary that he wanted her to write several checks. "My hand is a little unsteady," he said. "The checks are to be made out to my wife, my two daughters, to be given to them in the event of my death. I tell you this and want it kept in the strictest confidence."

"Of course, Mr. Warner."

When Harry told her the amount, the secretary was so startled she had to write the figures down on a separate piece of paper so she wouldn't make a mistake when writing the checks. Harry asked checks to be written to Rea for $3 million and to his daughters Doris and Betty for $1.5 million each. (To Betty's knowledge these checks never got distributed. The only explanation would be that Harry was worried Jack would play another trick to get the money back.)

After the secretary had made out the checks, Harry, with great sadness in his voice, said, "I'll never set foot on the studio grounds again."

Embarrassed, the secretary replied, "Oh, yes you will, Mr. Warner."

Harry took one last look around the office. It was a long time before he answered. "No, I won't."

Harry never returned to the Burbank studio. To further distance himself from the company's operation, he sold an additional 90,550 of shares of stock to Semenenko, which left him with 28,700 shares. He asked Doris and Betty whether they wanted to sell their stock and they went along with their father's wishes. When the sale was complete, Harry had made $8 million for all the work he had put into the studio over the years. Jack would continue to receive money from the company's assets. In December the Warners home office building in New York was sold for $2.5 million, an excellent profit (much of which went to Jack) from a property that had been in Warners' possession for thirty years. Construction began on a new building at 666 Fifth Avenue for the Semenenko syndicate. Betty always referred to the address as "Sick, sick, sick Fifth Avenue."

Shortly afterward, Harry, always the visionary, penned a story he titled "Initiative." In effect he was writing his own eulogy. It said in part:

> Sometimes a definite pattern of life, whatever its apparent advantages, can be a potent danger to man's advancement. . . . The sun will not wait, nor will the world. Empires, religions, and great industrial establishments alike have toppled into oblivion for lack of elasticity to move with changing times. And a man who holds to a set pattern, who fails to realize that a blend of initiative and tenacity is indispensable, wakes one morning to find himself out of step with an era which has long since passed him by. Hardening of the intellectual arteries is a dangerous element, both to its possessor and to humanity at large. Petrified trees maintain the status quo, but they build no houses.
>
> America was carved out of the wilderness by initiative—an initiative forged in lonely pioneer fires and tempered by the expanding hearths of industry. As long as it remains our heritage there can be no closed frontiers in all the Western World.

Six months after Harry had the secretary make out checks to his family, he became seriously ill, then suffered another stroke. And soon after, another one.

His daughter Betty recalled:

> Rea was under the impression that Harry knew what was going on even after he had several strokes. She'd talk to him as if he was normal and expect him to respond, asking him to do things: "Come on, Harry, let's go for a ride." "Harry, let's take a walk." There were brief moments when he would come around and seem to know what was being said. Watching television and eating were the only ways he kept in touch with life. He didn't like being in silence or alone.

On August 23, 1957, Harry and Rea celebrated their golden wedding anniversary with their family and a few close friends (a total of 125 attended) at the ranch in San Fernando. A dinner had been planned at the Beverly Hills Hotel for 300 but had to be canceled as Harry's condition worsened. Harry and Rea's three daughters, Doris, Betty, and Lita, were there, as well as eleven grandchildren; the youngest was Lita's two-and-a-half-year-old daughter, Vicki, and the oldest was Doris's son, Warner LeRoy, twenty-one, who had flown in from New York for the event. Jack Warner Jr. also attended. Harry was propped stiffly in an armchair near a window where he could watch the outdoor activities. Members of the family took turns being with him.

The youngest grandchildren were playing on the front lawn next to a party tent that had been set up for the celebration when, to everyone's horror, Jack Warner drove up. He jumped out of the car, all smiles, cracking jokes about Harry living with one woman for fifty years. He took a glass of champagne from a waiter and asked, "Where's H.M.? I'm bustin' my balls at the studio and he's livin' the good life!"

Then, seeing that everyone was staring at him, Jack gulped down the champagne, grabbed another glass, and glanced at the house's picture window. He saw his brother sitting there and with an embarrassed laugh said, "I guess I'd better pay my respects."

Even though I was only nine, I was fascinated to see this man, this mystery brother, whose name I had learned was a dirty word in the family. He was charismatic. His face was shiny, tanned, his hair slicked back . . . He seemed to glow, like he had just been polished. He was so animated he reminded me of a Looney Tunes character.

Other Voices

Jack Warner Jr.

I was sitting in the living room with Harry when to my surprise the door swung open to admit my father. He bounced into the room exuding good cheer and jollity, his unnaturally black hair and thin mustache glistening. He stepped close to his brother and tried to say something of little consequence, hoping Harry would perhaps notice him. Harry did notice . . . and he did the only thing he could still do. He closed his eyes tightly, shutting his brother from sight—and two big tears slowly rolled down his sunken cheeks. My father suddenly stopped speaking, stood stiffly in front of his brother, who had communicated in the only way left to him. Then, with his face suddenly gone bright red, my father turned and almost ran out of the silent room. I reached over to hold my uncle's hand for a while until the tears stopped.

I watched my great-uncle Jack's car speed down the driveway. Strangely, no one had said a word to him; no one had even said his name. It was as if he were already dead. I knew he had done something to Grandpa to hurt him, but I didn't know what. I only knew Grandpa had gotten very sick, had lost his spunk and spirit.

Worried about my grandpa, I walked into the living room and sat next to him, remembering the many times I had visited him on the ranch, ridden with him, talked with him . . . I looked out the window at the same spot he stared at—our spot—the place where we had spent every Sunday talking. Tears came to my eyes but I quickly wiped them away so he wouldn't notice.

I moved my chair closer to his side so I could put my arm around his shoulder, like he used to do to me. We sat there in the deepening silence for a long time.

Harry Warner died July 27, 1958. He was seventy-six. The cause of death was listed as a cerebral occlusion, but some said it was a broken heart. Funeral services were held at the Home of Peace Cemetery close to where his only son, Lewis, had been buried. Among the mourners were director Michael Curtiz and actors Edward G. Robinson and James Stewart. Rabbi Edgar Magnin, a longtime friend of the family, gave the eulogy, calling Harry a "plain, simple man who loved above all else being a farmer."

After the funeral, Rea, who was appalled that Jack had not attended, issued a malediction: "Harry didn't die; Jack killed him."

At the time, Jack was vacationing at his villa on Cap d'Antibes.

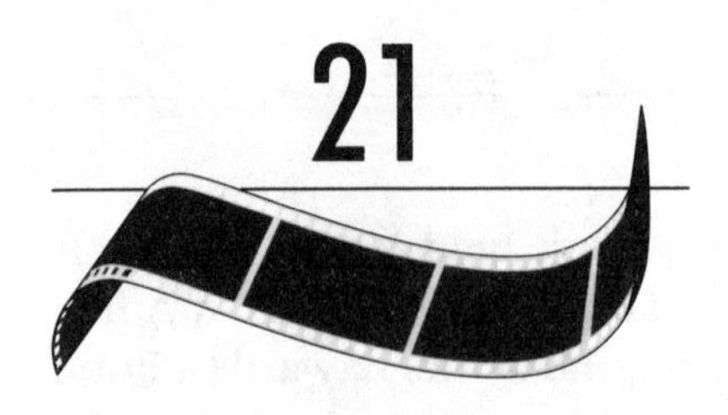

21

The End of a Dynasty

The telegram read:

FUNERAL FOR HARRY HELD UP PENDING YOUR ARRIVAL. PLEASE RESPOND.

Jack dropped the telegram onto the table. He paced the room for a while, then stood gazing at the view of the Mediterranean from his villa's picture window. Turning away, he called to his secretary. "Send a wire saying 'Unable to attend. Please pay my respects.'"

Inside the Gold Room of the Palm Beach Casino in Cannes was a baccarat table with comfortable chairs, on one of which was a brass plate engraved with the name Jack L. Warner. He had won this recognition as one of the more distinguished—and more frequent—players. Four days after Harry's funeral, August 4, 1958, was a big night for Jack. In a marathon six-hour session of baccarat he won a stack of chips totaling 2 million francs ($4,000). Staggering slightly—he was more tired than drunk, as he rarely had more than a glass or two of champagne, or a scotch on the rocks—he checked his watch: three o'clock in the morning. It was only a five-minute drive to his Villa Aujourd'hui in Juan les Pins, a quaint fishing village, lately taken over by tourists and American expatriates.

Cashing in his chips, he collected a wad of money, which he stuffed into his pocket. On the way out of the casino, eyelids heavy, he bowed deeply to everyone and tipped an imaginary top hat. "Bonsoir." "Bonsoir."

Someone said, "Arrivederci."

Jack cracked, "And a dirty river to you too."

He stepped out into the moonlight and into his waiting 1947 Alfa-Romeo with its right-hand drive, and a driver's door that had an irritating tendency to pop open. As he sped away from the casino, the door flew open and he cursed it.

He sailed down the winding Cornish Road with its two-foot-high curbstones, and as he passed the villa of his friend the Aga Khan, the Alfa-Romeo slowly drifted across the center line until it was moving at a high rate of speed on the far left side of the road.

Jack Warner was no doubt sound asleep when his car smashed almost head-on into a parked coal truck. In an agonizing screech of metal against metal the Alfa-Romeo flipped over and bounded down the road, a flaming mass. Jack Warner was pitched out through the right door and tossed forty feet from the burning car. He landed partly on the right front side of his head, his outstretched arms taking up much of the shock as he struck the pavement. Gravely injured, he lay there unconscious and bleeding. Behind him the coal truck burst into flames. The impact had trapped the driver behind the steering wheel, but the fire did not spread to the truck's cab.

A moment later a Volkswagen's headlights illuminated Jack. The tiny car, crammed with people who had just left the casino, skidded to a halt beside his crumpled body. They recognized the unconscious man in the road.

"Good God, it's Jack Warner!"

"Is he dead?"

"Sure as hell looks like it."

After calling for help at the nearest house and prying the truck door open and pulling the driver free, the rescuers did something in violation of all first-aid directives: They picked Jack up and jammed his limp body—pushing and twisting his arms and legs as though they were rubber bands—into the back seat of the Volkswagen and drove him to the hospital in Cannes, where the doctors, experienced in such matters thanks to the crazy drivers on the Riviera, would know what to do—as long as his spinal cord hadn't been damaged by the good Samaritans.

The only person from the family who was in France with Jack when the accident occurred was his daughter, Barbara, who was now married to Claude Terrail of the Tour d'Argent restaurant in Paris. Early in the morning Barbara called Bill Orr, husband of Jack's stepdaughter Joy, to tell him Jack was in intensive care and to ask him to inform Ann. Bill Orr did not notify Jack Warner Jr.

Other Voices

Jack Warner Jr.

I was eating an early breakfast when I heard the news on the radio. I phoned the studio at once, but there was not much they could tell me. My father could be dead—he could be alive. I rushed down to the studio and upon entering heard Jack's secretary, Bill Schaefer, talking to Ann on the phone: I could hear her voice coming through the receiver in a shrill excited tone. I asked to speak to her and the call was broken off. Later that day she flew to France, accompanied by Joy and her lawyer, Arnold Grant. Grant, an attorney from New York, seemed to me to be a strange choice for traveling companion but later turned out to be logical. (Arnold Grant was a young opportunist who had known Lewis, and after Lewis died, Grant had come to Harry for a job but had been turned down. Furious, Grant had told Harry, "I will ruin you, if it's the last thing I do.")

I knew that over the last ten years my father and Ann had lived almost separate lives. She isolated herself either in the house on Angelo Drive, in one of their Palm Springs homes, or behind the walls of the Scripps Institute. Her appearances in public were almost nonexistent. He took long holidays in the south of France or stayed in their two-floor penthouse apartment in New York's Sherry-Netherland Tower.

I decided to take my wife and go to Cannes, thinking, Had my father come back for Harry's funeral, he wouldn't have had the accident. I thought about that many times in the days that followed. If there's such a thing as divine retribution, then it proved out in this case.

It took two days before we were able to have our passports renewed and secure the necessary visas. After we landed at Nice our cabdriver drove us past the exact spot where my father's car had collided with the truck. We could plainly see the smudges and debris from the fire. We drove straight on to the Hospital Brussailes on a hill overlooking Cannes. It was near dawn and we were not admitted, so we went to the Carlton Hotel, where I tried to phone Ann at the villa, but she was not in.

Because of a combination of jet lag and deep concern, we were unable to sleep and went back to the hospital, where I was told my father was still in a critical state. He was unconscious and for a long time had been plunged into an ice-filled tub to hold down his temperature to avoid swelling of brain tissue, which might cause death.

We met with Ann at the hospital. Surprisingly, she invited my wife and me to the villa. There she said that when my father recovered and we were all in California again, we would be together and make a real start toward a happy relationship. "Good times are coming," Ann repeated several times as she waved good-bye at the door of the villa. During our conversation, I noted her attorney, Arnold Grant, standing nearby, arms crossed, a slight smile on his face.

When my wife and I appeared at the hospital the next day, we were told we could not see my father.

A week passed. I still hadn't been permitted to see my father. Finally the doctor told me he was out of danger, fully conscious, if a bit groggy, and I could see him for a minute.

My father lay on a bed, his face shrunken: He looked like someone I had never seen before, a stranger balanced between life and death. He stared at me with eyes I could still recognize and gave a weak smile. His left hand trembled as it reached out a few inches, fingers stretched toward me. I placed my hand in his and the fingers closed to hold me with a surprising grip. I felt much better, because up to the accident our relationship had greatly improved. I had seen him almost daily at the studio and he was very pleased with my work with the TV commercials and industrial films, especially the Bell Science series, of which he was proud.

The next day, he said my name in a very low voice. After I was with him for four minutes, the doctor asked me to leave, but my father clung to my hand and mumbled something that sounded like, "Let him stay." My wife stepped in for a few moments and held his hand. As we left, my father crinkled his face into a tiny smile and moved his hand in almost a wave.

My wife and I walked down the hospital corridor feeling greatly relieved. Ahead of us the double doors burst open and through them came Ann, her face twisted into a mask of rage. She stormed past us without a word. I tried to say how improved my father was, but she was already through the door, leaving an almost palpable trail of fire and smoke in her wake. I have never before or since seen so much hatred in one person's face.

That evening the doctor phoned me to say it was best I not see my father again, as his temperature had gone up seriously and he must not have visitors.

Bill Orr arrived several days after Jack Jr. When he saw Jack Sr. he was still on ice and barely conscious. Bill took Jack's hand and said, "How are you, Chief?" Jack replied, "Heart of gold and teeth to match," a saying he had used years before to describe an actor. Jack was conscious enough to realize his two front teeth were missing.

Other Voices

Jack Warner Jr.

After my wife and I left the hospital and returned to the hotel, I received a phone call from a French journalist (who worked for something like a French *National Enquirer,* I was soon to discover), asking if he could get into the hospital room and photograph my father. I told him absolutely not, that it would be an invasion of privacy. He pressed on, saying he would give me the camera so I could take some pictures. I told him my father was seriously ill and it was out of the question—and hung up.

A day later a story appeared in the French press which was translated for me. It said, in effect, that I had told the reporter my father was "near death." From this came a call from Al Goetz, my father's stockbroker, who was terribly upset that what I had said would depress Warners stock.

I told him the true story and he relaxed. I shook my head when I learned that the wire services had picked up on the story that "Jack Warner was not expected to survive," and even prepared obituaries, some of which were printed. Worst of all, Ann made it known to my father that I had committed a grievous sin.

I later heard from Doris, who had attended a party at which Ann and Arnold Grant had been guests, that Grant mentioned what "terrible things" young Jack and his wife had done to his father when "the Chief was lying so close to death in the hospital." Grant stated that when we visited my father, each of us had gone into his room and made some kind of horrible scene, had shouted at him in a manner which "caused his temperature to shoot up, very nearly ending his life." Looking at Ann, Grant added, "We had to bar him from the hospital and ask him to leave Cannes."

Recalling this period, Jack and Ann's daughter Barbara said:

> I never understood what went wrong between my mother and Jack Jr., as she never talked to me about their relationship. All I can think of—and this is my personal opinion—is that she thought Jack Jr. was going to do something terrible to her when my father died. It became an obsession. Perhaps it goes back to the time when he was a young man and sided with his mother, Irma, about the divorce. My mother made Jack Jr. into this big boogieman. She was so scared, she was irrational.

Jack struggled to survive. After four months' recuperation in France, he returned to Los Angeles. He joked with reporters: "I forgot to duck." By the end of 1958, after a few weeks' rest, he decided he was ready to come back to the Burbank studios. Before he could return, he had one unpleasant chore to accomplish.

He had to fire his son.

Other Voices

Jack Warner Jr.

Before I departed France, I left several letters for my father from both my wife and me and our two daughters, Betsy, now eight, and Debbie, just two. Although my father, by his own unbelievable choice, had yet to see either of his granddaughters, I felt his aloofness had to do with his fears of Ann's reaction. The letters were never answered, nor were the many to follow.

I went back to work at the studio, deep in shooting television films. I kept inquiring daily at my father's office about how he was doing. One bright Burbank morning I picked up the paper and read that my father had been flown back to California the day before and was at his home. I drove to the house at once—but the gates remained shut, and when I said my name the metallic voice through the speaker box told me, "No visitors!" I tried several more times, but the answer was always a cold "No visitors."

On December 30, 1958, just before the joyous New Year, I was summoned to my father's office. When I entered, Arnold Grant, whom I had last seen with Ann at the French villa, was seated at the side of the big black desk, behind which was my father's empty chair. On the other side of the room sat company attorney Roy Obringer, who was armed with some legal-looking papers. He was an old friend and at this moment looked terribly upset. Bill Schaefer, my father's secretary, followed me into the room, notebook and pencil in hand.

I seated myself facing the three men. Grant began to speak. "Jack, your father has given me a most unpleasant task to perform . . ."

At this point I cut in and asked what his exact function was, as he was not an officer of the company.

"Well," he responded, "you can call me an errand boy . . . I am to tell you that effective this date you are no longer affiliated with Warner Brothers."

The bluntness of his statement stunned me and I drew in my breath.

Grant looked at me stiffly. "You are *terminated* from your pos-tion in the television department. You will get six months' severance pay."

Still in a state of shock, I asked whether my father realized what he was doing. "I'm his son . . . an officer of this company. Doesn't he re-alize what this would do to the business? Doesn't he comprehend the consequences of this terrible act?"

"He doesn't care," Grant said, his voice like ice. "He's too emo-tionally involved to be near you. He wants you out of here—now."

Then Roy Obringer, looking embarrassed and unhappy, pushed forward several documents, which included my typed resignation as a vice president of the company I had been associated with for more years than any other person in the room.

"You really prepared yourself for this, didn't you?" I held the res-ignation letter by its edge. "How do you want me to tear it, lengthwise or crossways?" I tore it in two pieces and handed it back. I was so out-raged I could hardly breathe, then my mind started to clear. "Arnold, you're not an official of this company. You can't request my resigna-tion or order me off the lot. You have nothing to do with Warner Brothers."

"I am representing your father."

"Are you? Or are you Ann's lawyer?"

"I represent them both. He wants you off the lot."

"Why?"

"So he can come back to the studio."

"Why!"

He paused. "Every time he looks at you he sees Irma."

Every time he looks at you he sees Irma. I had heard it before, yet still found it hard to believe. The truth hit hard. I was family—but I was a part of his past. A past he wanted to forget.

I said, "I'm not resigning, I'm coming to work. I'll be here next Tuesday." Tuesday was the day after New Year's, 1959—a great way to start a year. I turned to leave the room and saw Ann's picture on a table. I should have picked it up and put it on my father's desk and said, "She's running the studio."

As I left, her eyes seemed to follow me out of the room.

Bill Schaefer recalled:

> Ann had an interesting power over Jack. She was incredibly attractive and was a fascinating creature when Jack met her. Due to her weight problem, and her disillusionment with Hollywood—everyone wanted something from you because of who you were—she went into seclusion. Jack had women other than Ann. Once Ann spoke to one of them: "You may be with him, but you aren't going to get him."

Agent Dick Dorso had his own opinions about Jack and Ann's relationship:

> In the early '40s, she treated Jack like a butler—no, more like a gardener. He was like a child with her. He fawned on her. He was obsessed with her. She used the right combination of love and disregard with him. He felt comfortable with that disregard. She was tough; she was a self-willed lady.

This from Bill Orr, husband of Ann's daughter Joy: "I loved to talk to her, because she read a lot—like books on gardening, Eastern philosophies. Yet it was hard to talk to her, because she knew it all. She was a very strong woman, but also a warm one."

My mother said, "Jack was putty in everyone's hands, but especially Ann's."

Other Voices

Jack Warner Jr.

I went home to my wife and told her about the meeting with Arnold Grant in my father's office. I had planned a big New Year's Eve party at the studio for the one hundred people who worked in my television division and decided to go ahead with it. I gave a talk complimenting everyone on a great year past—and toasted the one to come.

I watched the bowl games on Monday, then drove to work the following day. At the gate the studio police blocked my entry. They looked at me solemnly, almost in tears. These were all men I knew, fellow Masons; I was president of the Sam L. Warner Square Club. Police chief Blainey Matthews said, "This is the worst day of my life."

"Blainey, you don't have to explain to me. I will cooperate, but first I demand to go to my office."

He said, "No, I've been instructed not to allow you in the building."

"Well, let me explain something. I was a colonel commanding an Army Reserve unit and I have classified papers locked in my office. No one has a right to touch them but me." I started to walk around him.

He sidestepped and blocked my path. "I can't allow it."

"Then how are we going to resolve this?"

"Mr. Warner, they already changed the locks and took your name off the office door. I'll have to get someone who is cleared to get the papers and bring them to you in person."

Boom, just like that, it was over. I knew there was nothing I could do. I looked up and caught a glimpse of my assistant and others of my division looking out the window. I then noticed a man removing my name from the information board at the gate. I was becoming a nonperson. And I knew who was writing this miserable script: Ann.

Sadly, I got into my car and drove off. As I passed under the entrance, several men were hanging up a big banner. It said: WELCOME BACK, JACK.

Jack Jr. said he wrote a number of letters to his father after he was barred from the studio.

> I remember one of them in particular because I loved the ending. I said a lot of things to him about our relationship and how it had deteriorated and how I didn't want it this way, and at the end I said, "I want you to remember one thing. I am not your enemy. I am your son, Jack." He never answered the letter.

It was in the studio trophy room that Jack L. Warner performed the final act of banishing his son.

Other Voices

Jack Warner Jr.

During a visit several years after I left the studio, I entered the trophy room to discover that there was not a single sign to show that I existed.

I remembered the walls being covered with glass sheets to protect the photographs. For a long time, photos of family members were up there on the walls, including shots of me: one in a Boy Scout uniform; another in a group with my father and assorted presidents of the USA; both of us at the 1932 Olympics; me in a military academy uniform, and one at a movie premiere looking uncomfortable in my first tux.

Now there were no longer photographs of any of Jack L. Warner's brothers or sisters, except Sam, safely dead all these years. Where our pictures once had been were now shots of Ann, my father, and several frozen celebrities. In his office was only one picture, the portrait of Ann which had stared at me during that fateful meeting with Arnold Grant.

In 1964, five years after Jack Warner had expelled his son from the studio—and his life—he published an autobiography, *My First Hundred Years in Hollywood.* Written with a selective memory, the book is pure Jack Warner, splashed with anecdotes and corny humor. Although a great deal of copy was devoted to his masseur, Abdul Maljan, any reference to his antagonism to Harry or the 1956 sale was withheld. He also omitted to report that he had an ex-wife, Irma—and a son, Jack Warner Jr.

Jack did include his second wife in the book and describes his fateful meeting with Ann:

> I fist met Ann some thirty years ago. . . . I was unmarried; I was eligible; and I was a man of some substance in my business. So I was a natural target for unattached women who may or may have not been charming companions, and I was not interested in taking any of them out of circulation. . . . I forgot all about the rule books, or what day it was, or what picture we were making, and I asked her to marry me.

The book was ghostwritten by Dean Jennings (who would later publish several articles describing a most unhappy and unpleasant collaboration). Ann interpolated her own musings into the text, set off in italic print. The final page, "A postscript from one who has shared so much," said, in part: "He holds no malice for those who have done him wrong . . . These are the little things, the human things, that I love him for."

Jack L. Warner, his health slowly returning, was once again in charge of Warner Bros. There were still films to be made, and the Warners publicity factory ensured that Jack's name was displayed prominently in its press releases: "Jack Warner announced today . . ." "Jack Warner accepted the Oscar for . . ."

In 1960 *The Sundowners* was nominated for Best Picture but lost to United Artists' *The Apartment. Splendor in the Grass* won the studio's only Oscar in 1961 (for William Inge's screenplay). In 1962 *The Music Man* was nominated, but Columbia's *Lawrence of Arabia* walked away with the Best Picture award. *Gypsy,* adapted from the great Broadway show, was another musical highlight from Warners that year. At least Jack could report to his new partners that the studio was in the black: In each of the first three years of the '60s the studio's net profit was a little over $7 million.

The studio's top grossers in 1963 were *Days of Wine and Roses* and *Whatever Happened to Baby Jane?* The year's net profit was a dismal $3,738,000. The big one for 1964 was *My Fair Lady,* which did win the Best Picture Oscar, the first for Warners since *Casablanca* nineteen years before.

At the Academy Awards, Julie Andrews, whom Jack had passed over for the starring role in *My Fair Lady* in favor of Audrey Hepburn, accepted the Best Actress award for *Mary Poppins.* As Andrews held up the statuette, she looked at Jack Warner in his seat near the stage and said, "I want to express my gratitude to Jack Warner for making all this possible."

The next year, 1965, was a lackluster one, with *The Great Race* being the big moneymaker. By 1966, studio production was cut to thirteen films. One of them, *Who's Afraid of Virginia Woolf?,* caused an outcry from industry censors, who objected to such phrases from Edward Albee's play as "monkey nipples" and "angel tits." The Production Code Administration's powers had been severely reduced in 1952 when the Supreme Court ruled that the First Amendment extended to movies and provided "the same protected status held by newspapers, magazines and organized speech." By 1956 the Code was modified to allow controversial matter, but still drew the line to bar obscenity. By the mid-'60s the

Production Code had been weakened by the rapidly changing morality of the American public. In the spring of 1966, Jack J. Valenti, a special adviser to President Lyndon Johnson, took over the job as president of the Motion Picture Association of America. Valenti's first confrontation with a studio came when he reviewed *Virginia Woolf.*

He and his fellow Association members were appalled by the language of the on-screen screaming matches between stars Elizabeth Taylor and Richard Burton, particularly several "Screw you!"s and Burton's proposed parlor game called "Hump the Hostess." In a bartering session with Valenti in Jack Warner's office, a solution was reached by trading one word for another. Warner was allowed to retain "Hump the Hostess," in return for giving up one "screw." By the time it was over Warner was allowed to retain two "screws" and "Hump the Hostess." (After it was over, Valenti vowed he would never go through that nonsense again and introduced the rating system, which would allow producers greater freedom and warn parents about the content of movies.) *Virginia Woolf* was hugely successful, and was nominated for thirteen Academy Awards. Elizabeth Taylor, who did not attend Oscar night (even though Jack insisted), won for Best Actress.

By the mid-'60s, the movie mogul was an endangered species. Louis B. Mayer was dead; so was Harry Cohn. Darryl Zanuck had given up producing pictures and spent all his time in Europe. Sam Goldwyn had found making movies too great a strain and had filmed his last picture. There was only one ruler of a movie empire left: Jack Warner.

He sat alone on his throne.

But even Jack had decided that movies "weren't fun anymore." He told his studio confidant, Bill Orr, that it was too big a hassle to produce films since the heyday of the studio system had passed: "Casting a picture is a pain in the ass." He hated dealing with "all those fuckin' agents." Jack was seventy-four. It was time to relinquish power.

On November 14, 1966, a headline in a trade paper reported: JACK L. WARNER SELLS INTEREST IN WARNER BROS. Jack had announced plans to sell his 1.6 million shares to Seven Arts Productions for \$20 a share. The sale would earn him \$32 million and he figured to collect \$24 million after capital gains taxes. "Not bad for a butcher boy from Youngstown, Ohio," he bragged.

My father told me that he had dinner with Jack shortly after the sale was announced and mentioned to Jack that he thought he'd got a great price for the studio. Jack stopped eating and said, "Yeah, today I'm Jack L. Warner, but wait until tomorrow. I'll just be another rich Jew."

Richard Gully observed:

> Selling the studio was the greatest mistake of Jack's life. It made him miserable. The week after the sale, no one showed up for the Saturday and Sunday tennis parties. Jack was heartbroken. No one came because they no longer needed to seek favors from him. That's Hollywood politics. I once told him, "Jack, you're so alone in this world." He said, "Richard, if you have power, you can't have friends, because they always want something out of you."

Jack Warner was not out of the picture business; the agreement stipulated that he would remain an independent producer and the company would finance his pictures. While the deal was being settled, Jack made *Bonnie and Clyde.* Warren Beatty had argued for the production, saying that Warners had begun by making gangster movies. Jack had responded, "Sure, but we didn't have dames shooting guns." Jack also made the acclaimed thriller *Wait Until Dark* with Audrey Hepburn, produced by her husband, Mel Ferrer.

Then he decided on one last picture, a $14.5 million musical spectacular called *Camelot.*

Camelot!

Was it mere coincidence that my great-uncle Jack Warner chose this to be the last movie he made under the Warner Bros. shield? Or was he aware of the parallels in this powerful story to his own? I see the role of Guinevere as being played by Hollywood; Harry as King Arthur; Jack as Lancelot. Arthur's liking for Lancelot was immediate, full of respect for his abilities as a knight, even though Lancelot had knocked him off his horse jousting. Lancelot was totally in agreement with Arthur's concept of a new order, and became a cocreator of Arthur's court and the Round Table. King Arthur understood Lancelot's love and uncontrollable passion for Guinevere. He felt the same way.

The Camelot tale ended in turmoil, confusion, and betrayal.

Yet, looking at the bigger picture, the ideas and intentions behind Arthur's dreams lived on.

Yes, there was one brief shining moment called Camelot that lived on with those who heard the story. At the end of the film King Arthur secretly meets with his beloved Guinevere and Lancelot and says, "Revenge—the most worthless of causes. All we have been through . . . all for an idea . . . the fates must not have the last word."

Epilogue

Passions and Dreams

Imagine Jack Warner, alone in the screening room, watching the final cut of *Camelot.* The thoughts of the past, memories of his father, his brothers . . . these he has momentarily erased from his mind.

On the screen, King Arthur, his once proud Round Table in shambles, retreats to the woods, and finds his old teacher, Merlin. Arthur asks his wizened mentor, "What's the best thing for feeling sad? You told me once."

Merlin's answer: "The best thing for being sad is to *learn* something."

Arthur: "Learn something?"

Merlin: "It's one thing that never fails. You may grow old and trembling in your arteries, you may lie awake at night listening to the disorder of your pain, you may miss your father, your mother, your dog, your only love . . ."

Arthur whispers: "My love . . ."

Merlin: "There is only one thing for all of it—learn. Learn *why* the world wags, and what wags it."

Jack taps his fingers on the arm of his seat, and grumbles, "Merlin, you bastard, you know who you sound like? Yeah, good old Harry."

Jack continues to watch the story unfold, listening more intently as Arthur later speaks the line: "I can't remember all that Merlin taught me, but I do remember this—that happiness is a virtue. No one can be happy and wicked. Triumphant, perhaps, but not happy."

Jack Warner breathes heavily, but says nothing.

Then, as the color images from the screen are reflected in Jack's eyes, the scene changes and is shrouded in the gray light before dawn. A bugle blows to begin the battle. The sound of a breaking twig catches King Arthur's attention. It's a young boy who has heard the tales of the Round Table and has come to be a knight in King Arthur's court.

Arthur, revitalized with hope, knights the youth and orders him *not* to fight in the battle, but to grow up and grow old, and tell anyone who hasn't heard the story that there was once a fleeting wisp of glory called Camelot. As the boy runs behind the line to safety, King Arthur roars with the ecstasy of knowing he has won the battle before it has even begun.

"Yeah, run, boy!" Jack chuckles as the boy disappears into the fog and Arthur shouts, "The boy is one of what we all are! No less than a drop in the great sunlit sea. But it seems some of the drops sparkle. *Some of them do sparkle!*"

Jack's forehead creases into a frown as the *Camelot* theme builds behind Arthur's closing line. Over the music, Jack mumbles, "The schmuck. Some king. All he does is scream lines. I shoulda stuck with Burton—and Andrews. Yeah, why did I let Josh Logan con me into casting this English asshole, Harris, as Arthur? And Redgrave . . . scrawny broad . . . Franco Nero as Lancelot? He's an Italian, for chrissakes! *Hell, none of them could even sing!*" Jack shakes his head wearily. "Fourteen and a half million bucks . . ."

The screening room lights come up and a fast-talking corporation executive enters. "How are we coming, Jack? We gonna pop this puppy on schedule? Marketing is on hold, and foreign sales is breathing down my neck . . ."

Jack stares at the now blank screen. He contains his anger, and finally grins. "Uneasy lies the head that wears the toilet seat."

The executive looks at Jack with disdain and leaves.

Jack remains seated, letting the soft felt cushions envelop him like a warm soothing bath.

So many movies . . .

In his mind, images appear on the gray screen . . . The barrel of a gun aimed at the camera, a flash of smoke as it fires. Yes, *The Great Train*

Robbery. And Sam, his arm weary, cranking the Kinetoscope . . . A covered wagon, just one damn covered wagon, that's all we had. But it was enough for *Perils of the Plains* . . . Then *My Four Years in Germany.* Good-bye Poverty Row—hello Hollywood . . . Rin Tin Tin, the mortgage lifter, a diamond collar around his neck . . . *The Jazz Singer.* Jolson, you lucky bastard. You couldn't keep your damn mouth shut . . . *42nd Street.* You're gonna come back a star! . . . *The Life of Emile Zola.* Our first Best Picture. Okay, Harry, we'll do it. Sure, sure, whatever you say: enlighten, educate—but we'll also entertain! . . . *Mission to Moscow.* That one almost brought down the studio . . . *Casablanca.* Here's looking at you, kid . . . *A Star Is Born.* Garland, you were great . . . *Giant* . . . *Gypsy* . . . *My Fair Lady* . . .

And now, *Camelot* . . .

Jack turns his head and whispers as if someone is sitting next to him, "Harry, you would have liked *this* one . . ."

He chuckles, then breaks out into an uproarious laugh. As the sound of his laughter echoes through the room, a young man enters with a small tray. "Your coffee, Mr. Warner."

Jack starts to brush the youth aside, but something makes him stop. He stares at him for a moment, then buzzes the projectionist in the booth.

"Hey, kid, ya wanna see a movie?"

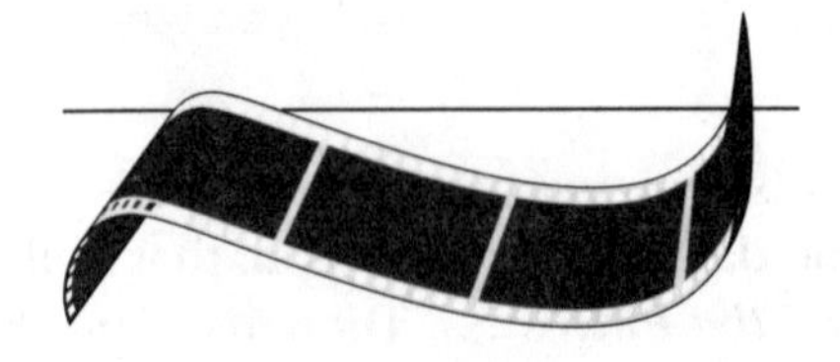

Fadeout

Jack Warner

Camelot, Jack L. Warner's last picture for Warner Bros., received scathing reviews. The critics thought the 178-minute film had done little to re-create the magic of the musical stage show. Although its general release attracted a fair-sized audience, it never made back its almost $15 million cost. Director Josh Logan's miscasting and his uninspired direction (when asked by critic Charles Champlin why the movie was poorly shot, Logan screamed, "Jack Warner made me film it on the back lot!") shattered Jack's dream of ending his career on a high note.

It was only a few months after Warner Bros. and Seven Arts merged as Warner Bros.-Seven Arts that the company became a target for takeover moves. The winner was Steven J. Ross, president of Kinney National Service Incorporated, a group that specialized in parking lots and funeral parlors. The combined corporation was named Warner Communications, which would become one of the largest entertainment concerns in the country.

On the last day of October 1969, Jack made one last tour of the lot in his Bentley and drove out the gate. There was no one to say good-bye.

Jack hated to relinquish his power, but even he realized the studio czar was obsolete.

For a while he set up an office in Century City near Los Angeles and produced a Broadway musical, *Jimmy,* based on the life of Jimmy Walker, the flamboyant mayor of New York. The theater critics were less than kind to this Hollywood interloper, and after opening night Jack was advised to close: "Get out while you can. It's like a vacuum cleaner. Everybody's in it to see what they can take out of your pockets." Jack answered, "You don't understand, I can't do that. Who would have lunch with me then?" Before Jack closed the show three months later, it had cost him $1.5 million. When friends said anything about the losses he sustained, he shrugged. "It's only a million." He produced two independent movies, *Dirty Little Billy,* a dismal remake of the stale story of Billy the Kid, and *1776,* from the Broadway musical. The latter was a strange choice for Jack, as he had always resisted any film set in the period of the American Revolution. Jack Jr. remembered his father saying, "Nobody wants to see a movie where the men wear knickers, silver buckles on their shoes, and wigs!" He was right; the noble project bombed at the box office.

At a Screen Producers Guild dinner, Jack lashed out at movie critics, calling them "downbeat bums." He then asked the assembled members to take up arms against the press. He ranted, "It's high time we strike back! No one strikes back except me." As he sat down, the master of ceremonies, George Jessel (who admittedly had not sent Jack any love notes since *The Jazz Singer*), shook his head and asked, "How the hell did you become the head of a great studio?"

Although most of Jack's mistresses had been glamour girls, in the later part of his life he shifted his attention to women who would give him companionship. Richard Gully remembers Jack's last mistress as a middle-aged woman, charming, soft-spoken, very likable, good-looking, but not glamorous. This was a very serious affair. He bought her a beautiful house in Palm Springs, and even put her son through college.

Barbara Warner, Jack and Ann's daughter, said:

> My father was explainable. He was what you saw. You knew what to expect. He had a great smile, and great fun in life. He liked pretty women and good food. He liked to travel, but my mother wouldn't travel with him. I think she was bored with his social scene at the casino and some of the hangers-on. She preferred being with writers, painters, and politicians. She retired into her life, and I think she was lonely and wanted companionship. She never acknowledged that he

had anybody [another woman], and he certainly let her know as little as he could. They were never legally separated, and I know they never wanted a divorce.

In 1972 Jack celebrated—as he called it—the "Big Eight-Oh." It was one of the few times he mentioned his age. When told that ex–movie mogul Joe Schenck had died at eighty-two, Jack intoned, "Too bad . . . a young man."

By 1973 Jack's health was beginning to fail, partly because of a fall he took while playing tennis. A new man had come to play and didn't realize he wasn't supposed to make Jack run for balls. It was the last time Jack played tennis. When he began to act disoriented, Ann kept him in the house and became his nurse, and even studied the use of different medicines. By Christmas 1977 he had suffered a stroke and was confined to a wheelchair. Ann shielded him from visitors and devoted all her time to him.

Other Voices

Barbara Warner

When my father was sick, my mother became the most amazing nurse, tending to all his needs. She took care of all his medicine; never left his side. Then he had another stroke, after which he lost his sight. This was terrible for a man who loved film.

I have a tape of the two of them talking to each other. I think she was taping music for him to listen to, and they both began to talk. Or, perhaps, she just wanted to tape his voice. He said, "I love you, darling," and she replied, "I love you. And you're going to get well."

Jack Warner died of edema on September 9, 1978, at the age of eighty-six. His estate was conservatively estimated at more than $15 million. His will stipulated how his mausoleum should be built at the Home of Peace Cemetery, and precisely who would have permission to enter during his long sleep. There was also the implied demand that it be as far as possible from where Harry, Rea, his parents, and others of the Warner family rested. It took Ann almost four years to have the mausoleum completed.

Jack Warner Jr. offers this final note about his father:

> To be thought well of by one's fellows is a basic human need, unless delicate controls and balances deep inside have succumbed to inner rot. My father had a deep desire to be liked, but at times perversely rushed to extremes to be unlikable, unlovable, and unbearable. While he had his admirers, most of them seemed driven by their self-centered needs and dependency on his good will to succeed, whatever the personal cost, silently accepting public humiliation if it meant their gains and security.
>
> With all his faults, Jack L. Warner loved producing motion pictures and the power that went into running a great studio and he should be given credit for bringing about the creation of films which even today demand respect as gleaming nuggets from the "Golden Age" of Hollywood, where making the deal now means more than making the film.
>
> I hear my father now in the executive dining room saying, "I'm not here to make friends—I'm here to run a studio and turn out movies to fill theater seats with behinds to make money for this company!"
>
> Perhaps Jack Warner wrote his own epitaph when he said: "I am what I am, and I will probably never change."

Ann Warner

Ann remained a recluse after Jack's death. The only time she ventured out of the house was for her husband's funeral. Jack Warner Jr. also attended.

Other Voices

Jack Warner Jr.

Barbara and I and our daughters were the first to arrive at the Wilshire Boulevard Temple. Betsy (the mother of our first grandson—the great-grandson my father never saw) and Debbie (who had never met her grandfather in all her twenty-one years, although unknown to me she had twice gone up to the closed bronze gates of his estate to introduce herself, but had been rebuffed) stood silently by our side. Betsy's would be the only tears I saw shed at the funeral, tears of frustration and anger. We went up the staircase to meet Rabbi Magnin in the upper hall just outside the chapel. I wondered why the services were to be held up here instead of in the temple, with its magnificent Warner Memorial Murals, a sweeping graphic monument picturing the Old Testament in tribute to a faith, a people—and a family. The murals had been given by the Warner brothers to honor their departed parents.

Rabbi Magnin had been told to make it very brief and I could see this troubled him, as there was much he would have liked to say. I stood with him a moment inside the chapel and looked at the blanket of red carnations completely hiding the casket, which in turn concealed the man inside—my father resting behind his last wall.

We watched from the upstairs window as a line of limousines drew up to the curb. The doors of the first one swung open and a man stepped out to assist a heavyset woman, dressed in black, as she slowly eased out and stood on the sidewalk. It was hard to believe this short, dark figure standing down there was the same woman I last saw twenty years before at the villa in France saying to us, "When this is over and Jack is better and we're all back in California, we'll get together, all of us, and have a wonderful party!" And now here we were, all of us, together again at last . . . only it wasn't a party.

Very slowly they came up the stairs, their feet making soft shuffling sounds on the marble. Rabbi Magnin, standing next to me in his sweeping black robes, stared as their heads came into view. He nudged me and whispered, "Jack—which one is she? Which one is she? I've never seen her."

I indicated the black shape in the middle of the small crowd, the old woman even I had difficulty recognizing. She was surrounded by her two daughters, Barbara and Joy, a son-in-law, cousins and strangers, who like a dark moving wall masked but could not conceal the woman. For a moment they slowed and Ann glanced sideways toward me and my family. I saw a flash of almost childlike innocence in her aged face that quickly changed to the tight-lipped look of a woman who had achieved everything she had set out to do.

Ann continued living at the Angelo Drive estate. Richard Gully said, "After Jack died, Ann would have loved to leave Hollywood and go to France to live in the villa. But she couldn't because of her weight. She was afraid of being seen. The Beverly Hills house was like a prison to her, except in this case, the walls and gate served the purpose of keeping people out. She loved to stay up late and read and watch movies on television. One of her favorites was *The Sailor Who Fell from Grace with the Sea*."

Ann's diabetes got progressively worse, and she died from it (and complications, pneumonia and kidney failure) on March 8, 1990. She was eighty-one.

Her daughter Barbara told a reporter her mother died peacefully and that in death her face had an incredibly serene beauty. A few days after Ann's passing, a few relatives and close associates were invited to the mansion for an occasion of remembrance. In attendance were Barbara's husband Cy Howard, Ann's daughter Joy with her husband, Bill Orr, and two of their children, Gregory and Diane. Also in attendance were Ann's attorney, her business manager, her secretary, and several nurses who took care of her during the last months of her life. She did not have a funeral and was cremated.

In 1991 entertainment executive David Geffen purchased the Angelo Drive property and furnishings for an estimated $47.5 million. The sale broke records for the price of a single-family residence. Of all the Hollywood estates created in the '20s and '30s, Jack L. Warner's was the only one to remain intact with its original mansion and extensive grounds.

Shortly after Ann Warner's death, Richard Gully wrote an article for *Beverly Hills* magazine titled "A Tribute to the Last of Hollywood's Royalty—Ann Warner." He ended it by saying: "During the fun of Hollywood's golden era with all its colorful movie moguls and world-famous, unforgettable movie stars, Ann Warner was royalty. She looked like a queen; she lived like a queen. God bless her."

Irma Warner

Irma and her husband, Al Rogell, who had been a top director in silent films, directing such stars as Tom Mix, lived out their lives in Beverly Hills. Irma loved to compose and play music on the piano. Her other hobbies were horse racing and gambling, and she and Al owned several racehorses. She remained on friendly terms with Rea Warner. Irma died of colon cancer in 1985. Rogell died in 1990.

Jack Warner Jr.

Jack became an independent film producer and with a partner formed the Warner-Schor Company to develop projects for theatrical and television distribution. In 1976 he was the executive producer and director of the syndicated TV series "Eddie Kantar on Bridge."

In 1982 his novel, *Bijou Dream*, a story about a bombastic studio chief named Hamilton J. Robbins, of Robbins International Studios, was published by Crown. Although a work of fiction, the story of "Ham" Robbins, his older brother Maurie, and Ham's second wife Nita was written from the author's experiences. The book was issued as a paperback in 1988 under the title *The Dream Factory*.

Jack Jr. now lives near Los Angeles with his wife Barbara. The Warners have two children, Betsy and Debbie.

Joy Paige

Joy, Ann's daughter by her first husband, had three children with Bill Orr. Joy currently lives in Los Angeles. Bill Orr has a home in Palm Springs.

Barbara Warner

After attending Sarah Lawrence College, Barbara moved to Europe, where she met Claude Terrail, the owner of the Paris restaurant the Tour d'Argent. When Barbara was twenty-one, they married, and in 1959 they had a daughter, Anne. When this marriage was over, she met and married a French composer, Raymond LeSenechal, who died unexpectedly. She moved back to Los Angeles in the early '70s, and in 1977 married Cy Howard, a Hollywood screen and television writer, producer, and director. Howard died in May 1993 after a long illness. Barbara is on the board of the New York Theatre Workshop, and resides in New York and Los Angeles.

Albert "Abe" Warner

On November 26, 1967, the same day *Camelot* was released, Abe Warner died peacefully in his home in Miami Beach. He was eighty-three.

His step-grandson John Steel recalls:

> Abe passed away very suddenly. He was living in a condo on Miami Beach and he had just gone to the track as he had every day for the last several years. He must have had a pretty good day, because he walked in and told Bessie, who was bedridden at the time, that he had won a few hundred bucks. He went into his little den, sat down in the chair in front of his television—and had a stroke, and died. The circuits blew up and he was dead.
>
> We had a small service for Abe in Miami Beach, then flew his body to New York and had a large funeral in Temple Emanu-El. I saw Jack there and he had this incredible smile on his face. I remember seeing Jack arrive with a lot of pomp and circumstance—with this smile. He couldn't get it off his face.

Jack Warner Jr. also attended Abe's funeral.

Other Voices

Jack Warner Jr.

I thought a great deal of Abe Warner, whose exterior was gruff—but who had a warm heart and love for everyone. I went to the funeral and was pleased to see my father enter the temple and seat himself in the front row facing his brother's casket. As the seat next to him was empty, I sat next to him. I thought of an earlier meeting when he said that we only seemed to see each other at funerals. As the rabbi spoke about what a fine family man Albert Warner had been all his life, his acts of charity and his wonderful relationships with others, I could feel some tension from my father, but perhaps that was normal in the presence of death.

I rode in his car to the cemetery and afterward we both wound up sitting at a table in his Sherry-Netherland apartment. The view was breathtaking: snow in Central Park . . . glistening boulevards and cross streets . . . New York on a sparkling winter evening. He poured glasses of Chivas Regal for us both. It was a little wake for Uncle Abe and we wished he could be there to drink with us. The warmth I felt did not come from the scotch. I was very happy to be with my father. It was a good, solid talk. Not a father-son loving conversation, but a nice talk.

The next morning I changed my plane reservations and booked a seat next to him. As his chauffeur was driving us to the airport, my father leaned over to me and said, "Will you quit writing me those fucking letters."

As the plane crossed the Rockies I turned and asked if I could drive him to his house. He said, "No, I made arrangements." I noticed that the closer we got to Los Angeles, the more reserved, the more distant he became, and he moved over away from me as far as he could. I thought, he's scared. *The closer he gets to Ann, the deeper the chill sets in.*

The Los Angeles city lights covered the whole sky with their glow as we passed over the freeway coming in for the landing. He was still far over in his seat as we touched down.

"Good-bye, drive carefully," he said as I rose from my seat.

"Are you sure I can't drive you home? My car's right here."

A quick answer in monotone: "No . . . good-bye . . . good-bye." He sat in the plane until I'd left and I watched as he came down the ladder and got into a waiting car. As I walked away there was a chill in the evening air and it felt like snow in Los Angeles.

Abe Warner was buried not in Miami Beach, the city he had loved and where he had spent the last two decades of his life, but in Brooklyn, New York. Harry Warner had built a mausoleum, a beautiful four-tiered structure, for himself there but decided he didn't want to be buried there. He gave it to his brother. Albert and his second wife had buried their first spouses, Bessie Krieger Warner, and Jonas Siegal, there and arranged it so they would be laid to rest between them. Bessie Levy Siegal and Abe had been married forty-six years. She died of influenza. Although she and Abe did not have children, Bessie had a son from her first marriage, Arthur Steel, and two grandsons, John and Lewis.

Rea Warner

After Harry died, Rea moved into a condominium on Wilshire Boulevard near Beverly Hills. She loved baseball and was an avid Dodger fan. She continued to travel, and kept up her social activities with her women friends in Los Angeles. She died of colon cancer in 1969 at age eighty-six.

Doris Warner

Doris's union with Mervyn LeRoy lasted eleven years and produced two children: Warner and Linda.

Doris was married to Charles Vidor from 1945 until his death in 1959. They had two sons together, Brian and Quentin. Vidor was directing the film *Song Without End* in Vienna, Austria, when he suffered a fatal heart attack, the same night Doris had returned to New York for her daughter's college graduation.

After Vidor's death, Doris was a studio executive at United Artists. In 1962 she moved back to New York to be closer to her children, and while there helped to establish the Phoenix House drug rehabilitation center and the Beaumont Theatre.

In the fall of 1963, Doris became severely ill with colon cancer. At the time, Billy Rose, a successful financier, became interested in her and pursued her until she agreed to marry him in 1964. They were divorced six weeks later, and Rose died shortly after.

Because of her physical problems, she had too many doctors who prescribed various addictive painkillers, all of which took their toll. On August 28, 1978, she died from complications of intestinal surgery. Doris was sixty-five.

Betty Warner

Betty remained married to Milton Sperling for twenty-four years. They had three daughters and a son.

Long before it was fashionable for a woman to select a role other than wife and mother, Betty found a way to express her passion for painting and sculpting. She also took night courses at UCLA in philosophy, comparative religion, and literature. At thirty she began art school and has continued taking classes throughout the years.

It was always a surprise to come home and see what Mom was up to. Whatever else was going on, she was to be found in the living room, and later in the garage-studio, painting or welding huge pieces of scrap metal or weaving yarn into something wonderfully aesthetic to experience. I remember the joy she once had from receiving a wrecked and crumpled car as a gift. The next thing I knew, this gorgeous, shiny, powerful, winged horse standing on its haunches—a phoenix rising from the ashes—was gracing our garden. Magic. My mom performed magic.

Not only that, she managed to be a real supporter of the arts by not only collecting it, but opening a gallery in Santa Barbara, California, and New York to help other artists show and sell their work. Both galleries were devoted to the promotion of fine crafts, such as woodworking, ceramics, glass blowing, and weaving.

To me she was a superwoman as I witnessed her ability to juggle her time so that she could also be involved with social and political issues.

In 1964, Betty married Stanley Sheinbaum, an economist who at the time was a senior fellow at one of the first think tanks, the Center for the Study of Democratic Institutions. Sheinbaum ran for Congress in Santa Barbara in 1966 and 1968 with Betty at his side. In 1973, he became chairman of the American Civil Liberties Union Foundation of Southern California, and later was a regent of the University of California.

My stepfather Stan was at the forefront of so many admirable involvements that it was and remains difficult to keep up with him. He is always into something mysterious and exciting that's happening on national and international fronts. He's a consummate networker and worker dedicated to human rights, justice, education, politics and world affairs.

Following the Rodney King beating, Stan became president of the Los Angeles Board of Police Commissioners and was instrumental in repairing the damage of that incident.

Milton Sperling

Betty's first husband, Milton Sperling, worked for Warners for ten years as an independent producer, and—as he liked to say—"a referee between Jack and Harry." Sperling wrote, produced, and often ended up directing films that Humphrey Bogart, Anthony Quinn, and many other famous actors took part in. He was responsible for producing and writing forty-seven feature films and over half a dozen television movies. In 1955 Sperling wrote and produced *The Court Martial of Billy Mitchell,* starring Gary Cooper and Rod Steiger. The movie was nominated for the Best Original Screenplay Oscar. In 1958 Sperling produced *Marjorie Morningstar,* starring Natalie Wood and Gene Kelly.

My father was inquisitive and fascinated by everything going on around him; in touch with world affairs, and well versed in just about any subject. He genuinely loved his work. He was a master storyteller. With his sense of humor, he was always able to lighten up any situation. It was because of his interest and advice that I continued with my passion for telling the Warner Bros. story.

He loved to consult and encourage people to do their art. Toward the end of his life, he was asked to teach a special advanced scriptwriting class at Northridge University and UCLA. He was actively involved in writing up until the day he lost the battle against cancer. He died in August 1988, surrounded by his family, and his wife, Margit, to whom he had been married for almost twenty-five years.

Lina Basquette Warner

Lina, Sam Warner's widow, is now in her eighth decade. The thick dark hair that entranced moviegoers in De Mille's *Godless Girl* is now a silver gray, and with her dark tan and sparkling eyes she looks many years younger.

For thirty years she raised and handled Great Danes. In 1971 and 1973 she was named All-Breed Handler of the Year by *Kennel Review* magazine. Today she is a well-known American Kennel Club Dog Show judge. Since 1975, Lina has lived alone in an apartment carved out of an old mansion atop a hill in Wheeling, West Virginia, sixty miles south of Pittsburgh.

Although she had not made a movie since the 1943 picture *A Night for Crime,* in 1989 she was cast by filmmaker Danny Boyd in *Paradise Park,* in which she played Nada, an aging grandmother who dreams God is coming to grant a wish to residents of an Appalachian trailer park.

"It was easy to get back into acting," Lina said. "I was a little apprehensive, but once I did it, it was simple. There weren't any browbeating studio executives making everyone nervous. It was really a lot of fun." When asked if she'd do it again, she replied, "Oh, if the right part came along I might take a look at it. But I mean, how many parts are there for old ladies, really!"

In 1990, her account of the first forty years of her life, *Lina, De Mille's Godless Girl,* was published by Denlinger's. She is presently working on the sequel and hopes that someday the story of her life will be filmed.

Lita Warner

Sam's daughter and Dr. Nathan Hiatt had three children. When she married Mort Heller, an investment counselor, in 1960, he brought his two children into the family. They now live in Aspen, Colorado, where they are very active in the community. Lita is involved in the Music Association of Aspen as well as Dance Aspen.

The Warner Sisters

The three Warner sisters lived far from the limelight. They were almost completely dependent on Harry's financial support, and behaved as if they were "poor relatives," which in a sense they were, compared to the lifestyles of Harry and Abe, and especially Jack. Even after they were married, their husbands looked to Harry for job security. Harry was, in effect, the sisters' father, husband, and brother—the ultimate patriarch. "The girls' husbands didn't relish their roles and it made for uncomfortable marriages," Jack Jr. says. "Because of this, the sisters were jealous, clannish, and critical of others."

They all died of colon cancer in the late '50s, within four years of each other.

Annie Warner Robbins

Annie, the oldest daughter, is remembered by the family as assuming the role of the good Jewish daughter, helping her mother run the house, do the cooking, the laundry, and all the backbreaking labor of preindustrial domestic immigrant life. She was small and plump, with a pleasant personality.

Annie married Dave Robbins of Youngstown. The marriage did not produce any children. Late in life, after becoming a widow, she moved to California to live near her parents. Jack Jr. remembers Annie as "a silent and pale persona, not very noticeable, or wanting to be, the sister who was easily lost in a group photograph." She died March 20, 1958, a few months before her beloved brother, Harry, died. She was seventy-nine.

Sadie Warner Halper

Sadie married Louis J. Halper of Youngstown and they had two children, Sam and Evelyn. Betty Warner remembers Sadie as being "short and dumpy—the traditional yenta-type." She was the family gossip, and openly expressed her hatred of Jack for the way he had treated his first wife, Irma. After her husband died Sadie turned increasingly bitter about Jack's lifestyle. When Jack was in bed following his near-fatal accident in France, she visited him, saying that God had punished him for not returning for his brother's funeral. Jack had her ejected and never spoke to her again. When she died a year later, on September 7, 1959, at age sixty-four, he refused to go to her funeral.

Rose Warner Charnas

Rose married Harry Charnas of Youngstown and they adopted one child, Milton. She was tall and big-boned and was thought of by the family as being a bit of a rebel. Jack Warner Jr. remembers her as being the "cream, the diamond, the genuine article beside which the others paled. I felt good in knowing there was a Rose Warner Charnas. She filled in most of the black holes in the family and gave off a kind of golden light."

Rose died on December 26, 1955, at the age of sixty-five.

Warner Family Tree

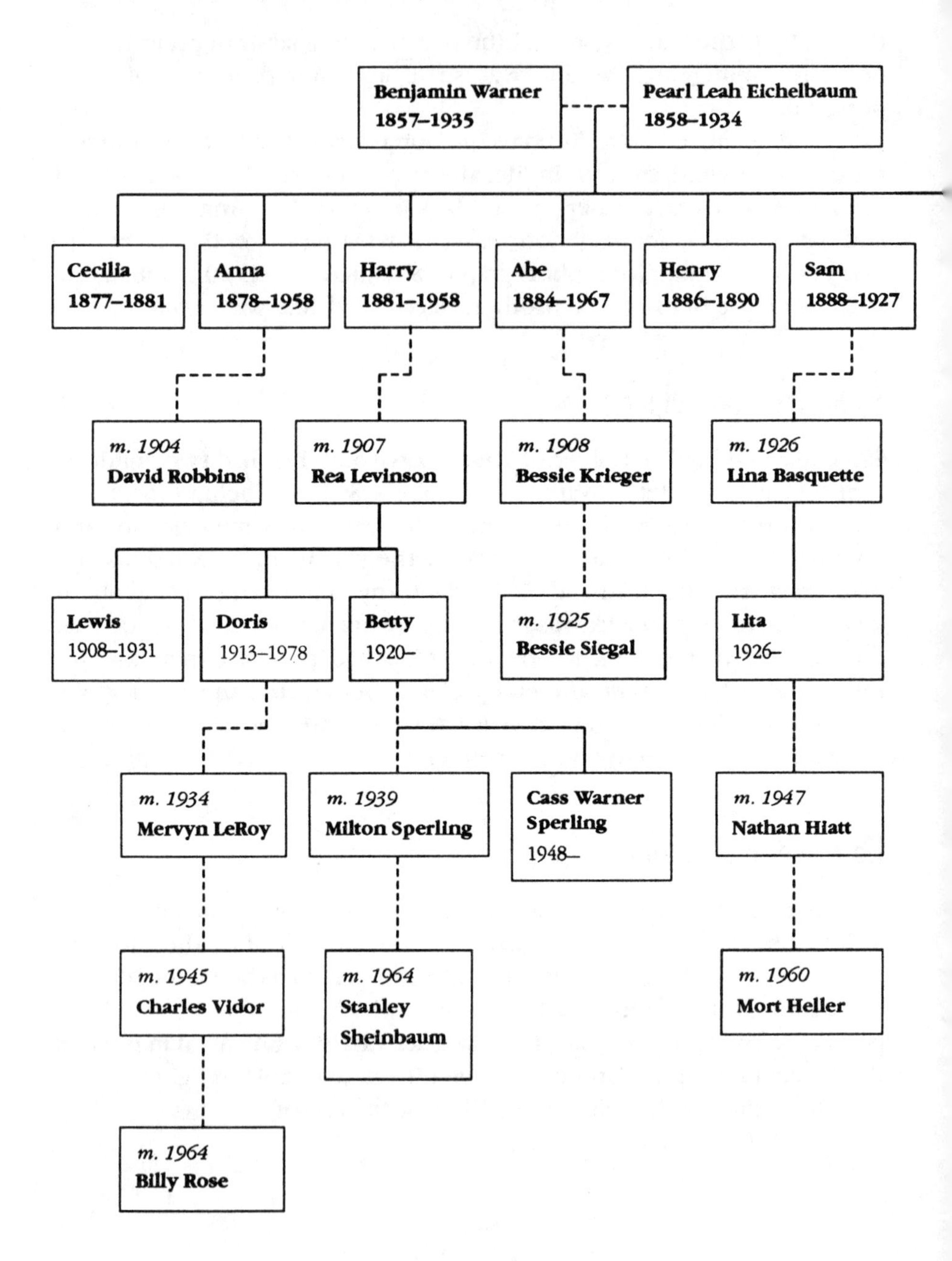

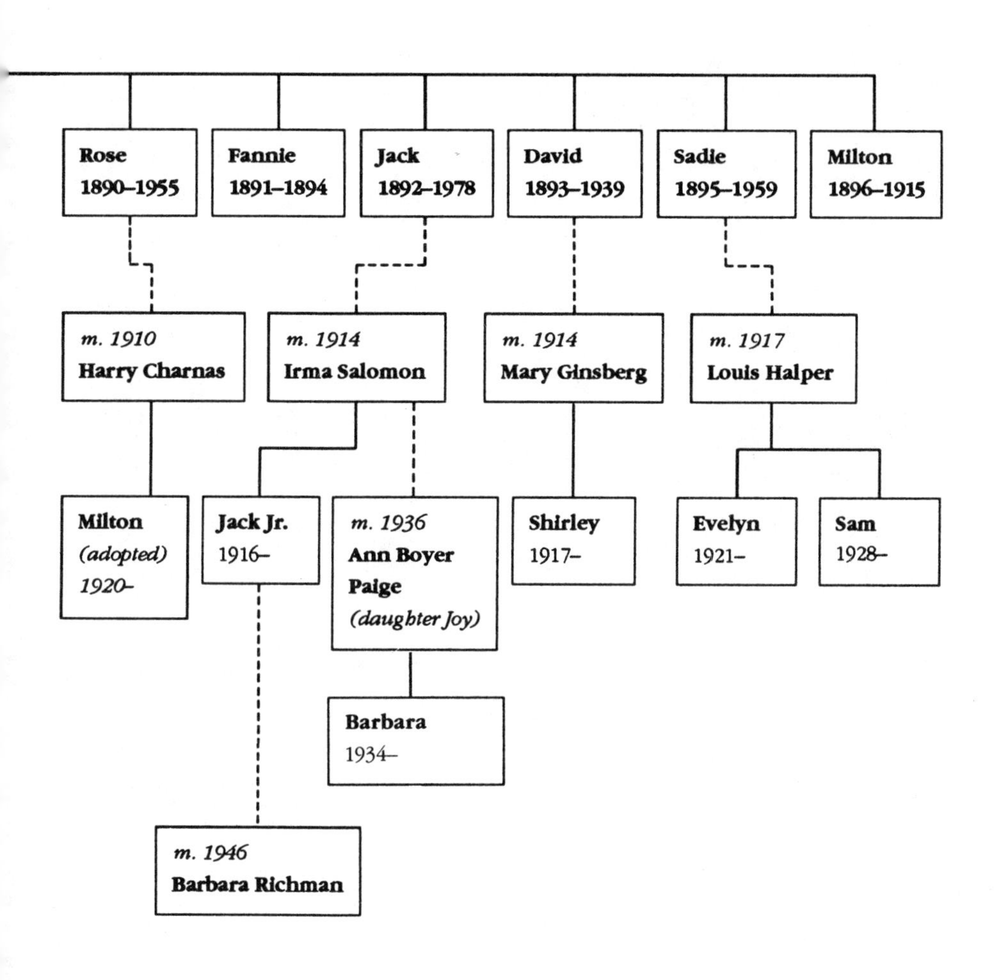

Rose
1890–1955
Fannie
1891–1894
Jack
1892–1978
David
1893–1939
Sadie
1895–1959
Milton
1896–1915
m. 1910
Harry Charnas
m. 1914
Irma Salomon
m. 1914
Mary Ginsberg
m. 1917
Louis Halper
Milton
(adopted)
1920–
Jack Jr.
1916–
m. 1936
Ann Boyer
Paige
(daughter Joy)
Shirley
1917–
Evelyn
1921–
Sam
1928–
Barbara
1934–
m. 1946
Barbara Richman

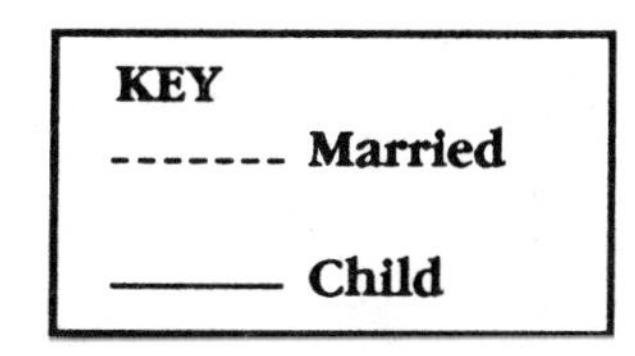

KEY
Married
Child

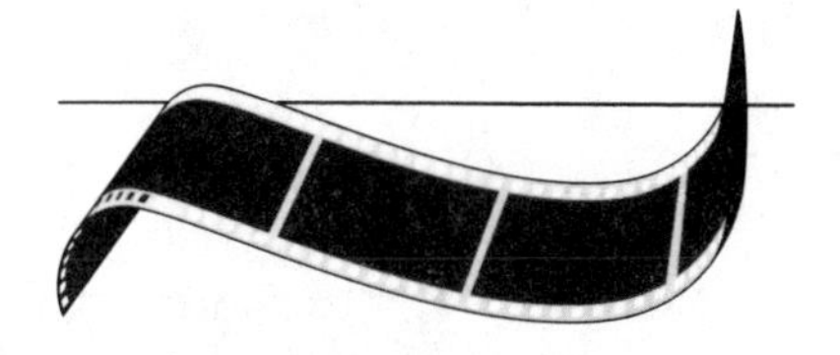

Index

A

C

D

E

F

G

H

I

J

K

N

S

T

U

V

W

Y

Z